Studies in Computational Intelligence

Volume 526

Series editor

Janusz Kacprzyk, Polish Academy of Sciences, Warsaw, Poland
e-mail: kacprzyk@ibspan.waw.pl

For further volumes:
http://www.springer.com/series/7092

About this Series

The series "Studies in Computational Intelligence" (SCI) publishes new developments and advances in the various areas of computational intelligence—quickly and with a high quality. The intent is to cover the theory, applications, and design methods of computational intelligence, as embedded in the fields of engineering, computer science, physics and life sciences, as well as the methodologies behind them. The series contains monographs, lecture notes and edited volumes in computational intelligence spanning the areas of neural networks, connectionist systems, genetic algorithms, evolutionary computation, artificial intelligence, cellular automata, self-organizing systems, soft computing, fuzzy systems, and hybrid intelligent systems. Of particular value to both the contributors and the readership are the short publication timeframe and the world-wide distribution, which enable both wide and rapid dissemination of research output.

Witold Pedrycz · Shyi-Ming Chen
Editors

Social Networks: A Framework of Computational Intelligence

 Springer

Editors
Witold Pedrycz
Department of Electrical and Computer
 Engineering
University of Alberta
Edmonton
Canada

Department of Electrical and Computer
 Engineering, Faculty of Engineering
King Abdulaziz University
Jeddah
Saudi Arabia

and

Systems Research Institute, Polish Academy
 of Sciences
Warsaw
Poland

Shyi-Ming Chen
Department of Computer Science
 and Information Engineering
National Taiwan University of Science
 and Technology
Taipei
Taiwan

ISSN 1860-949X   ISSN 1860-9503   (electronic)
ISBN 978-3-319-37487-1 ISBN 978-3-319-02993-1   (eBook)
DOI 10.1007/978-3-319-02993-1
Springer Cham Heidelberg New York Dordrecht London

Printed on acid-free paper

Springer is part of Springer Science+Business Media (www.springer.com)

Preface

A social network is an architecture made up of a set of actors (such as individuals or organizations) and a web of linkages between these actors. The social network perspective provides a clear way of analyzing the structure of whole social entities and identifying essential links among various nodes of the network. The study of these structures uses social network analysis to identify local and global patterns, locate influential entities, and examine complex network dynamics.

Indisputably, social networks and their analysis form an inherently interdisciplinary academic field, which has emerged from social psychology, sociology, statistics, and graph theory and attracts a great deal of attention from computer sciences and engineering. A few lines about the history could be worth bringing here. G. Simmel authored early structural theories in sociology emphasizing the dynamics of triads and "web of group affiliations." Moreno is credited with developing the first sociograms in the 1930s to study interpersonal relationships. These approaches were mathematically formalized in the 1950s and theories and methods of social networks became pervasive in the social and behavioral sciences. Social network analysis is now one of the major paradigms in numerous disciplines. Together with other complex networks, social networks contribute to the emergence of the discipline of network science.

Generally speaking, social networks are self-organizing, emergent, and complex structures. Globally, coherent patterns appear as a result of some local interactions of the elements that make up the system. These patterns become more apparent as network size increases. However, a global network analysis of, for example, all interpersonal relationships in the world is not feasible and is likely to contain so much information that its relevance becomes questionable. Practical limitations come also with available computing power. The nuances of a local system may be lost in a large network analysis, hence the quality of information may be more important than its scale for understanding network properties. Thus, social networks are analyzed at the scale relevant to the research objectives. In other words, in one way or another, advanced analysis and synthesis environments need to engage concepts of information granularity. Although levels of analysis are not necessarily mutually exclusive, there are three general levels into which networks may fall, namely micro-level, meso-level, and macro-level.

In light of the inherent human-centric facet of social networks, the principles and practice of Computational Intelligence have been poised to play a vital role in the analysis, design, and interpretation of the architectures and functioning of social networks. As a matter of fact, it has already exhibited a visible position there. In particular, we capitalize on the important facets of learning, structural design, and interpretability along with human-centricity, where all of these aspects are vigorously supported by the leading technologies of Computational Intelligence.

Taking into account the synergy among neurocomputing, fuzzy sets, and evolutionary optimization it is worth stressing that the same synergy plays a pivotal role in the analysis and synthesis of social networks. Fuzzy sets and Granular Computing bring a highly desirable feature of transparency of models of social networks and offer insights into their dynamics and interrelationships among key functional substructures. Evolutionary optimization and population-based optimization are of relevance in the context of the design of adaptive social networks, especially when being concerned with their optimization of underlying topologies and refinements of their hierarchical structures.

The ultimate objectives of the proposed edited volume is to provide the reader with an updated, in-depth material on the conceptually appealing and practically sound information technology of Computational Intelligence in social networks analysis, synthesis, and evaluation.

The volume involves studies devoted to key issues of social networks including community structure detection in networks, online social networks, knowledge growth and evaluation, and diversity of collaboration mechanisms. The book engages a wealth of methods of Computational Intelligence along with well-known techniques of linear programming Formal Concept Analysis, machine learning, and agent modeling. Human-centricity is of paramount relevance and this facet manifests in many ways including personalized semantics, trust metric, and personal knowledge management; just to highlight a few of these aspects. The contributors to this volume report on various essential applications including cyber attacks detection, building enterprise social networks, business intelligence, and forming collaboration schemes.

An overall concise characterization of the objectives of the edited volume can be presented by identifying several focal points:

- Systematic exposure of the concepts, design methodology, and detailed algorithms. The structure of the volume adheres to the top-down strategy starting with the concepts and motivation and then proceeding with the detailed design that materializes in specific algorithms and representative applications.
- Individual chapters with clearly delineated agenda and well-defined focus and additional reading material available via carefully selected references.
- A wealth of carefully structured and organized illustrative material. The volume includes a series of brief illustrative numeric experiments, detailed schemes, and more advanced problems.
- Self-containment. The intent is to provide a material that is self-contained and, if necessary, augment some parts with a step-by-step explanation of more advanced concepts supported by a significant amount of illustrative numeric material and some application scenarios.

Given the theme of this undertaking, this book is aimed at a broad audience of researchers and practitioners. Owing to the nature of the material being covered and the way it is organized, the volume will appeal to well-established communities including those active in various disciplines in which social networks, their analysis, and optimization are of genuine relevance. Those involved in operations research, management, various branches of engineering, and economics will benefit from the exposure to the subject matter.

Considering the way in which the edited volume is structured, this book may serve as a highly useful reference material for graduate students and senior undergraduate students in courses such as those on decision-making, Internet engineering, Computational Intelligence, management, operations research, and knowledge-based systems.

We would like to take this opportunity to express our sincere thanks to the authors for sharing the results of their innovative research and delivering their insights into the area. The reviewers deserve our thanks for their constructive and timely input. We greatly appreciate the continuous support and encouragement coming from the Editor-in-Chief, Prof. Janusz Kacprzyk whose leadership and vision makes this book series a unique vehicle to disseminate the most recent, highly relevant, and far-fetching publications in the domain of Computational Intelligence and its various applications.

We hope that the readers will find this volume of genuine interest and the research reported here will help foster further progress in research, education, and numerous practical endeavors.

Witold Pedrycz
Shyi-Ming Chen

Contents

Detecting Community Structures in Networks Using a Linear-Programming Based Approach: a Review

William Y. C. Chen, Andreas Dress and Winking Q. Yu

Abstract We give an account of an approach to community-structure detection in networks using linear programming: given a finite simple graph G, we assign penalties for manipulating this graph by either deleting or adding edges, and then consider the problem of turning G, by performing these two operations, at minimal total cost into a graph that represents a *community structure*, i.e., that is a disjoint union of complete subgraphs. We show that this minimization problem can be reformulated (and solved!) in terms of a one-parameter family of linear-programming problems relative to which some kind of a "second-order phase transition" can be observed, and we demonstrate by example that this interpretation provides a viable alternative for dealing with the much studied task of detecting community structures in networks. And by reporting our discussions with a leading ecologist, we demonstrate how our approach can be used to analyse food webs and to support the elucidation of their "global" implications.

Keywords Networks · Graph · Graph manipulation · Community structure · Zachary's karate club · Food web · Modularity · GN algorithm · Linear programming (LP) · Integer linear programming (ILP) · One-parameter families of

W. Y. C. Chen
Center for Applied Mathematics, Tianjin University, Tianjin 300072,
People's Republic of China
e-mail: chenyc@tju.edu.cn

A. Dress (✉)
CAS-MPG Partner Institute and Key Lab for Computational Biology, Shanghai,
People's Republic of China
e-mail: dress@picb.ac.cn

A. Dress
Max-Planck-Institute for Mathematics in the Sciences, Leipzig, Germany

W. Q. Yu
Center for Combinatorics, Nankai University, Tianjin 300071,
People's Republic of China
e-mail: winking2002@163.com

W. Pedrycz and S.-M. Chen (eds.), *Social Networks: A Framework of Computational Intelligence*, Studies in Computational Intelligence 526, DOI: 10.1007/978-3-319-02993-1_1, © Springer International Publishing Switzerland 2014

1

LP-problems · Second-order phase transitions for parametrized LP-problems · Perturbation (of community data)

1 Introduction

The study of networks has attracted a lot of attention ever since the current network hype began with the proclamation of *scale-free* and *small-world* networks as constituting important new and universally applicable paradigms of interaction schemes observed in real-world systems, and suggesting fundamentally new basic laws governing important processes addressed in the natural and the social sciences (see e.g. [2, 5, 12, 24, 30, 33, 34] for reviews from 1998 to 2010). They are studied in many fields, from social research (e.g. the investigation of the World-Wide Web [1, 19] or patterns of scientific collaboration [20] and citation networks [27]) to ecology [35] and molecular biology (regulatory [10], protein [17], and metabolic networks [16]). There are now many thousands of papers that mention the term "network" in their title, and many more mention this term in their abstract.

Why are networks currently that popular? Apparently, this is due to the close connection between networks and dynamical systems: networks often are nothing but *snapshots* of dynamical systems while dynamical systems can often be viewed as networks *in action*. Consequently, there is some good hope that proper network analysis can help to elucidate important traits of a system's dynamics.

Standard methods of dynamical systems analysis require a lot of detailed input information about the *mechanisms* of interaction between the various agents participating in the system's activity, as well as the respective inter- and reaction rates. Given such information, a lot of detailed information about the dynamics of actual processes can then be deduced by solving the resulting (ordinary and/or partial differential) equations or by mimicking the interaction schemes by computer simulation. However, in many of the rather complex dynamical systems currently of interest in biology and, even more so, in the social sciences, such input information is simply not available.

So, what can be done if all that is (more or less) known are the system's *agents* and which agents are closely related to–or "interact" with–each other? On a *very abstract* level, such knowledge can best be described by the system's *interaction graph* consisting of

(i) A finite set V whose elements represent the single agents of the system, and

(ii) A subset E of the set $\binom{V}{2}$ of all 2-subsets $\{u, v\}$ of V consisting of those pairs $\{u, v\} \in \binom{V}{2}$ of distinct agents $u, v \in V$ that one believes to be, one way or the other, closely related to each other.

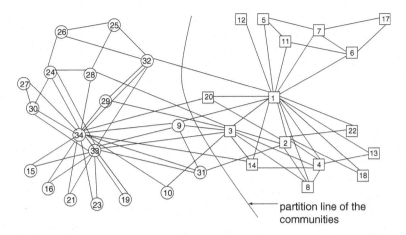

Fig. 1 Zachary's "Karate club friendship network" and the partition line separating the members of the two new clubs into which the original one broke up [36]

In other words, one represents the system's *interaction topology* in terms of a *simple graph* with *vertex* or *node set V* and *edge set E*.

A famous example of such an interaction graph is Zachary's "friendship network" [37] for which Google finds 12,90,000 quotations in about a quarter of a second (cf. Fig. 1). The data were collected in 1977 by Wayne Zachary from the members of a former university karate club. Each node represents one of the 34 members of this club, and two members of the club are connected by an edge whenever both asserted to be on friendly terms with each other. When Zachary collected the data, the club had, due to some internal struggle, broken up into two, and the partition line in Fig. 1 separates the members of the two clubs that resulted from this breakup:

Given a system's interaction graph, one currently much studied task is that of detecting this system's *community structure*, that is, of partitioning the system's agents appropriately into a collection of disjoint *communities*, with each community consisting of agents that appear to strongly interact with each other and less strongly with the agents in the other communities. For instance, community detection procedures aim for predicting the partition line from the topology of Zachary's "friendship network".

A quantitative measure of fit of a proposed community structure called "modularity" was suggested in [22] while Radicchi et al. defined "weak" and "strong" communities [26]. For other definitions, see [32] and [11]. Many other approaches and algorithms to detect "community structures" have also been proposed. The so-called "computer science approaches" (spectral bisection [25] and the Kernighan-Lin [18] algorithm) only bisect graphs. In contrast, the sociologists' key idea is essentially that of *hierarchical clustering* [29] (also called the *agglomerative* method). Recently, more efficient approaches have been found.

Girvan and Newman proposed an algorithm in 2002 [13] that has later been dubbed the "GN algorithm" [26]. Unlike the *agglomerative* algorithms, the GN algorithm begins with the entire original network, and deletes one edge at a time. It works very well for many networks, and it does not presuppose the number of communities into which a network should best be divided. Yet, its complexity is relatively high, i.e., it is $O(m^2n)$, where m is the number of edges and n is the number of vertices or "nodes" of the graph in question. Tyler et al. [31] employed the GN algorithm and gave a Monte Carlo estimate of the "betweenness parameter" that is crucial for the GN algorithm; their method seems to be much faster. Also based on deleting edges step by step, Radicchi et al. [26] proposed an algorithm in 2004 that detects the local property of each edge with complexity $O(m^4/n^2)$. And in the same year, Newman [21] proposed a rather fast algorithm based on the modularity defined in [22] whose complexity is about $O(mn)$.

Another algorithm, of complexity $O(n^3 \lg n)$, was proposed by Wu and Huberman [36]. It is motivated by properties of resistor networks and avoids edge cutting. Clauset, Newman, and Moore [8] improved the algorithm of Newman [21], and obtained an even faster algorithm, dubbed the CNM-algorithm, of still lower complexity. Reichardt et al. [28] found a method for the identification of fuzzy communities. And Bagrow and Bollt [3] developed a method of complexity $O(n^3)$ while Clauset developed an efficient method for finding local community structures [9]. Note that most of these methods require to artificially fix some parameters and, thus, do not provide just one, purely "intrinsic" result.

Here, we want to report on a proposal to deal with this task that is based on linear programming and depends on just one parameter, only, for which an apparently optimal value can be determined quite easily. More specifically, we assign penalties for manipulating a given graph G by either deleting or adding edges, and then consider the problem of turning G, by performing these two operations, at minimal total costs into a graph that *represents* a *community structure*, i.e., that is a union of disjoint complete subgraphs. We show that this minimization problem can be reformulated—and solved—in terms of a one-parameter family of integer linear-programming problems, and we demonstrate by example that this interpretation provides a viable alternative for dealing with the much studied task of detecting community structures in networks.

Our report closely follows our joint publications from 2006 [6, 7].

2 Forming Cliques

Note first that, given a finite set V, there is a canonical one-to-one correspondence between *set partitions* of V on the one hand and simple graphs with node set V that form a disjoint union of cliques (that is, that are a disjoint union of complete subgraphs) on the other. For short, any such graph will henceforth be dubbed a *target graph*.

A very straight forward fact that can therefore be used for describing community structures in terms of a simple graph had been published by Grötschel and Wakabayashi in 1989 [14] and 1990 [15], viz., the fact that there is an obvious purely local characterization of target graphs: Indeed, it was observed in these chapters that a graph $T = (V, F)$ with vertex set V and edge set $F \subseteq \binom{V}{2}$ is a target graph if and only if any two distinct vertices that are connected to a common neighbour are themselves connected by an edge, that is, if and only if $u, v, w \in V$, $\{u, v\}, \{v, w\} \in F$, and $u \neq w$ implies $\{u, w\} \in F$. Remarkably however, it seems that this simple fact that was the starting point for our investigations published in [6, 7], had been completely ignored in the context of community structure detection in all previous contributions dealing with this problem mentioned above.

Yet, it allows us to reformulate this problem as a "zero-one integer linear programming" problem as follows: Using a standard bookkeeping device, one may describe the edge set of a graph $G = (V, E)$ with vertex set V and edge set $E \subseteq \binom{V}{2}$ in terms of the associated *indicator function*, i.e. the map

$$\chi_G : \binom{V}{2} \rightarrow \{0, 1\} : \{u, v\} \mapsto \begin{cases} 1 \text{ in case } \{u, v\} \in E, \\ 0 \text{ otherwise.} \end{cases}$$

Clearly, using this function, a graph $T = (V, F)$ is easily seen to be a target network if and only if the linear inequality

$$\chi_T(uv) + \chi_T(vw) - \chi_T(uw) \leq 1 \tag{1}$$

holds, for any three distinct nodes $u, v, w \in V$, for the indicator function χ_T of T.

In other words, using the indicator function allows us to simply reformulate a geometric-combinatorial fact in purely algebraic-numerical terms.

Consequently, all that still needs to be done is to find out how, given a network $G = (V, E)$ as above, we can append and eliminate edges in a *most parsimonious* way so that the resulting network becomes a target network.

The most simple way to measure the deviation of the original network $G = (V, E)$ from a given target network $T = (V, F)$ is, of course, the total number

$$SW(G \rightarrow T) := |E - F| + |F - E|$$

of edges that need to be *switched* (i.e., of appended or eliminated) to obtain T from G, a number that is easily seen to coincide with

$$\sum_{\{u,v\} \in E} (1 - \chi_T(uv)) + \sum_{\{u,v\} \in \binom{V}{2} - E} \chi_T(uv)$$

giving rise to a *penalty function* that is apparently an affine linear function of the indicator function χ_T.

So, to obtain a target network T from the input graph G by switching *as few* of its edges as possible, we can follow the approach worked out so excellently by Grötschel and Wakabayashi, and use integer linear programming (ILP) relative to the constraints given by (1) to find an integer-valued map $\chi : \binom{V}{2} \to \{0, 1\}$ that satisfies these constraints and minimizes the penalty function $\sum_{\{u,v\}\in E} (1 - \chi(uv)) + \sum_{\{u,v\}\in\binom{V}{2}-E} \chi(uv)$ and, thus, is the indicator function of a target network T that satisfies our "least-deviation requirement" by minimizing $SW(G \to T)$.

However, ILP can also easily accommodate much more complex penalty functions: We are allowed to specify, for every 2-subset $\{u, v\} \in \binom{V}{2}$ of V, an arbitrary positive or negative number $L_{\text{a priori}}(uv)$ registering an *a priori* measure for the likelihood of the pair u, v being contained in the same community within the community structure we want to detect, and then use ILP to determine that target network T whose indicator function minimizes the resulting objective function

$$L(T) := \sum_{\{u,v\}\in\binom{V}{2}} \chi_T(uv) L_{\text{a priori}}(uv).$$

Note that the numbers $L_{\text{a priori}}(uv)$ could be derived from the overall local or global graph structure as well as from any additional information we may have been provided with. In particular, it may be tempting to experiment with the various edge parameters used in the work by Newman and others referred to above.

In our work, we have been using the "CPLEX" software package to investigate this approach, experimenting, just for a start, with a parameterized *a priori* penalty function of the form

$$L_{\text{a priori}}(uv) := \begin{cases} -s & \text{if} \{u, v\} \in E \text{ holds,} \\ 1 & \text{otherwise,} \end{cases}$$

or, favouring edges that are incident with at least one vertex of high degree (or *hub*, as vertices of high degree have also been dubbed),

$$L_{\text{a priori}}^{\text{deg}}(uv) := \begin{cases} -s(\deg_G(u) + \deg_G(v)) & \text{if} \{u, v\} \in E \text{ holds,} \\ 2(|V| - 1) - \deg_G(u) - \deg_G(v) & \text{otherwise.} \end{cases}$$

Here, $\deg_G(x)$ denotes, for any node x in a graph $G = (V, E)$, of course the number of edges that are incident with that node, and s is a positive real number that we introduced for appropriately *calibrating* our objective function allowing us to give distinct weights to deletions versus insertions (Figs. 2 and 3).

Fig. 2 The result for the karate club for $L_{\text{a priori}}^{\text{deg}}$

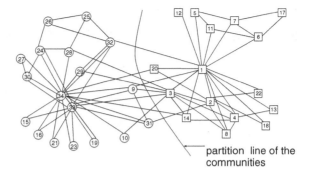

partition line of the communities

Fig. 3 The result for the karate club for $L_{\text{a priori}}$ yielding four communities

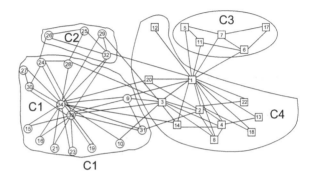

For Zachary's network, our algorithm reproduces exactly the same partition line that represents the real-world situation for the edge-dependent penalties, yet only for $s_0 := 38.8$, a much higher value than we ever expected, and it yields four communities for the edge-independent penalties for a similar value of s.

Yet, what is more remarkable, one does not even need to run the ILP algorithm for many times and different s-values to find the "correct" s-value 38.8, and even less does one have to know the "correct" result beforehand to find it. Instead, looking at many examples, we could observe that—in each and every case—some rather unexpected, yet very distinctive type of *second-order phase transition* regarding the s-parameter took place at some critical and easily recognized s-value (see also Figs. 4, 5, and 6): Increasing the control parameter s from 1 to larger and larger values, the running time of the ILP problem becomes shorter and shorter until a value, say, s^* is found for which

(i) The running time of the associated ILP problem is approximately that of the corresponding *relaxed LP problem*.[1]
(ii) The solutions of both problems coincide, i.e., the relaxed problem has an integral solution. And

---

[1] That is, the LP problem for which all requirements regarding the "integrality" of the solution vector have been dropped and only the requirement $\chi_T(uv) \in [0, 1]$ is kept.

Fig. 4 The computation time
as a function of s for
Zachary's network

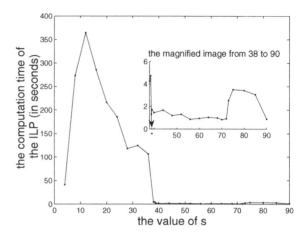

Fig. 5 The ILP/LP
computation time ratio as a
function of s for Zachary's
network

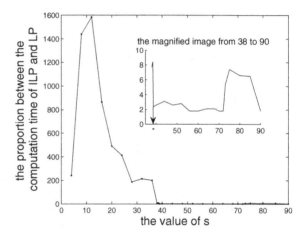

(iii) The community structure resulting in case $s = s^*$ has, so far, consistently
turned out to basically coincide with *that* community structure that is considered to be the "correct" one by the respective experts.

This is demonstrated in Figs. 4, 5, and 6 below for Zachary's network, but it has
been observed also in all other cases we investigated.

In consequence, we propose now to use the following variant of our algorithm:

(1) Start with s := 1.
(2) Run the relaxed LP program for larger and larger values of s.
(3) Stop when, for the first time, you'll find an integer solution.
(4) And accept the associated target graph as the community structure you search
    for.

Fig. 6 The ratio of the
optimal values of the
objective function obtained,
respectively, for the LP and
the ILP problem, as a
function of s for Zachary's
network

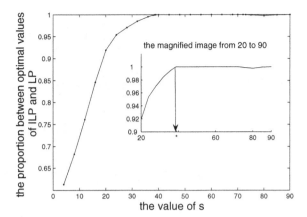

Fig. 7 The Chesapeake food
web and its three
communities obtained for
both, degree-dependent and
degree-independent penalties

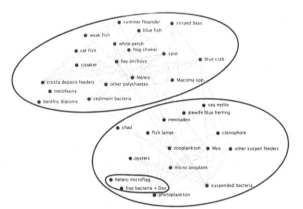

3 The Chesapeake Bay Food Web

A second example is the celebrated "Chesapeake Bay Food Web" (see [4] and [13]).

The Chesapeake Bay is a large estuary on the east coast of the United States.
The *Chesapeake Bay Food Web* relates marine organisms living in this bay. It was
compiled by *David Baird* and *Robert Ulanowicz* [4]. Following [13], we use 33
taxa that represent the ecosystem's most prominent species, from phytoplankton
and heterotrophic microflagellates to oysters, herring, white perch, blue fish, and
striped bass. Some of these species (represented by squares in Fig. 8) are *pelagic*,
i.e., they live and feed near the surface or in the open water column of coastal,
ocean, and lake waters, but not on the bottom of the sea or the lake. In contrast,
benthos live in the "benthic zone", the ecological region at the lowest level of a
body of water such as the bottom of an ocean or a lake, including the sediment
surface and even some sub-surface layers. They are represented by squares in

Fig. 8 An artistic animation
of this food web by Peter
Serocka that the PICB used
for the cover page of its
Research Report for its first
years 2005–2007

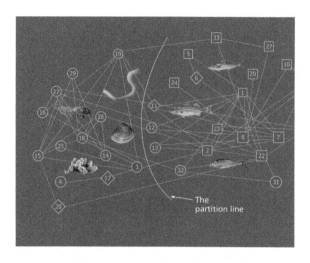

Fig. 8. The edges indicate *trophic* relationships (i.e., who eats whom). The network and our result (using both, the edge-dependent and edge-independent penalties) are represented in Fig. 7. We also compare, in Fig. 9 below, our result with that of the GN-algorithm. And we asked Robert Ulanowicz for comments. He wrote back:

Dear Andreas and Winking,

I finally had the chance to look at the groupings that you sent to me. To begin with, I noticed only a single transposition difference between your grouping and that by Girvan and Newman. Namely, you group blue crab (19) in the second group and blue fish (30) in the first. GN does the opposite.

My judgement is that both groupings are quite good, but yours probably wins out by a hair.

Please notice that the groupings are according to where the organisms "feed", not where they are located. Hence, three of the "mistakes" you indicated are not mistakes at all by my reckoning. That mya (12), oysters (13) and other suspension feeders (11) live on the bottom is only incidental. They are all filter feeders and take their nourishment from the water column.

In terms of feeding, they belong with the pelagic organisms.

On the other hand, spot (27), white perch (28), croaker (25), catfish (29) and hogchoker (26) are technically nektonic, but they all derive their nourishment primarily from the benthos and can logically be placed in the second category.

As for the single discrepancy between the two methods, I would judge that blue crabs (19) belong decidedly in the benthic feeding group, as your method detected.

Blue fish (30) feed mostly on other nekton, but ultimately derive most of their sustenance from the benthos.

In fact, in the paper I sent you on Oct. 10, we note how the indirect diet of striped bass (33) differs from that of blue fish (30) because the former derives most of its sustenance from the pelagic domain, whereas the latter comes ultimately (but not directly) from the benthos.

Hence, blue fish is a "borderline" species, and GN do not err gravely by placing it among the benthic feeders. Their bigger error is in placing blue crab (19) among the pelagic feeders.

Fig. 9 A comparison between the GN-algorithm and the LP-method applied to the Chesapeake Bay food web

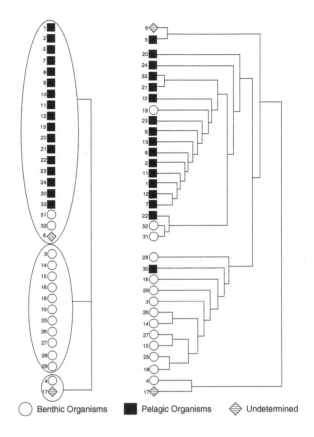

So by placing blue crab correctly among the benthic feeders, you win by a slight edge. :)
I do hope these observations have been helpful.
I am impressed by the power of your grouping algorithm.
Sincerely, Bob.

Taking into account that, when we started this work, we did not even know the terms "pelagic" and "benthic" and, therefore, had even less an idea of the—in hindsight, of course, quite obvious–ecological relevance of separating these two types of species, we were, in turn, impressed by the amount of direct and indirect information that David Baird and Robert Ulanowicz had incorporated into their Chesapeake bay food web and that could be extracted from it.

To us, it seemed also remarkable that we did not even had to use the fact that trophic relationships are, by their very nature, "directed" relationships and, therefore, would probably much better be represented by a directed graph telling us whether, in case a is connected by an edge with b, it is a who eats b or b who eats a. However, for separating the two groups of species, one apparently does not even need to know about this specific information.

4 Further Examples

We also investigated two further food webs dubbed "Marshes and Sloughs" and "Cypress" that we found on Robert Ulanowicz's project webpage. They represent two ecological systems in Florida Bay (USA).

For the first one representing trophic relationships between 63 species of animals and plants, we again used both, the degree-dependent and the degree-independent penalties, and found that, once again, both yield the same, quite satisfying result. One of the two communities we obtained gathers almost all of the species that live in the brackish water (fishes and water plants) or are directly related to the wetland (amphibians etc.) including also some birds that take species living in the wetland as food. The second community contains all the mammals that live in the much drier "upland" habitat.

For "Cypress", there are–after abandoning "Vultures" because of their isolated position in the network—63 species. The two penalty functions yield slightly distinct results, yet both identify three communities. Again, the first two communities obtained for either one of the two penalty functions gather wetland and upland species, respectively,—the latter one including most mammals. And the third one is the same for both penalty functions. It contains three species including Squirrels.

We again asked Bob and his colleague C. Bondavalli for expert comments on our result, and their judgement confirmed that, in principle, the classification according to "wetland" and "upland" habitats was right. For "Marshes and Sloughs", the only error was grouping "Lizards" with the brackwater species as they are not really related to the wetland zone and should better be put into the "upland" group. So, there is essentially only one "error" for those 63 organisms.

For "Cypress", the separation into wetland and upland habitats for both penalty functions seemed to be essentially correct. The two experts confirmed that, if we restrict our attention to the first two communities, there are three "errors" for the degree-dependent, and eight for the degree-independent penalty function. Yet, some species could be assigned to both groups, and a strict partition of the wetland and upland vegetation also seemed problematic. In addition, we are not quite sure why the third three-element community containing squirrels, vine, and cypress appeared for both types of penalty functions except that squirrels surely like to live in a habitat containing cypress trees, and vine may also prefer to grow there. For simplicity, here we only present, in Fig. 10, the result obtained for the degree-independent penalty function.

A further example can be found in [23]. In this work, Pocklington et al. studied a network containing 101 proteins and detected communities with the modified GN-algorithm [22]. There were thirteen communities in their result. Among them, three were much bigger than the others, and two of the three had unstable vertices exhibiting more external than internal connections; and four other communities were "unstable" because each exhibited more exterior than internal edges; and only the last six communities were "good".

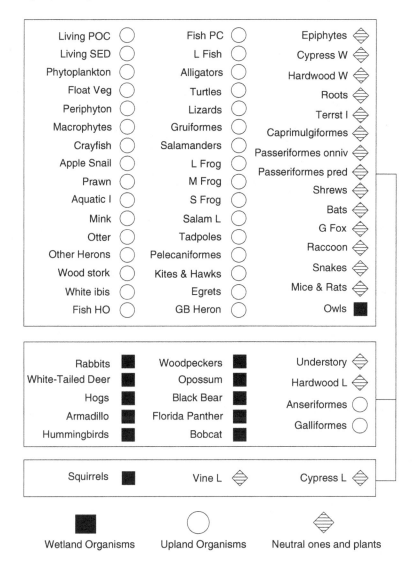

Fig. 10 The result for the Cypress food web for the degree-independent penalty function

Using the degree-dependent penalties, we detected seven communities. One of them contained the three biggest original communities and, in addition, some further proteins; five of them were, up to at most one protein, identical with five of the six good communities found by Pocklington et al. And to each of the communities we found, one could clearly assign specific biological functions as indicated in Fig. 11.

And using the degree-independent penalties, we detected six communities. Four of them were identical with communities found by Pocklington et al. while the

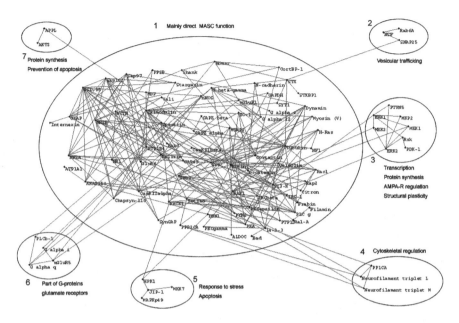

Fig. 11 The seven communities detected in Pocklington's protein network using the degree-dependent penalties, and their specific biological function

Fig. 12 A comparison between the (averaged) maintenance ratio obtained for our perturbed data using our "LP" method and the CNM algorithm

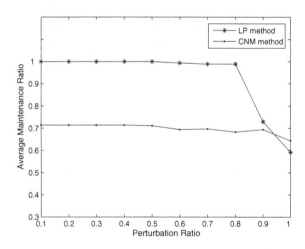

other two were unions of some of these communities. Remarkably, one could again assign specific biology functions to all of the communities we found, revealing some additional aspects of interest besides those suggested already in [23]. For further details, see [7] (Fig. 12).

5 Some Perturbation Experiments Using Artificially Created Data Sets

Finally, here are some results that we obtained studying the "reconstructability" of a given target graph $T = (V, E)$ from graphs G obtained from T by systematically *perturbing* T:

We choose the disjoint union of 4 cliques containing $12, 9, 8, 6$ nodes, respectively,—and, hence, altogether $66 + 36 + 28 + 15 = 145$ edges—as our target graph T.

To explore how well this graph is reconstructed from the perturbed graphs G, we considered randomly generated graphs $H = (V, F)$ whose edge sets were, in each instance, supposed to represent all the edges that were to be switched, and formed the graph $G = T \Delta H := (V, (E - F) \cup (F - E))$. Applying our algorithm to G, yielded a target graph $T' = (V, E')$ for which we considered the resulting *maintenance ratio* relative to T, i.e., the quotient $\frac{Min(T' \Rightarrow T)}{|V|}$ of the (minimal) number $Min(T' \Rightarrow T)$ of nodes that have to be moved from one clique to another one to obtain T from T' and the number $|V|$ of all nodes in V (which happened to be 35).

Table 1 lists the average maintenance ratio obtained for (ten) randomly chosen graphs $H = (V, F)$ for various pregiven numbers $|F|$ of switched edges or, equivalently, for various given *perturbation ratios*, that is, the quotients $|F|/|E| = |F|/145$. In addition, we used the simpler degree-independent penalty function $L_{a\ priori}$ in these investigations.

Note that a perturbation ratio of 1 does, of course, not mean that all 145 edges of T were replaced by others, but that we chose arbitrarily 145 or 24 % of the altogether $\binom{35}{2} = 595$ edges one can draw between 35 nodes.

Table 1 The perturbation and the maintenance ratio

Number of switched edges	Perturbation ratio	Maintenance ratio
15	0.103	1
29	0.200	1
44	0.303	1
51	0.352	1
53	0.366	1
55	0.379	1
58	0.400	1
73	0.503	1
87	0.600	0.99
102	0.703	1
116	0.800	0.98
131	0.903	0.63
145	1	0.36

We also compared these numbers with those obtained using the CNM algorithm:

As the LP-method turned out to having maintenance ratio almost 1 even up to a perturbation ratio of 80 % and, thus, a much higher maintenance ratio than the CNM algorithm that seemed to never go beyond 0.7, we asked Aaron Clauset for comments. On August 27, 2006, he wrote back:

> *Dear Andreas,*
>
> *If your work is going to be presented in the computer science community, it would be most correct to call our approach a heuristic; the physics community is less picky, so there, it would be an algorithm.*
>
> *In either case, it would be most correct to say that it is a very fast greedy approach to the maximum-modularity problem, and that it returns demonstrably good results for many real-world networks.*
>
> *You could also note that it is not an optimal greedy algorithm, and that it is known to return poor results for certain pathological networks that appear to be unlike those we see in the real world. It doesn't really matter how big (or small) the pathological network is, the greedy approach will always give you a bad answer.*
>
> *So, the most accurate characterization would be to say that, as the size of the network you want to cluster increases, you have progressively fewer choices of algorithms for maximizing the modularity (or any other of those nice parameters used for community-structure detection) because most of them take time at least $O(n^2)$, and we want our algorithms to return an answer after a reasonable amount of time.*
>
> *For the very largest networks, say on the order of hundreds of thousands or millions of nodes, basically only our greedy algorithm will return an answer to you within this constraint.*

So, the CNM-algorithm is clearly an algorithm that is tuned towards obtaining reasonably good results even for very large network for which, for instance, our method would completely break down, and not for producing almost optimal results for every case. It might also be that our perturbed networks are sort of "pathological" from the CNM-algorithms perspective even though they were, of course, not designed on purpose to behave like that.

More striking though seems to be that this algorithm does produce solutions with a reasonably good maintenance ratio of approximately 0.7 almost independently of the perturbation ratio even for perturbation ratios for which the LP-method begins to become rather poor.

6 Directions for Future Work

Our results lead to various problems and questions that deserve to be investigated further, and suggest several tasks that deserve to be pursued in the future:

(1) Optimizing the use of the CPLEX program (now freely available on the web),
(2) Studying a larger range of objective functions and trying to determine those that seem to be particularly *appropriate* for a specific task, including objective

functions related to edge-weighted networks and asymmetric ones representing directed networks,

(3) Trying to better understand the influence exercised by, and in particular the apparent *phase transition* behaviour of, the control parameter s,

(4) Analysing the *landscape* defined on the set of target graphs by a given (s-parametrized) objective function using stochastic models from statistical physics and, in particular, the *entropy* concept,

(5) Trying to understand also the apparent *phase transition* behaviour of the maintenance ratio relative to the perturbation ratio,

(6) Developing fast approximative variants for *large-scale* applications,

(7) Creating a data base containing the results obtained by applying the LP and other algorithms to *real-world* data gathered from the existing literature.

Acknowledgments The authors are grateful to R.Ulanowicz and C. Bondavalli for their expert comments on our results and many helpful suggestions, and to M. Briesemann for his help on the data packing.

This work was financially supported by the Max Planck Society of Germany, the 973 Project on Mathematical Mechanization, the Ministry of Education, the Ministry of Science and Technology, and the National Science Foundation of China.

References

1. Albert, R., Jeong, H., Barabási, A.-L.: Diameter of the World-Wide Web. Nature **401**, 130–131 (1999)
2. Amaral, L.A.N., Scala, A., Barthélémy, M., Stanley, H.E.: Classes of samll-world networks. Proc. Natl. Acad. Sci. U.S.A. **97**, 11149–11152 (2000)
3. Bagrow, J.P., Bollt, E.M.: Local method for detecting communities. Phys. Rev. E **72**, 046108 (2005)
4. Baird, D., Ulanowicz, R.E.: The seasonal dynamics of the Chesapeake Bay ecosystem. Ecol. Monogr. **59**, 329–364 (1989)
5. Barabási, A.-L., Albert, R.: Emergence of scaling in random networks. Science **286**, 509–512 (1999)
6. Chen, W.Y.C., Dress, A.W.M., Yu,W.Q.: Community structures of networks. In: IET Systems Biology, Proceedings Mathematics Aspects of Computer and Information Sciences (MACIS 2006), Beijing, China, and Mathematics computer science vol. 1, pp. 441–457 (2008)
7. Chen, W.Y.C., Dress, A.W.M., Yu,W.Q.: Checking the reliability of a linear-programming based approach towards detecting community structures in networks, Proceedings International Conference on Computational Systems Biology (ICCSB 2006), Shanghai, China, and IET System Biology, vol. 5, pp. 286–291 (2007)
8. Clauset. A., Newman M.E.J., Moore, C.: Finding community structure in very large networks. Phys. Rev. E. **69**, 026113 (2004)
9. Clauset. A.: Finding local community structure in networks. Phys. Rev. E. **72,** 026132 (2005)

10. Davidson, E., et al.: A genomic regulatory network for development. Science **295**, 1669–1678 (2002)
11. Flake, G.W., Lawrence, S.R., Giles, C.L., Coetzee, F.M.: Self-organization and identification of web communities. IEEE Comput. **35**, 66–71 (2002)
12. Fortunao, S.: Community detection in graphs. Phys. Rep. **486**(3–5), 75–174 (2010). doi:10.1016/j.physrep.2009.11.002
13. Girvan, M., Newman, M.E.J.: Community structure in social and biological networks. Proc. Natl. Acad. Sci. USA **99**, 7821–7826 (2002)
14. Grötschel, M., Wakabayashi, Y.: A cutting plane algorithm for a clustering problem. Math. Program. **45**, 59–96 (1989)
15. Grötschel, M., Wakabayashi, Y.: Facets of the clique partitioning polytope. Math. Program. **47**, 367–387 (1990)
16. Jeong, H., Tombor, B., Albert, R., Oltvai, Z.N., Barabási, A.-L.: The large-scale organization of metabolic networks. Nature **407**, 651–654 (2000)
17. Jeong, H., Mason, S.P., Barabási, A.-L., Oltvai, Z.N.: Lethality and centrality in protein networks. Nature **411**, 41–42 (2001)
18. Kernighan, B.W., Lin, S.: An efficient heuristic procedure for partitioning graphs. Bell Syst. Tech. J. **49**, 291–307 (1970)
19. Kleinberg, J., Lawrence, S.: The structure of the Web. Science **294**, 1849–1850 (2001)
20. Newman, M.E.: The structure of scientific collaboration networks. Proc. Natl. Acad. Sci. U.S.A. **98**, 404–409 (2001)
21. Newman, M.E.: Fast algorithm for detecting community structure in networks. Phys. Rev. E **69**, 066133 (2004)
22. Newman, M.E.J., Girvan, M.: Finding and evaluating community structure in networks. Phys. Rev. E **69**, 026113 (2004)
23. Pocklington, A., Cumiskey, M., Armstrong, J., Grant, S.: The proteomes of neurotransmitter receptor complexes form modular networks with distributed functionality underlying plasticity and behaviour. Mol. Syst. Biol. (2006). doi: 10.1038/msb4100041
24. Porter, M.A., Onnela, J.-P., Mucha, P.J.: Communities in networks. Not. Amer. Math. Soc. **56**(9), 1082–1097, 1164–1166 (2009)
25. Pothen, A., Simon, H., Liou, K.-P.: Partitioning sparse matrices with eigenvectors of graphs. SIAM J. Matrix Anal. Appl. **11**, 430–452 (1990)
26. Radicchi, F., Castellano, C., Cecconi, F., Loreto, V., Parisi, D.: Defining and idetifying communities in networks. Proc. Natl. Acad. Sci. U.S.A. **101**, 2658–2663 (2004)
27. Redner, S.: How popular is your paper? an empirical study of the citation distribution. Eur. Phys. J. B **4**, 131–134 (1998)
28. Reichardt, J., Bornholdt, S.: Detecting fuzzy community structures in complex networks with a potts model. Phys. Rev. Lett. **93**, 218701 (2004)
29. Scott, J.: Social Network Analysis: A Handbook., 2nd edn. Sage, London, (2000)
30. Strogatz, S.H.: Exploring complex networks. Nature **410**, 268–276 (2001)
31. Tyler, J.R., Wilkinson, D.M., Huberman, B.A.: Email as spectroscopy: automated discovery of community structure within organizations. In: Huysman, M., Wenger, E., Wulf. V. (eds.) Proceedings of the first international conference on communities and technologies, Kluwer, Dordrecht (2003)
32. Wasserman, S., Faust, K.: Social Network Analysis. Cambridge University Press, Cambridge (1994)
33. Watts, D.J., Strogatz, S.H.: Collective dynamics of 'small world' networks. Nature **393**, 440–442 (1998)
34. Watts, D.J.: Small Worlds. Princeton University Press, Princeton (1999)

35. Williams, R.J., Martinez, N.D.: Simple rules yield complex food webs. Nature **404**, 180–183 (2000)
36. Wu, F., Huberman, B.A.: Finding communities in linear time: a physics approach. Eur. Phys. J. B **38**, 331–338 (2004)
37. Zachary, W.W.: An information flow model for conflict and fisson in small groups. J. Anthropol. Res. **33**, 452–473 (1977)

Personalization of Social Networks: Adaptive Semantic Layer Approach

Alexander Ryjov

Abstract This work describes the idea of an adaptive semantic layer for large-scale databases, allowing to effectively handle a large amount of information. This effect is reached by providing an opportunity to search information on the basis of generalized concepts, or in other words, linguistic descriptions. These concepts are formulated by the user in natural language, and modelled by fuzzy sets, defined on the universe of the significances of the characteristics of the data base objects. After adjustment of user's concepts based on search results, we have "personalized semantics" for all terms which particular person uses for communications with data base or social networks (for example, "young person" will be different for teenager and for old person; "good restaurant" will be different for people with different income, age, etc.).

Keywords Personalization · Adaptive semantic layer · Fuzzy linguistic scales · Measure of fuzziness · Loss of information and information noise for fuzzy data

1 Motivation

Social Networks (SN) is one of the striking phenomena of the last decade. This is one of the dynamic and fast-growing segments of Information and Communications Technologies, which will largely determine the landscape of the industry in the coming decades [1]. SN is very diverse, and we do not have their universally recognized classification for now. They differ in size (ranging from mini networks—for example, corporate or project network to wide area networks such as Facebook, Linked In), the breadth of expertise (from highly specialized—for

A. Ryjov (✉)
Department of Mechanics and Mathematics, Chair MaTIS, Moscow State University, Moscow 119899, Russia
e-mail: alexander.ryjov@gmail.com

W. Pedrycz and S.-M. Chen (eds.), *Social Networks: A Framework of Computational Intelligence*, Studies in Computational Intelligence 526, DOI: 10.1007/978-3-319-02993-1_2, © Springer International Publishing Switzerland 2014

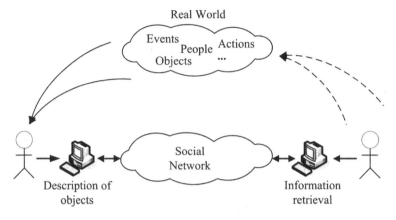

Fig. 1 SN as an information model of real world

example, the Berkeley Initiative in Soft Computing, to general ones—for example, Facebook), life cycle (writing of this book was supported narrow specialized mini networks who "will die" on the day the book will be published; Facebook can accompany a person throughout his conscious life), and many other parameters. Tasks that are important for networks of the particular type may be out of interest to other types of networks.

For sufficiently large networks (most of the participants do not know each other) important is the task of finding the "right" items—partners, restaurants, theaters, things—depending on the "specialization" of the network. Some of these objects are described by the "natural" values (for example, gender, age, price, average bill); other ones—through rubricates, classifiers, category and other tools.

SN is an information model of the real world (Fig. 1). All important from user's point of view objects from a real world are presented in SN as an "information image". We describe the real objects, search their images in SN, and use the results for operation in a real world.

From technical point of view, social networks could be interpreted as a collection of interrelated data bases. The present data processing technology only allows us to search information using concepts (words, symbols, figures) which are present in the data base descriptions of objects. This leads to difficulties in those situations where the information we need is not expressed unequivocally in the language of significances of attributes of object descriptions. The translation of such queries towards the latter search language tends to deform their meaning, and, hence, to reduce the efficiency and the quality of using these data bases.

One of the important properties of the information we need which distinguishes it from the information in the data base, is the fuzziness of the concepts of the user. The user, like any human being, thinks in qualitative categories [13-15], whereas the data base information is basically clear (sharp, non-fuzzy). This is of one of the main problems in the "translation" of user's need of information towards the data base query.

We can illustrate this with the following example.

Example 1 Choosing a car.

We consider the situation of a costumer choosing an car from a data base (electronic catalogue), which contains the following information:

- Price
- Model
- Year of issue
- Fuel consumption.

For the formulation of a request in such a data base, the user is forced to present his query in the language of concrete, definite numbers and models. If our user has in mind a very specific car, for which all the above-mentioned attributes have definite values, and if he has decided that other cars (similar, or for instance a little more economical) are not suitable, the standard technology of data bases solves all his problems. But, he wants to buy a car which is "not very expensive", "prestigious", "economical", and "not old", the formulation of such a query to the data base is not a very simple task.

A similar example can be given for the task of choosing a flat or other real estate property using a data base, containing the description of particular flats in a city. The processing of queries of the type "comfortable, not very expensive flat, close to a park, and not very far from an underground station" is thus possible. Applications of this approach for political problems (for example, monitoring and evaluation of State's nuclear activities [5], department of safeguards, International Atomic Energy Agency) have been described in [10].

In general, it can be stated that the problem described above is very important when using automated (electronic) catalogues of goods and services, i.e., data bases, containing particular information about particular objects of the same general type. Especially this way could be effective for several tasks in big data analysis—the next frontier for innovation, competition, and productivity which was introduced by McKinsey Global Institute in May 2011 [1] and was supported by leading companies [2–4].

2 The Concept of an Adaptive Semantic Layer

The structure of an adaptive semantic layer is shown in Fig. 2. Here we use data base as a simplified model of SN. The idea of the adaptive semantic layer is to provide by user an interface which allows:

- define user's concepts;
- search an information by this concepts;
- adjustment of user's concepts based on search results.

Fig. 2 The structure of an
adaptive semantic layer

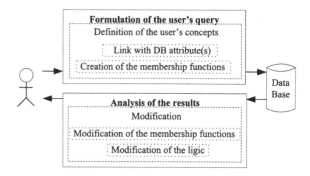

2.1 Definition of the User's Concepts

The work of user begins with a linguistic description of the objects he wants to find
in the data base. If the system does not recognise or know this linguistic
description, control is transmitted to the program block for the construction of
membership functions.[1] If, on the other hand, the system recognises the description, it will retrieve the membership function, associated with it. The user can then,
in case of disagreement, edit the membership function, and a new membership
function will now be associated with this user, reflecting his "view" (interpretation) of the description. The membership function editor is based on the principle
of cognitive graphics and does not require a specialised knowledge of computers.

We can mark out two situations: metric ($U \subseteq R^1$) and non-metric ($U \not\subseteq R^1$)
univerums. For both cases we have well-defined methods which were tested in a
number of applications of fuzzy models. We can provide the following continu-
ation of example 1:

Example 2
(a) Metric universum: $U = $ [\$10,000, \$50,000]; user's concept $A = $ "not-very-
    expensive car" is presented in Fig. 3. We can also use fuzzy clustering
    methods (for instance, Fuzzy C-Means) for building membership functions
    (Fig. 3).
(b) Non-metric universum: U is a set of all cars' models from the data base; user's
    concept B $ = $ "sport cars for city" is presented in Fig. 4. Using the same way,
    we can define "cars for hunting", "cars for farmers", "cars for young girls",
    etc.

Here in right column we have all the cars; green box is collection of cars
definitely belongs to user's concept; red box is collection of cars definitely not
belongs to user's concept; yellow box is a set of cars which partially belongs to
user's concept. We start from empty boxes (all models are in the right column) and
split all the models to these three boxes. We can order elements from yellow box

---

[1] Following [15], we associate semantics of the terms (words) with membership functions.

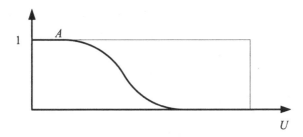

Fig. 3 Example of semantic of user's concept for metric universum

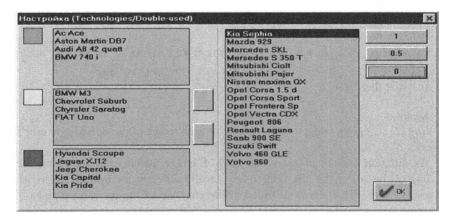

Fig. 4 Example of semantic of user's concept for non-metric universum

according to belongings to user's concept and define membership function as linear one.

After the formulation of the linguistic description, and the association of the membership functions with this description, an information search in the data base can be effected.

2.2 The Information Retrieval Algorithm

The information retrieval algorithm consists in calculating for each record in data base the degree of satisfaction to the formulated request: from 1 (total satisfaction) to 0 (total non-satisfaction). The result of the search is an ordering of the records in the data base on the basis of the degree of satisfaction to the request.

Notations:

- i ($1 \leq i \leq N$)—index of data base attributes;
- U_i—domain of attribute i;

- $A^i = \{A^i_1, \ldots, A^i_{ni}\}$—user's concepts defined on i-attribute $(ni \geq 0)$;
- $a^i_{nk}(u_i) = \mu_{A^i_{nk}}(u_i)$—membership function of nk–concept of i-attribute $(nk \leq ni, u_i \in U_i)$;
- $Q = \langle A^1_{k1} \circ A^2_{k2} \circ \cdots \circ A^N_{kN} \rangle$, where $ki \leq ni$, $A^i_{ki} \in A^i \cup \emptyset (1 \leq i \leq N)$; $\circ \in \{$and, or, not$\}$—is user's concepts based database query.

Algorithm:

1. $r = 0$ (r—index of records in database; $r \leq R$);
2. $r = r + 1$;
3. $i = 0$ (i—index of record's attribute in database);
4. $i = i + 1$;
5. if $A^i(.) \in Q$ then calculate $a^i_{nk}(u_i)$;
6. if $i < N$ then goto 4;
7. calculate $\mu_Q(r) = a^1_{k1} \bullet a^2_{k2} \bullet \cdots \bullet a^N_{kN}$, where $\bullet$ is: t-norm if "$\circ$" in Q is "and"; t-conorm if "$\circ$" in Q is "or"; $1 - a^i_{nk}(u_i)$ if "$\circ$" in Q is "not";
8. if $r < R$ then goto 2.

Result:

$\mu_Q(1), \ldots, \mu_Q(R)$—degree of belonging of each records from database to user's query.

As different classes of users can have different membership functions, the results of the search for the same query can be different for different users, or classes of users. This allows us to have different "views" on the same data base.

2.3 Adjustment of User's Concepts Based on Search Results

It is obvious enough that different users (classes of users) can have different formalization of the concepts (different membership functions). For example, concept "expensive" for student and for businessman can be different. How can we make our interface "personalized"?

In general terms, if we allow using uncertainty at the point of "entry" of the system, we have to provide tools for manipulation of uncertainty at the "output".

We can propose two ways to adjust or tune the interface. First way is adjustment of membership function, second one is tuning of t-notms and t-conorms. The following example can explain this idea.

Example 3

(a) Adjustment of the membership functions is shown in Fig. 5.
    Here *more*, *less*—directions of modification; *a bit*, *not so far*, *a far*,...—volume ("power") of modificators.
    This approach is described in [9].

Fig. 5 Adjustment of the semantic of user's concept A

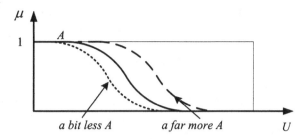

a bit less A   *a far more A*

(b) Adjustment of the logic (*t*-norms and *t*-conorms).

We can use parametric representation of *t*-norms and *t*-conorms like

$$T_\lambda(a,b) = \frac{\mu_a \times \mu_b}{\lambda + (1-\lambda)(\mu_a + \mu_b - \mu_a \times \mu_b)}$$

and use genetic algorithms for choosing the best value of λ.

This approach is described in detail in [6].

3 Optimization of Semantic Layer

It is assumed that the person describes the properties of real objects in the form of linguistic values. The subjective degree of convenience of such a description depends on the selection and the composition of such linguistic values. Let us explain this on a model example.

Example 4 Let it be required to evaluate the height of a man. Let us consider two extreme situations.

Situation 1   It is permitted to use only two values: "small" and "high".

Situation 2   It is permitted to use many values: "very small", "not very high",..., "not small and not high",..., "very high".

Situation 1 is inconvenient. In fact, for many men both the permitted values may be unsuitable and, in describing them, we select between two "bad" values.

Situation 2 is also inconvenient. In fact, in describing height of men, several of the permitted values may be suitable. We again experience a problem but now due to the fact that we are forced to select between two or more "good" values. Could a set of linguistic values be optimal in this case?

In SN, one object may be described by different persons. Therefore it is desirable to have assurances that the different participants of the SN describe one and the same object in the most "uniform" way.

On the basis of the above we may formulate the first problem as follows:

Problem 1 Is it possible, taking into account certain features of the man's perception of objects of the real world and their description, to formulate a rule for selection of the optimum set of values of characteristics on the basis of which these objects may be described? Two optimality criteria are possible:

Criterion 1. We regard as optimum those sets of values through whose use man experiences the minimum uncertainty in describing objects.

Criterion 2. If the object is described by a certain number of user's, then we regard as optimum those sets of values which provide the minimum degree of divergence of the descriptions.

This problem is studied in Sect. 3.1. It is shown that we can formulate a method of selecting the optimum set of values of qualitative indications. Moreover, it is shown that such a method is stable, i.e., the natural small errors that may occur in constructing the membership functions do not have a significant influence on the selection of the optimum set of values. The sets which are optimal according to criteria 1 and 2 coincide.

What gives us the optimal set of values of qualitative attributes for information retrieval in a SN (see Fig. 1)? In this connection the following problem arises.

Problem 2 Is it possible to define the indices of quality of information retrieval in fuzzy (linguistic) databases and to formulate a rule for the selection of such a set of linguistic values, use of which would provide the maximum indices of quality of information retrieval?

This problem is studied in Sect. 3.2. It is shown that it is possible to introduce indices of the quality of information retrieval in fuzzy (linguistic) databases and to formalize them. It is shown that it is possible to formulate a method of selecting the optimum set of values of qualitative indications which provides the maximum quality indices of information retrieval. Moreover, it is shown that such a method is also stable.

3.1 Description of Objects for Social Networks

The model of an estimating of real object's properties by a person as the procedure of measuring in Fuzzy Linguistic Scale (FLS) has been analyzed at first time in [11] and described in details in [7]. The set of scale values of some FLS is a collection of fuzzy sets defined on the same universum.

Let us consider t fuzzy variables with the names $a_1, a_2, \ldots, a_t$, specified in one universal set (Fig. 6). We shall call such set the scale values set of a FLS.

Let us introduce a system of limitations for the membership functions of the fuzzy variables comprising s_t. For the sake of simplicity, we shall designate the membership function a_j as μ_j. We shall consider that:

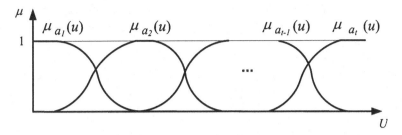

Fig. 6 The scale values set of a FLS

1. $\forall \mu_j (1 \leq j \leq t) \, \exists \, U_j^1 \neq \emptyset$, where $U_j^1 = \{u \in U : \mu_j(u) = 1\}$, U_j^1 is an interval or a point;

2. $\forall j (1 \leq j \leq t) \, \mu_j$ does not decrease on the left of U_j^1 and does not increase on the right of U_j^1 (since, according to 1, U_j^1 is an interval or a point, the concepts "on the left" and "on the right" are determined unambiguously).

   Requirements 1 and 2 are quite natural for membership functions of concepts forming a scale values set of a FLS. In fact, the first one signifies that, for any concept used in the universal set, there exists at least one object which is standard for the given concept. If there are many such standards, they are positioned in a series and are not "scattered" around the universe. The second requirement signifies that, if the objects are "similar" in the metrics sense in a universal set, they are also "similar" in the sense of FLS.

   Henceforth, we shall need to use the characteristic functions as well as the membership functions, and so we shall need to fulfil the following technical condition:

3. $\forall j (1 \leq j \leq t) \, \mu_j$ has not more than two points of discontinuity of the first kind.

   For simplicity, let us designate the requirements 1–3 as L.

   Let us also introduce a system of limitations for the sets of membership functions of fuzzy variables comprising s_t. Thus, we may consider that:

4. $\forall u \in U \, \exists \, j (1 \leq j \leq t) : \mu_j(u) > 0$;

5. $\forall u \in U \sum_{j=1}^{t} \mu_j(u) = 1$.

Requirements 4 and 5 also have quite a natural interpretation. Requirement 4, designated the *completeness* requirement, signifies that for any object from the universal set there exists at least one concept of FLS to which it may belong. This means that in our scale values set there are no "holes". Requirement 5, designated the *orthogonally* requirement, signifies that we do not permit the use of semantically similar concepts or synonyms, and we require sufficient distinction of the concepts used. Note also that this requirements is often fulfilled or not fulfilled depending on the method used for constructing the membership functions of the concepts forming the scale values set of a FLS [12].

For simplicity, we shall designate requirements 4 and 5 as G.

We shall term the FLS with scale values set consisting of fuzzy variables, the membership functions of which satisfy the requirements 1–3, and their populations the requirements 4 and 5, a *complete orthogonal FLS* and denote it *G(L)*.

As can be seen from example 4, the different FLS have a different degree of internal uncertainty. Is it possible to measure this degree of uncertainty? For complete orthogonal FLS the answer to this question is yes.

To prove this fact and derive a corresponding formula, we need to introduce a series of additional concepts.

Let there be a certain population of t membership functions $s_t \in G(L)$. Let $s_t = \{\mu_1, \mu_2, \ldots, \mu_t\}$. Let us designate the population of t characteristic functions $\hat{s}_t = \{h_1, h_2, \ldots, h_t\}$ as *the most similar population of characteristic functions*, if $\forall j (1 \leq j \leq t)$

$$h_i(u) = \begin{cases} 1, & \text{if } \mu_i(u) = 1 \\ 0, & \text{otherwise} \end{cases} \tag{1}$$

It is not difficult to see that, if the complete orthogonal FLS consists not of membership functions but of characteristic functions, then no uncertainty will arise when describing objects in it. The expert unambiguously chooses the term a_j, if the object is in the corresponding region of the universal set. Some experts describe one and the same object with one and the same term. This situation may be illustrated as follows. Let us assume that we have scales of a certain accuracy and we have the opportunity to weigh a certain material. Moreover, we have agreed that, if the weight of the material falls within a certain range, it belongs to one of the categories. Then we shall have the situation accurately described. The problem lies in the fact that for our task there are no such scales nor do we have the opportunity to weigh on them the objects of interest to us.

However we can assume that, of the two FLS, the one having the least uncertainty will be that which is most "similar" to the space consisting of the populations of characteristic functions. In mathematics, distance can be a degree of similarity. Is it possible to introduce distance among FLS? For complete orthogonal FLS it is possible.

First of all, note that the set of functions L is a subset of integrable functions on an interval, so we can enter the distance in L, for example,

$$\rho(f, g) = \int_U |f(u) - g(u)| du, f \in L, g \in L.$$

Lemma   1 Let $s_t \in G(L)$, $s_t' \in G(L)$; $s_t = \{\mu_1(u), \mu_2(u), \ldots, \mu_t(u)\}$, $s_t' = \{\mu_1'(u), \mu_2'(u), \ldots, \mu_t'(u)\}$; $\rho(f, g)$—some   distance   in L.   Then $d(s_t, s_t') = \sum_{j=1}^{t} \rho\left(\mu_j, \mu_j'\right)$ is a distance in *G(L)*.

The semantic statements formulated by us above may be formalized as follows.

Let $s_t \in G(L)$. For the measure of uncertainty of s_t we shall take the value of the functional $\xi(s_t)$, determined by the elements of $G(L)$ and assuming the values in $[0,1]$ (i.e. $\xi(s_t) : G(L) \to [0, 1]$), satisfying the following conditions (axioms):

A1. $\xi(s_t) = 0$, if s_t is a set of characteristic functions;

A2. Let $s_t, s'_{t'} \in G(L)$, t and t' may be equal or not equal to each other. Then $\xi(s_t) \leq \xi(s'_{t'})$, if $d(s_t, \hat{s}_t) \leq d(s'_{t'}, \hat{s}'_{t'})$.

(Let us recall that $\hat{s}_t$ is the set of characteristic functions determined by (1) closest to s_t.).

Do such functional exist? The answer to this question is given by the following Theorem [12].

Theorem 1 *(Theorem of existence).* Let $s_t \in G(L)$. Then the functional

$$\xi(s_t) = \frac{1}{|U|} \int_U f\left(\mu_{i_1^*}(u) - \mu_{i_2^*}(u)\right) du, \qquad (2)$$

where

$$\mu_{i_1^*}(u) = \max_{1 \leq j \leq t} \mu_j(u), \; \mu_{i_2^*}(u) = \max_{1 \leq j \leq t, j \neq i_1^*} \mu_j(u), \qquad (3)$$

f satisfies the following conditions:

F1. $f(0) = 1$, $f(1) = 0$;

F2. f does not increase—is a measure of uncertainty of st, i.e. satisfies the axioms A1 and A2.

There are many functional satisfying the conditions of Theorem 1. They are described in sufficient detail in [12]. The simplest of them is the functional in which the function f is linear. It is not difficult to see that conditions F1 and F2 are satisfied by the sole linear function $f(x) = 1 - x$. Substituting it in (2), we obtain the following simplest measure of uncertainty of the complete orthogonal FLS:

$$\xi(s_t) = \frac{1}{|U|} \int_U \left(1 - \left(\mu_{i_1^*}(u) - \mu_{i_2^*}(u)\right)\right) du, \qquad (4)$$

where $\mu_{i_1^*}(u)$, $\mu_{i_2^*}(u)$ are determined by the relations (3).

We can provide the following interpretation of (4). We consider the process of description by person of a real objects. We do not have any uncertainty in the process of a linguistic description of an object which possessing a "physical" significance of the attribute u_1 (Fig. 7).

We attribute it to term a_1 with total reliance. We can to repeat these statement about an object which have "physical" significance of attribute u_5. We choose the term a_3 for its description without fluctuations. We begin to test the difficulties of choosing of a suitable linguistic significance in the description of an object,

Fig. 7 Interpretation of
degree of fuzziness of a FLS

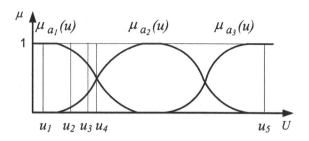

possessing the physical significance of attribute u_2. These difficulties grow (u_3) and reach the maximal significance for an objects, possessing the physical significance of attribute u_4: for such objects both linguistic significances (a_1 and a_2) are equal. If we consider the significance of the integrand function

$$\eta(s_t u) = 1 - \left(\mu_{i_1^*}(u) - \mu_{i_2^*}(u) \right)$$

in these points, we can say, that

$$0 = \eta(s_t, u_5) = \eta(s_t, u_1) < \eta(s_t, u_2) < \eta(s_t, u_3) < \eta(s_t, u_4) = 1$$

Thus, the value of the integral (4) is possible to be interpret as an average human doubts degree while describing some real object.

It is also proved that the functional has natural and good properties for fuzziness degree. In particular the following Theorems then hold [12].

Let us define the following subset of function set L:

$\bar{L}$ is a set of functions from L, which are part-linear and linear on $\bar{U} = \{u \in U : \forall j (1 \leq j \leq t)\, 0 < \mu_j(u) < 1\}$;

$\hat{L}$ is a set of functions from L, which are part-linear on U (including $\bar{U}$).

Theorem 2 Let $s_t \in G(\bar{L})$. Then $\xi(s_t) = \frac{d}{2|U|}$, where $\mathrm{d} = |\bar{U}|$.

Theorem 3 Let $s_t \in G(\hat{L})$. Then $\xi(s_t) = c\frac{d}{|U|}$, where $c < 1$, $c = \mathrm{Const}$.

Adjustment of the membership functions (Sect. 2.3) means theirs displacements to left or right according to directions of modification (*more* or *less*). What will be with our measure of uncertainty after these displacements?

Let g is some biunique function, which is defined on U. This function is induced transformation of some FLS $s_t \in G(L)$ on universum U to FLS $g(s_t)$ on universum U', where $U' = g(U) = \{u' : u' = g(u), u \in U\}$. The above induction is defined by following way: $g(s_t)$ is a set of membership functions $\{\mu_1'(u'), \mu_2'(u'), \ldots, \mu_t'(u')\}$, where $\mu_j'(u') = \mu_j'(g(u)) = \mu_j(g^{-1}(u')) = \mu_j(u), 1 \leq j \leq t$. The following example illustrates this definition.

Fig. 8 Presentation of $G^\delta(L)$

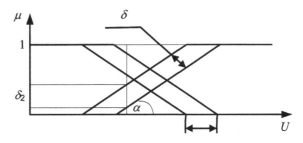

Fig. 9 FLS with minimal degrees of fuzziness

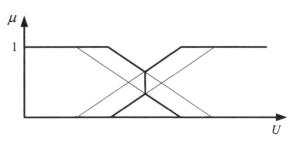

Example 5 Let $s_t \in G(L)$, U is universum s_t, and g is expansion (compression) of universum U. In this case, $g(s_t)$ is a set of functions produced from s_t by the same expansion (compression).

The following Theorem is hold [12].

Theorem 4 Let $s_t \in G(L)$, U is universum s_t, g is some linear function defined on U, and $\xi(s_t) \neq 0$. Then $\xi(s_t) = \xi(g(s_t))$.

Inasmuch as displacement is a subcase of linear function, our measure of uncertainty will be not change during adjustment of the membership functions.

Finally, we present the results of the analysis of our model, when the membership functions which are members of the given collection of fuzzy sets, are not given with absolute precision, but with some maximal inaccuracy δ (Fig. 8). Let us call this particular situation the δ-model and denote it by $G^\delta(L)$.

In this situation we can calculate the top ($\bar{\xi}(s_t)$) and the bottom ($\underline{\xi}(s_t)$) valuations of the degree of fuzziness.

FLS with minimal and maximal degrees of fuzziness for simple case $t = 2$ is shown on Figs. 9 and 10 correspondingly.

The following Theorem is hold [12].

Theorem 5 Let $s_t \in G^\delta(\bar{L})$. Then $\underline{\xi}(s_t) = \frac{d(1-\delta_2)^2}{2|U|}$, $\bar{\xi}(s_t) = \frac{d(1+2\delta_2)}{2|U|}$, where $d = |\bar{U}|$.

By comparing the results of the Theorem 2 and 5, we see that for small significances δ, the main laws of our model are preserved. Therefore, we can use our technique of estimation of the degree of fuzziness in practical tasks, since we have shown it to be stable.

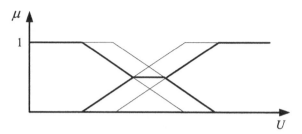

Fig. 10 FLS with maximal degrees of fuzziness

Based on these results we can propose the following method for selection of the optimum set of values of characteristics:

1. All the "reasonable" sets of linguistic values are formulated;
2. Each of such sets is represented in the form of $G(L)$;
3. For each set the measure of uncertainty (4) is calculated;
4. As the optimum set minimizing both the uncertainty in the description of objects and the degree of divergence of opinions of users we select the one, the uncertainty of which is minimal.

Following this method, we may describe objects with minimum possible uncertainty, i.e. guarantee that different users will describe the objects for SN in the most possible unified manner (see Criterion 2 in Problem 1). It means that the number of situations when one real object has more than one image in SN, or different real objects have the same image in SN, will be minimal. Accordingly, we will have a maximal possible adequacy of the SN as a model of real world from this point of view. Stability of the measure of uncertainty (Theorem 5) allows us to use this method in practical applications. We also will have an optimal set of values of attributes after adjustments (Sect. 2.3) we need for personalization (see Theorem 4).

3.2 Modeling of Information Retrieval in Social Networks

SN is an information model of the real world (Fig. 1). The quality of this model is expressed, in particular, through parameters of the information retrieval. If the database containing the linguistic descriptions of objects of a subject area allows to carry out qualitative and effective search of the relevant information then the system will work also qualitatively and effectively.

As well as in Sect. 3.1, we shall consider that the set of the linguistic meanings can be submitted as $G(L)$.

In our study of the process of information searches in data bases whose objects have a linguistic description, we introduced the concepts of loss of information ($\Pi_X(U)$) and of information noise ($H_X(U)$). These concepts apply to information searches in these data bases, whose attributes have a set of significances X, which

Fig. 11 Simple case $t = 2$

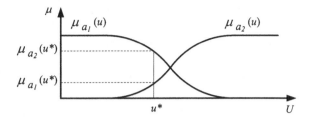

are modelled by the fuzzy sets in s_t. The meaning of these concepts can informally be described as follows [8]. While interacting with the system, a user formulates his query for objects satisfying certain linguistic characteristics, and gets an answer according to his search request. If he knows the real (not the linguistic) values of the characteristics, he would probably delete some of the objects returned by the system (information noise), and he would probably add some others from the data base, not returned by the system (information losses). Information noise and information losses have their origin in the fuzziness of the linguistic descriptions of the characteristics.

These concepts can be formalized as follows.

Let's consider the case $t = 2$ (Fig. 11). Let's fix the number $u^* \varepsilon U$ and introduce following denotes:

- $N(u^*)$ is the number of objects, the descriptions of which are stored in the data base, that possess a real (physical, not linguistic) significance equal to $N(u^*)$;
- N^{user}—the number of users of the system.

Then

- $N_{a_1}(u^*) = \mu_{a_1}(u^*)N(u^*)$—the number of data base descriptions, which have real meaning of some characteristic equal $N(u^*)$ and is described by source of information as a_1;
- $N_{a_2}(u^*) = \mu_{a_2}(u^*)N(u^*)$—the number of the objects, which are described as a_2;
- $N_{a_1}^{user}(u^*) = \mu_{a_1}(u^*)N^{user}$— the number of the system's users who believe that $N(u^*)$ is a_1;
- $N_{a_2}^{user}(u^*) = \mu_{a_2}(u^*)N^{user}$—the number of the users who believe that $N(u^*)$ is a_2.

That's why under the request "To find all objects which have a meaning of an attribute, equal a_1" (let's designate it as $\langle I(O) = a_1 \rangle$) the user gets $N_{a_1}(u^*)$ descriptions of objects with real meaning of search characteristic is equal to $N(u^*)$. Under these circumstances $N_{a_1}^{user}(u^*)$ users do not get $N_{a_2}(u^*)$ object descriptions (they carry loses). It goes about descriptions of objects which have the meaning of characteristic equal $N(u^*)$, but described by sources as a_2. By analogy the rest $N_{a_2}^{user}(u^*)$ users get noise ("unnecessary" descriptions in the volume of given $N_{a_1}(u^*)$ descriptions).

Average individual loses for users in the point $N(u^*)$ under the request are equal

$$\pi_{a_1}(u^*) = \frac{1}{N^{user}} N^{user}_{a_1}(u^*) \times N_{a_2}(u^*) = \mu_{a_1}(u^*)\mu_{a_2}(u^*)N(u^*) \tag{5}$$

By analogy average individual noises in the point $N(u^*)$

$$h_{a_1}(u^*) = \frac{1}{N^{user}} N^{user}_{a_2}(u^*) \times N_{a_1}(u^*) = \mu_{a_1}(u^*)\mu_{a_2}(u^*)N(u^*) \tag{6}$$

Average individual information loses and noises, given under analyzed request ($\Pi_{a_1}(U)$ and $H_{a_1}(U)$ accordingly) are naturally defined as

$$\Pi_{a_1}(U) = \frac{1}{|U|}\int_U \pi_{a_1}(u)du, \ H_{a_1}(U) = \frac{1}{|U|}\int_U h_{a_1}(u)du$$

It's obvious that

$$\Pi_{a_1}(u^*) = H_{a_1}(u^*) = \frac{1}{|U|}\int_U \mu_{a_1}(u)\mu_{a_2}(u)N(u)du \tag{7}$$

By analogy for the request $\langle I(O) = a_2 \rangle$ or from symmetry considerations we can get that in this case average loses and noises are equal ($\Pi_{a_2}(U) = H_{a_2}(U)$) too and are equal the right part of (7). Under information loses and noises appearing during some actions with characteristic which has the set of significance $X = \{a_1, a_2\}$ (($\Pi_X(U)$ and $H_X(U)$) and we naturally understand

$$\Pi_X(U) = p_1\Pi_{a_1}(U) + p_2\Pi_{a_2}(U), H_X(U) = p_1 H_{a_1}(U) + p_2 H_{a_2}(U),$$

where $p_i (i = 1, 2)$—the probability of some request offering in some i—meaning of the characteristic.

It's obvious that as $p_1 + p_2 = 1$, then

$$\Pi_X(U) = H_X(U) = \frac{1}{|U|}\int_U \mu_{a_1}(u)\mu_{a_2}(u)N(u)du \tag{8}$$

Let's consider general case: t—meanings of the retrieval attribute. We can generalize the formula (8) in case of t meanings of the retrieval attribute the following way [9]:

$$\Pi_X(U) = H_X(U) = \frac{1}{|U|}\sum_{j=1}^{t-1}(p_j + p_{j+1})\int_U \mu_{a_j}(u)\mu_{a_{j+1}}(u)N(u)du, \tag{9}$$

where $X = \{a_1, \ldots, a_t\}$, $p_i(i = 1, 2, \ldots, t)$—the probability of some request offering in some i—meaning of the characteristic.

Theorem 6 Let $s_t \in G\ (\bar{L})$, $N(u) = N = Const$ and $p_j = \frac{1}{t}$ $(j = 1, \ldots, t)$. Then $\Pi_X(U) = H_X(U) = \frac{ND}{3t|U|}$, where $D = |\bar{U}|$.

Corollary 1 Let the restrictions of the Theorem 6 are true. Then $\Pi_X(U) = H_X(U) = \frac{2N}{3t} \xi(s_t)$.

For proof of the Corollary is enough to compare Theorems 2 and 6.

We can generalize Corollary 1 for $s_t \in G(L)$. The following Theorem is hold.

Theorem 7 Let $s_t \in G(L)$, $N(u) = N = Const$ and $p_j = \frac{1}{t}$ $(j = 1, \ldots, t)$. Then $\Pi_X(U) = H_X(U) = \frac{c}{t} \xi(s_t)$, where c is a constant with depends from N only.

The proof of Theorems 6 and 7 are given in [9].

This Theorems showing that the volumes of losses of the information and of information noise arising by search of the information in a SN are coordinated with a degree of uncertainty of the description of objects. It means that describing objects by an optimum way (with minimization of degree of uncertainty) we provide also optimum search of the information in SN.

By analogue with Sect. 3.1, we can construct the top $\left(\overline{\Pi}_X(U)\overline{H}_X(U),)\right)$ and bottom $\left(\underline{\Pi}_X(U), \underline{H}_X(U)\right)$ valuations of the $\Pi_X(U)$ and $H_X(U)$.

The following Theorems and Corollaries are hold [9].

Theorem 8 Let $X = \{a_1, \ldots, a_t\}$, $s_t \in G^\delta(\bar{L})$, $N(u) = N = Const$ and $p_j = \frac{1}{t}$ $(j = 1, \ldots, t)$. Then

$$\underline{\Pi}_X(U) = \underline{H}_X(U) = \frac{ND(1 - \delta_2)^3}{3t|U|}, \tag{10}$$

where $D = |\bar{U}|$.

Corollary 2 Let the restrictions of the Theorem 8 are true. Then

$$\underline{\Pi}_X(U) = \underline{H}_X(U) = \frac{2N}{3t}(1 - \delta_2)\underline{\xi}(s_t) \tag{11}$$

Theorem 9 Let $X = \{a_1, \ldots, a_t\}$, $s_t \in G^\delta(\bar{L})$, $N(u) = N = Const$ and $p_j = \frac{1}{t}$ $(j = 1, \ldots, t)$. Then

$$\overline{\Pi}_X(U) = \overline{H}_X(U) = \frac{ND(1 - \delta_2)^3}{3t|U|} + \frac{2ND\delta_2}{t|U|}, \tag{12}$$

where $D = |\bar{U}|$.

Corollary 3 Let the restrictions of the theorem 9 are true. Then

$$\overline{\Pi}_X(U) = \overline{H}_X(U) = \frac{2N}{t(1+2\delta)} \left[\frac{(1-\delta_2)^3}{3} + 2\delta_2 \right] \overline{\xi}(s_t) \qquad (13)$$

By comparing the results of the Theorem 6, 8 and 9 or the Corollary 1, 2, and 3, we see that for small significances δ, the main laws of our model of information retrieval are preserved. Therefore, we can use our technique of estimation of the degree of uncertainty and our model of information retrieval in fuzzy (linguistic) data bases in practical tasks, since we have shown it to be stable.

4 Conclusion Remarks

There is no SN without social beings. They are different, because they are human beings. It means that we have to take into consideration human's perception of objects of the real world and manner of their description for SN.

Here we have focused on two important from our point of view issues:

- How we can make our SN more personal, i.e. different for different users?
- How we can make our SN more optimal, i.e. more adequate as a model of a real world we would like to operate in?

This chapter describes only general properties of SN as a model of real world. I do hope that these ideas and results will allow building a maximal comfortable SN for the participants.

Taking an opportunity, I would like to express my thanks to Professor Witold Pedrycz and Professor Shy-Ming Chen for the great idea to prepare this book, valuable remarks, and help in preparation the text.

References

1. Big data: The next frontier for innovation, competition, and productivity. McKinsey Global Institute. http://www.mckinsey.com/insights/mgi/research/technology_and_innovation/big_data_the_next_frontier_for_innovation. (May 2011)
2. IBM. Bringing Big Data to the Enterprise. http://www-01.ibm.com/software/data/bigdata/. (2011)
3. Microsoft Big Data. http://www.microsoft.com/sqlserver/en/us/solutions-technologies/business-intelligence/big-data.aspx. (2012)
4. Oracle and Big Data. Big Data for the Enterprise. http://www.oracle.com/us/technologies/big-data/index.html. (2012)
5. Ryjov, A., Belenki, A., Hooper, R., Pouchkarev, V., Fattah, A., Zadeh, L.A.: Development of an Intelligent System for Monitoring and Evaluation of Peaceful Nuclear Activities (DISNA). IAEA, STR-310, p. 122. Vienna (1998)

6. Ryjov, A., Feodorova, M.: Genetic algorithms in selection of adequate aggregation operators for information monitoring systems. In: Proceedings of V Russian conference on Neurocomputers and its Applications, Moscow, pp. 267–270, February 1999
7. Ryjov, A.: Fuzzy linguistic scales: definition. Properties and applications. In: Reznik, L., Kreinovich, V. (eds.) Soft Computing in Measurement and Information Acquisition, pp. 23–38. Springer, New York (2003)
8. Ryjov, A.: Modeling and optimization of information retrieval for Perception-Based Information. In: Zanzotto, F.; Tsumoto, S.; Taatgen, N.; Yao, Y.Y (eds.) Proceedings of Brain Informatics. International Conference, BI 2012, (Dec. 2012) DOI =http://link.springer.com/chapter/10.1007/978-3-642-35139-6_14 (2012)
9. Ryjov, A.: Models of Information Retrieval in Fuzzy Environment, p. 96. Publishing house of center of applied research, department of mechanics and mathematics, Moscow University Publishing, Moscow (2004)
10. Ryjov, A.: On application of a linguistic modeling approach in information collection for future evaluation. In: Book of Extended Synopses, International Seminar on Integrated Information Systems, Vienna, IAEA-SR-212, pp. 30–34 (2000)
11. Ryjov, A.: The degree of fuzziness of fuzzy descriptions. In: Krushinsky, L.V., Yablonsky, S.V., Lupanov, O.B. (eds.) Mathematical Cybernetics and its Application to Biology, pp. 60–77. Moscow University Publishing, Moscow (1987)
12. Ryjov, A.: The Principles of Fuzzy Set Theory and Measurement of Fuzziness, p. 116. Dialog-MSU, Moscow (1998)
13. Zadeh, L.A.: Computing with words—principal concepts and ideas. Stud. Fuzziness Soft. Comput. Springer, ISBN **277**, 978-3-642-27472-5 (2012)
14. Zadeh, L.A.: From computing with numbers to computing with words—from manipulation of measurements to manipulation of perceptions. Int. J. Appl. Math. Comput. Sci. **12**(3), 307–324 (2002)
15. Zadeh, L.A.: The concept of a linguistic variable and its application to approximate reasoning. Part 1, 2, 3. Inform.Sci. **8**, 199–249; **8**, 301–357; **9**, 43–80 (1975)

Social Network and Formal Concept Analysis

Stanislav Krajči

Abstract In this contribution we present possible using of Formal Concept Analysis, a special method of relational data analysis, in the Social Network research. Firstly, we recall basic information about the classical Ganter and Wille's version of Formal Concept Analysis, and about our one-sided version of it. Then we give information about our experiment with a social network of students from one school class. Each pupil has characterized his/her relationships to all schoolmates by value from the given range. Then we use one-sided fuzzy Formal Concept Analysis and especially modified Rice-Siff's algorithm to form clusters. In the end we interpret the results, i.e. interesting groups of students which are viewed by their schoolmates in a similar way, as groups of friends.

Keywords Formal concept analysis · Concept lattice · Fuzzy · Clustering

1 Introduction

A social network can be characterized as a system of mutual relationships between people of some group. Such network is often expressed as an (oriented) graph—the set of vertices (representing persons) which are connected (or not connected) by (oriented) edges. In the simplest case the presence of each such edge means just the existence of the relationship examined from the person corresponding to the beginning of the edge to the person at the end vertex of this edge. Note that this graph is oriented in general, because the relationship to another person need not be

S. Krajči (✉)
UPJŠ Košice, Košice, Slovakia
e-mail: stanislav.krajci@upjs.sk

W. Pedrycz and S.-M. Chen (eds.), *Social Networks: A Framework*
of Computational Intelligence, Studies in Computational Intelligence 526,
DOI: 10.1007/978-3-319-02993-1_3, © Springer International Publishing Switzerland 2014

reciprocal. This graph can be represented by a classical relation (in the mathematical sense): If B is a set meaning group of the people examined, then our relationship is some subset of $B \times B$. If this relationship is expressed in the form of a table, the set B means rows of it (as objects) and columns of it (as attributes) too. Each value in this table means evaluation of single relationship of column-person to row-person.

So this is an ideal situation for applying some relational data method to gain some hidden information in our table. In this contribution we concentrate on using *Formal Concept Analysis*. As a result we have expected some clusters of persons which are considered by other people as similar. These clusters of people can be interpreted as groups of friends which we show in our experiment with real data. Information obtained this way can be helpful to a manager of such social network to know its details and consequently manage it better. In this context we can notice e.g. the paper [12] (dealing with visualization of a social network in the form of a concept lattice) or the survey [9] (describing different data-mining techniques in the social network research).

Our paper is organized as follows: Its theoretical part gives basic information on Formal Concept Analysis: In the Sect. 2, recall the classical version of this method working in a binary (yes/no) context given in [6] and illustrated by a toy example. In the Sect. 3 the problem of non-binary values is introduced and some its solutions are pointed. One of them, called one-sided fuzzy concept lattice is described in the Sect. 4, again motivated and illustrated by a simple example. Because the number of clusters found by this method is often too huge it is (at least in our case) necessary to reduce it. One of such reduction is Rice-Siff algorithm is introduced in the Sect. 5. This is shown in general first and the it is applied to our example from the previous section. The practical part of the paper in the Sect. 6 describes our experiment on the small social network of some school class where we use Formal Concept Analysis for obtaining clusters of pupils which are naturally interpreted by a domain expert in the Sect. 7.

2 Formal Concept Analysis

In this section we briefly recall basic notions of Formal Concept Analysis, respectable data-mining method invented by Ganter and Wille [6] and based on algebraic theory of Galois connections.

As an input it need one rectangle table of data (so-called formal context), some clusters of similar objects are expected as an output. For example, let we have this toy example—a table speaking about boys' sport preferences: Each cross in the table means that the boy from the appropriate row likes the sport from the appropriate column:

	Football	Tennis	Basketball	Golf	Hockey
John		×		×	
Frank	×		×		×
Michael	×				×
Paul		×		×	
Bob			×		

In general, an *formal context* is defined as a triple $\langle A, B, R \rangle$ where A and B are non-empty sets and R is a relation on them (i.e. $R \subseteq A \times B$). Elements of A are called *attributes*, elements of B are called *objects*, and R is called their *incidency relation*.

In our example, the attributes are the sports from the set

$$A = \{\text{football, tennis, basketball, golf, hockey}\}$$

and the objects are the boys from the set

$$B = \{\text{John, Frank, Michael, Paul, Bob}\}.$$

Their incidency relation is

$$R = \{\langle \text{football, Frank} \rangle, \langle \text{football, Michael} \rangle, \langle \text{tennis, John} \rangle, \langle \text{tennis, Paul} \rangle,$$
$$\langle \text{basketball, Frank} \rangle, \langle \text{basketball, Bob} \rangle, \langle \text{golf, John} \rangle, \langle \text{golf, Paul} \rangle,$$
$$\langle \text{hockey, Frank} \rangle \langle \text{hockey, Michael} \rangle \}.$$

In this context these two natural and mutually symmetric mappings—$\nearrow$: $\mathfrak{P}(B) \to \mathfrak{P}(A)$ and $\swarrow$: $\mathfrak{P}(A) \to \mathfrak{P}(B)$ (by $\mathfrak{P}(M)$ we understand the power set of the set M)—can be defined in the following way:

- If $X \subseteq B$ then

$$X^{\nearrow} = \{a \in A : (\forall b \in X), \langle a, b \rangle \in R\}.$$

- If $Y \subseteq A$ then

$$Y^{\swarrow} = \{b \in B : (\forall a \in Y), \langle a, b \rangle \in R\}.$$

Hence, the set $X^{\nearrow}$ is the collection of all attributes common to all objects from X, for example $\{\text{Frank, Michael}\}^{\nearrow} = \{\text{football, hockey}\}$ or $\{\text{Frank}\}^{\nearrow} = \{\text{football, basketball, hockey}\}$.

Similarly, $Y^{\swarrow}$ is the set of all object sharing all attributes from the set Y, e.g. $\{\text{football, hockey}\}^{\swarrow} = \{\text{Frank, Michael}\}$ or $\{\text{football, basketball}\}^{\swarrow} = \{\text{Frank}\}$.

It is easy to see that these two mappings form a so-called *Galois connection* which is common notion for these four properties:

1a If $X_1 \subseteq X_2 \subseteq B$ then $X_1^\nearrow \supseteq X_2^\nearrow$,
1b If $Y_1 \subseteq Y_2 \subseteq A$ then $Y_1^\swarrow \supseteq Y_2^\swarrow$,
2a If $X \subseteq B$ then $X \subseteq (X^\nearrow)^\swarrow$,
2b If $Y \subseteq A$ then $Y \subseteq (Y^\swarrow)^\nearrow$,

or, equivalently, $X^\nearrow \subseteq Y$ if and only if $Y^\swarrow \subseteq X$ for each $X \subseteq B$ and $Y \subseteq A$.

Note that attributes can be understood as some conditions which are (or are not) fulfilled by the objects. Then the property 1a says a natural thing: more objects share less common fulfilled condition. The property 1b says that more conditions generated less objects which fulfill them. The property 2a emphatizes that each group of objects fulfills thir common own conditions (but maybe they are some other objects which fulfill them too) and dually for the property 2b.

These four properties easily imply that the composition cl of $\nearrow$ a $\swarrow$ fulfills the following:

1 If $X_1 \subseteq X_2 \subseteq B$ then $\mathrm{cl}(X_1) \subseteq \mathrm{cl}(X_2)$.
2 If $X \subseteq B$ then $X \subseteq \mathrm{cl}(X)$.
3 If $X \subseteq B$ then $\mathrm{cl}(\mathrm{cl}(X)) = \mathrm{cl}(X)$.

Of course, dual three properties follows for the second composition of our mappings.

In our example we have e.g. $\mathrm{cl}(\{\text{Frank}, \text{Michael}\}) = \{\text{Frank}, \text{Michael}\}$ or $\mathrm{cl}(\{\text{Michael}\}) = \{\text{Frank}, \text{Michael}\}$. Especially, $\mathrm{cl}(B) = B$ (this holds in general) and $\mathrm{cl}(\emptyset) = \emptyset$ (this does not hold always). So, there is the equality in the formula $X \subseteq \mathrm{cl}(X)$ for some X's but for other X's not. We will be interesting in the former, so-called *fixpoints* of the function cl:

If $X \subseteq B$ a $Y \subseteq A$ are such that $X^\nearrow = Y$ and $Y^\swarrow = X$ then the pair $\langle X, Y \rangle$ will be called *(formal) concept*. The set X is called the *extent* of this concept and the set Y its *intent*.

Obviously, $\langle X, Y \rangle$ is the concept if and only if $X = \mathrm{cl}(X)$ and $Y = X^\nearrow$. In our case, there are concepts $\langle \{\text{Frank}, \text{Michael}\}, \{\text{football}, \text{hockey}\} \rangle$, $\langle \emptyset, A \rangle$, and $\langle B, \emptyset \rangle$.

Concepts can be ordered in the natural way: If $\langle X_1, Y_1 \rangle$ and $\langle X_2, Y_2 \rangle$ are concepts then $\langle X_1, Y_1 \rangle \leq \langle X_2, Y_2 \rangle$ whenever $X_1 \subseteq X_2$ (or, equivalently, $Y_1 \supseteq Y_2$). The order set $\langle \{ \langle X, Y \rangle : X^\nearrow = Y, Y^\swarrow = X \}, \leq \rangle$ is called a *concept lattice* and it can be denoted by $CL(A, B, R)$.

In our case we have the following poset (all name of boys and sports are abbreviated to their initials):

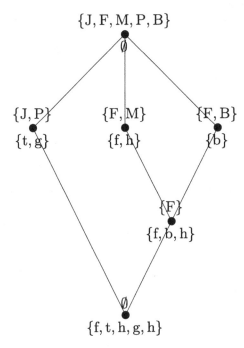

The intent {tennis, golf} of one of these concepts means that the boys from its extent, namely John and Paul likes individual sports. Similarly, the extent of the concept with the intent {football, hockey, basketball} consists in the boys which like contact sports.

As the following theorem says, a concept lattice is really a lattice, moreover complete:

Theorem 1 The Basic Theorem on Concept Lattices [6]

(1) *A concept lattice* $\mathrm{CL}(A, B, R)$ *is a complete lattice in which*

$$\bigwedge_{i \in I}, \langle X_i, Y_i \rangle = \left\langle \bigcap_{i \in I} X_i, \left(\bigcup_{i \in I} Y_i \right)^{\prime\prime} \right\rangle$$

and

$$\bigvee_{i \in I}, \langle X_i, Y_i \rangle = \left\langle \left(\bigcup_{i \in I} X_i \right)^{\prime\prime}, \bigcap_{i \in I} Y_i \right\rangle.$$

2) *A complete lattice* $\langle V, \preceq \rangle$ *is isomorphic to* $\mathrm{CL}(A, B, R)$ *if and only if there are mappings* $\alpha : A \to V$ *and* $\beta : B \to V$ *such that*

    1a. $\alpha[A]$ *je infimum-dense (i.e. each element of V is the infimum of some its subset).*

1b. $\beta[B]$ *je supremum-dense (i.e. each element of V is the supremum of some its subset).*

2. *For each $a \in A$ and $b \in B$,*

$$\alpha(a) \succeq \beta(b) \quad \text{if and only if} \quad \langle a, b \rangle \in R.$$

Moreover, $\langle V, \preceq \rangle$ is isomorphic to $\mathrm{cl}(V, V, \leq)$.

Note that input table can be transpose and then the roles of objects and attributes are interchanged. (However, what are objects and what are attributes is the matter of interpretation.) Because of mutual symmetry (or duality) of all definitions, it can be proved easily the concept lattice $CL(B, A, R^{\mathrm{T}})$ is dually isomorphic to original $CL(A, B, R)$.

If the extent of some concept is known, its intent can be computed easily (the inverse is true, of course, too). Hence, it is very often to understand the extent only as a concept. Hence, all such extents (as the subsets of B) are output clusters of this method. Moreover, they are ordered by the inclusion in some graphical structure.

3 Non-Binary Context

The classical formal concept analysis works on Yes/No relationships. But there are situations where two values do not describe an examined relationship fully, where more than two values for characterization of the quality of relationship are used. The natural question arise how to work with attributes with such non-binary values.

Some solution is given by Ganter and Wille in [6] already, namely a *scaling*: Each many-valued attribute is decomposed to its binary subattributes each of them corresponding to possible value. The problem is that the number of attributes arises very much.

There is large group of alternative solutions named fuzzy formal concept analysis. Fuzzy conceptual lattice was introduced by Umbreit-Pollandt (see e.g. [10]) and independently by Bělohlávek (e.g. [1] or [2]). They both work with fuzzy sets of objects and fuzzy sets of attributes computing the truth value of Ganter-Wille conditions in fuzzy logic. A comprehensive survey of fuzzifications of Formal Concept Analysis is done in Bělohlávek and Vychodil's paper [4].

In this contribution we use so-called *one-sided fuzzy concept lattice* given in our paper [7] which works exactly with crisp object subsets and fuzzy attribute subsets. This approach can be seen as substantially equivalent to the above-mentioned scaling or as its "quicker" version. (Note that similar approach was used by Ben Yahia and Jaoua in [5] and by Bělohlávek et al. in [3].)

4 One-Sided Fuzzy Concept Lattice

Let us recall basic notions from the one-sided fuzzy concept theory as it was stated in the [7]:

Let A and B be non-empty (finite) sets and let R be now *fuzzy relation* on their Cartesian product, i.e. a function $R : A \times B \to [0, 1]$. This relation can be understood as a table with rows and columns corresponding to objects and attributes respectively. Then the value $R(a, b)$ expresses the degree to which the object b carries the attribute a.

For (toy) example, in the following table we have relationships of our five boys to five school subjects. The greater number in data field, the better progress of the boy in the appropriate row to the subject in the appropriate column, e.g. John is excelent (1) in math but his knowledge of biology is rather bad (0.2).

	Math	Physics	Biology	English	History
John	1	0.8	0.2	0.3	0.5
Frank	0.8	1	0.2	0.6	0.9
Michael	0.2	0.3	0.2	0.3	0.4
Paul	0.4	0.7	0.1	0.2	0.3
Bob	1	0.9	0.3	0.2	0.4

Define a mapping $\uparrow: \mathfrak{P}(B) \to [0, 1]^A$ which assigns to every crisp set X of objects a fuzzy set $\uparrow(X)$ of attributes, a value in a point $a \in A$ of which is

$$\uparrow(X)(a) = \inf\{R(a, b) : b \in X\},$$

i.e. this function assigns to every attribute the greatest value s.t. all objects from X have this attribute at least in such grade. E.g.,

$\uparrow(\{\text{John, Frank, Bob}\})$

$= \{\langle\text{math}, \inf\{1, 0.8, 1\}\rangle, \langle\text{physics}, \inf\{0.8, 1, 0.9\}\rangle, \langle\text{biology}, \inf\{0.2, 0.2, 0.3\}\rangle,$

$\quad \langle\text{English}, \inf\{0.3, 0.6, 0.2\}\rangle, \langle\text{history}, \inf\{0.5, 0.9, 0.4\}\rangle\}$

$= \{\langle\text{math}, 0.8\rangle, \langle\text{physics}, 0.8\rangle, \langle\text{biology}, 0.2\rangle, \langle\text{English}, 0.2\rangle, \langle\text{history}, 0.4\rangle\}.$

Conversely, define a mapping $\downarrow: [0, 1]^A \to \mathfrak{P}(B)$, which assigns to every function $f : A \to [0, 1]$ a set

$$\downarrow(f) = \{b \in B : (\forall a \in A)R(a, b) \geq f(a)\},$$

i.e. all objects having all attributes at least in the degree set by the function f (in other words, the function of their fuzzy-membership to objects dominates over f). E.g.,

$$\downarrow(\langle \text{math}, 0.8\rangle, \langle \text{physics}, 0.8\rangle, \langle \text{biology}, 0.2\rangle, \langle \text{English}, 0.2\rangle, \langle \text{history}, 0.4\rangle)$$
$$= \{\text{John, Frank, Bob}\}$$

because all five values of John, Frank, and Bob fulfill this prescription.

It is easy to see that these mappings have the following properties:

- If $X_1 \subseteq X_2$ then $\uparrow(X_1) \geq \uparrow(X_2)$.
- If $f_1 \leq f_2$ then $\downarrow(f_1) \supseteq \downarrow(f_2)$.
- $X \subseteq \downarrow(\uparrow(X))$.
- $f \leq \uparrow(\downarrow(f))$.

By $f_1 \leq f_2$ we understand that for all elements x from a domain (common for both f_1 and f_2) is $f_1(x) \leq f_2(x)$.

These properties are equivalent to the assertion that for each $X \subseteq B$ and $f \in [0,1]^A$ holds

$$f \leq \uparrow(X) \text{ iff } X \subseteq \downarrow(f)$$

(or, equivalently, the quadruple $\langle \uparrow, \downarrow, \subseteq, \leq \rangle$ is a Galois connection).

Note that in a crisp case, i.e. if the range of an R is (at most) the two-element set $\{0,1\}$, $R(a,b) = 1$ means that the attribute a *is* an attribute of the object b, and conversely $R(a,b) = 0$ means that the a *is not* an attribute of the b. In this sense this approach is a generalization of the classical approach proposed by Ganter and Wille.

Now define a mapping cl : $\mathfrak{P}(B) \to \mathfrak{P}(B)$ as the composition of the mappings $\downarrow$ and $\uparrow$: for every $X \subseteq B$ set $\text{cl}(X) = \downarrow(\uparrow(X))$. E.g., as we have seen,

$$\text{cl}(\{\text{John, Frank, Bob}\}) = \{\text{John, Frank, Bob}\}.$$

It is easy to see from previous properties that cl is a closure operator, i.e. that the following conditions are fulfilled:

- $X \subseteq \text{cl}(X)$.
- If $X_1 \subseteq X_2$ then $\text{cl}(X_1) \subseteq \text{cl}(X_2)$.
- $\text{cl}(X) = \text{cl}(\text{cl}(X))$.

As in a crisp case an important role is played by such sets X of objects for which $X = \text{cl}(X)$ (because in such case $f = \uparrow(X)$ iff $X = \downarrow(f)$). Such pair $\langle X, \uparrow(X)\rangle$ is called a *one-sided fuzzy concept* (because X is a crisp set of objects and $\uparrow(X)$ is a fuzzy set of attributes). Then the X is the *extent* of this concept and the corresponding fuzzy set $\uparrow(X)$ is its *intent*. Because of possibility of reciprocal deriving of both coordinates of concept it can be considered only one of them, e.g. the first one. Then the set $\{X \in \mathfrak{P}(B) : X = \text{cl}(X)\}$ ordered by inclusion is a (complete) lattice, operation of which are defined as following: $X_1 \wedge X_2 = X_1 \cap X_2$ and $X_1 \vee X_2 = \mathrm{cl}(X_1 \cup X_2)$. This lattice is called a *one-sided fuzzy concept lattice*.

In our example we obtain the following lattice with 12 concepts. For simplicity, the boys and subjects are abbreviated to their initials:

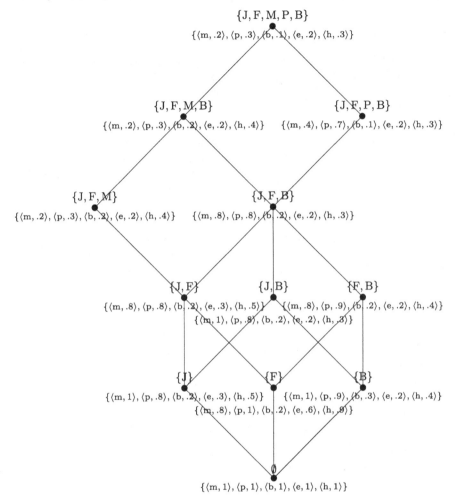

For example, the concept with the extent $\{J, B\}$ has the intent $\{\langle m, 1\rangle, \langle p, 0.8\rangle, \langle b, 0.2\rangle\langle e, 0.2\rangle, \langle h, 0.3\rangle\}$ which means that conditions are concentrate very strongly (1) to math and strongly to physics (0.8). Hence, the extent of this concept can be interpreted as the group of pupils oriented very strongly to math and rather good in physics.

If we weaken the condition to math from "very strongly" (1) to "strongly" (.8) Frank will be add to our group, so we obtain the group the concept with the extent $\{J, F, B\}$ of boys which are rather good in math and in physics.

Each of these concepts can be interpreted in similar way.

5 Rice-Siff Algorithm

The using of (crisp or fuzzy) formal concept analysis has usually one big disadvantage (similarly as other cluster methods)—a large number of clusters found. (Even our toy example 5×5 produces 12 concepts.) Therefore there are some methods for reducing this ugly property, some of them are done by Bělohlávek and his group (see e.g. [4]). They are based either on using some additional information about objects or attribute (e.g. their ordering or some other relation) or on reducing the cardinality of the value range using the so-called *truth-stressers*. But in our example we have no additional formalizable knowledge of such type about our data. Also we do not want to reduce our rather small numbers of values as that would mean that it was unnecessary to give pupils so many alternatives. (Moreover, we have tried to make an experiment in this direction when our seven possible values were naturally reduced to two as an possible answer to the question of sympathy: "+" (3, 2, and 1) and "−" (−3, −2, −1 and 0), but the number of concepts was still rather big, about 2000.)

If our clusters are to be useful for user's consideration in some way, they have to be rather small—in the number of them and/or in their cardinalities. There is one possible answer to this task, which is still based on formal concept analysis but it uses certain metric properties. It is given in [7] and we recall it shortly:

Define the function $\rho : \mathfrak{P}(B) \times \mathfrak{P}(B) \to \mathbb{R}$ in the following way: If $X_1, X_2 \subseteq B$, then

$$\rho(X_1, X_2) = 1 - \frac{\sum\limits_{a \in A} \min\{\uparrow(X_1)(a), \uparrow(X_2)(a)\}}{\sum\limits_{a \in A} \max\{\uparrow(X_1)(a), \uparrow(X_2)(a)\}}.$$

It can be proved that this function is a metric on the set $\{X \subseteq B : \mathrm{cl}(X) = X\}$ of all extents.

Now take the hierarchical clustering algorithm from [11] and simply replace their metric ρ' by our ρ. Then our algorithm can be expressed by the following pseudo-code:

```
input A, B, R
C ← D ← {cl({b}) : b ∈ B}
while (|D| > 1) do
{
    m ← min{ρ(X₁, X₂) : X₁, X₂ ∈ D, X₁ ≠ X₂}
    ε ← {⟨X₁, X₂⟩ ∈ D × D : ρ(X₁, X₂) = m}
    V ← {X ∈ D : (∃Y ∈ D)⟨X, Y⟩ ∈ ε}
    N ← {cl(X₁ ∪ X₂) : ⟨X₁, X₂⟩ ∈ ε}
    D ← (D ∖ V) ∪ N
    C ← C ∪ N
}
output C
```

The variable $\mathcal{D}$ is the set of clusters-concepts in an actual iteration, the $\mathcal{C}$ is the union of all iterations of $\mathcal{D}$ till now. The m is the minimal distance of pairs from $\mathcal{D}$. Then the set $\mathcal{E}$ contains "edges", such pairs of clusters from $\mathcal{D}$ which distance is this m and the variable $\mathcal{V}$ contains "vertices", the ends of such edges. Elements of the set $\mathcal{N}$ are new clusters. A new iteration of the $\mathcal{D}$ replaces clusters from $\mathcal{V}$ by ones from $\mathcal{N}$.

Loosely speaking, in each step we glue two clusters which are the nearest (in the sense of our metric)—throw them and add the closure of their union. Although this algorithm is exponential in principle, in the case that there is exactly one pair of the closest clusters in each step (what is clearly true for the most real data) we obtain about $2|B|$ clusters.

input: A, B, R from the table, $\overline{w} = \langle 1, 1, 1, 1, 1 \rangle$
$\mathcal{D}_0 = \{\{J\}, \{F\}, \{J, F, M\}, \{J, F, P, B\}, \{B\}\}$
$\mathcal{C}_0 = \{\{J\}, \{F\}, \{J, F, M\}, \{J, F, P, B\}, \{B\}\}$
$m_1 = \frac{2}{15}$
$\mathcal{E}_1 = \{\langle\{J\}, \{B\}\rangle, \langle\{B\}, \{J\}\rangle\}$
$\mathcal{V}_1 = \{\{J\}, \{B\}\}$
$\mathcal{N}_1 = \{\{J, B\}\}$
$\mathcal{D}_1 = \{\{J, B\}, \{F\}, \{J, F, M\}, \{J, F, P, B\}\}$
$\mathcal{C}_1 = \{\{J\}, \{F\}, \{J, F, M\}, \{J, F, P, B\}, \{B\}, \{J, B\}\}$
$m_2 = \frac{14}{37}$
$\mathcal{E}_2 = \{\langle\{J, B\}, \{F\}\rangle, \langle\{F\}, \{J, B\}\rangle\}$
$\mathcal{V}_2 = \{\{J, B\}, \{F\}\}$
$\mathcal{N}_2 = \{\{J, F, B\}\}$
$\mathcal{D}_2 = \{\{J, F, B\}, \{J, F, M\}, \{J, F, P, B\}\}$
$\mathcal{C}_2 = \{\{J\}, \{F\}, \{J, F, M\}, \{J, F, P, B\}, \{B\}, \{J, B\}, \{J, F, B\}\}$
$m_3 = \frac{7}{24}$
$\mathcal{E}_3 = \{\langle\{J, F, B\}, \{J, F, P, B\}\rangle, \langle\{J, F, P, B\}, \{J, F, B\}\rangle\}$
$\mathcal{V}_3 = \{\{J, F, B\}, \{J, F, P, B\}\}$
$\mathcal{N}_3 = \{\{J, F, P, B\}\}$
$\mathcal{D}_3 = \{\{J, F, M\}, \{J, F, P, B\}\}$
$\mathcal{C}_3 = \{\{J\}, \{F\}, \{J, F, M\}, \{J, F, P, B\}, \{B\}, \{J, B\}, \{J, F, B\}\} = \mathcal{C}_2$
$m_4 = \frac{7}{20}$
$\mathcal{E}_4 = \{\langle\{J, F, M\}, \{J, F, P, B\}\rangle, \langle\{J, F, P, B\}, \{J, F, M\}\rangle\}$
$\mathcal{V}_4 = \{\{J, F, M\}, \{J, F, P, B\}\}$
$\mathcal{N}_4 = \{\{J, F, P, B\}\}$
$\mathcal{D}_4 = \{\{J, F, M, P, B\}\}$
$\mathcal{C}_4 = \{\{J\}, \{F\}, \{J, F, M\}, \{J, F, P, B\}, \{B\}, \{J, B\}, \{J, F, B\}, \{J, F, M, P, B\}\}$
output $\mathcal{C}_4$

If values of all used variables in the k-th iteration of the algorithm are marked by the index k, we have in our example:

We have reduced the number of concepts from 12 to 8 in this way: (the extents of) the concepts $\emptyset$, $\{J, F\}$, $\{F, B\}$, and $\{J, F, M, B\}$ from our one-sided fuzzy conceptual lattice are omitted, because they are not clusters in this meaning.

6 Experiment

Our experiment from [8] was conducted in one eight-year secondary grammar school in Košice. We selected one consistent and closed group of pupils, which consists of 34 students (11 girls and 23 boys), all aged approximately 12. Their class has name *Second*, what means that they (more exactly the most of them) have known each other for more than 1 year. It is a class where our expert (and co-author of [8]) is very familiar with as she is the class teacher. Of course, she is not able fully understand all 34.33 mutual relationships between her pupils.

Each student was asked to express his/her relationship to each of his/her schoolmates by these seven values (notice that this number is in accordance with (the version of) well known Likert's scale):

- 3—he/she is my very good friend
- 2—he/she is my friend
- 1—I tend him/her positively
- 0—I tend him/her neutrally
- −1—I tend him/her negatively
- −2—I do not like him/her
- −3—I dislike him/her.

It means the values are graded, i.e. greater value means better relationship. Naturally, positive values express positive relationship, negative values negative ones, the zero is neutral. It should be said that these students have very good knowledge of meaning of negative numbers (moreover, this class specializes in mathematics), hence they have no problem to understand the task. They were promised that they will know some partial results. Therefore they cooperated seriously, in good working atmosphere, in no time stress and with no unnecessary fun.

Results are in the following table. The column of each value means the pupil who gives the evaluation, the row corresponds to the pupil who is being evaluated. It has to be said that values in the diagonal (i.e. self-evaluations of students) were uniformly changed to the maximal value (some of them used this value, some used

Pupil	1	2	3	4	5	6	7	8	9	10	11	12	13	14	15	16	17
1 Boy #1	3	−2	−1	−1	3	0	−1	0	3	0	−2	2	−2	1	1	0	0
2 Boy #2	1	3	3	0	0	0	3	3	0	−1	3	0	3	3	2	−1	2
3 Boy #3	0	3	3	2	−1	−2	3	3	−1	3	3	−1	2	3	2	−2	1
4 Boy #4	−1	−1	0	3	−1	−2	−2	2	0	0	−1	2	−1	−3	−1	−2	−2
5 Boy #5	3	0	2	−1	3	0	−3	0	2	0	0	2	0	2	2	0	1
6 Girl #1	−1	−1	−1	−1	−2	3	3	0	−1	0	3	0	−2	−1	−1	3	1
7 Boy #6	0	3	3	−1	1	0	3	3	−3	0	3	0	3	3	2	0	1
8 Boy #7	1	2	3	−1	3	0	2	3	−1	−1	3	−1	2	3	2	0	2
9 Boy #8	3	1	2	−1	3	−1	−1	−1	3	−1	−1	2	−3	−1	2	−1	2
10 Boy #9	0	0	3	−2	0	−2	0	2	−1	3	0	−1	−2	3	−1	−1	2
11 Boy #10	−1	3	3	−1	−1	0	3	3	−3	0	3	0	3	3	3	0	2
12 Boy #11	1	0	1	2	2	0	−1	0	0	0	3	3	0	−2	−1	−2	2
13 Boy #12	−1	3	3	0	−1	−2	3	3	−3	−1	3	1	3	3	2	−2	3
14 Boy #13	−1	3	2	−1	−1	−3	−1	3	−1	0	3	0	2	3	2	−2	3
15 Boy #14	−2	2	3	−1	−1	0	−1	2	0	−1	3	0	2	2	3	−1	2
16 Girl #2	−1	0	−2	0	−1	3	0	−1	−1	0	0	0	−2	−3	−1	3	3
17 Boy #15	−3	−1	0	−3	−3	−1	−1	0	−3	−1	0	0	−3	−3	3	−2	2
18 Girl #3	−1	1	−1	0	1	2	1	−1	−1	−2	3	0	1	−1	3	2	3
19 Boy #16	3	0	−1	−1	3	0	−1	−1	0	0	0	0	−1	−3	2	0	3
20 Boy #17	0	−2	−1	−1	1	0	−1	−1	−1	−1	3	3	−3	−1	2	0	2
21 Boy #18	3	1	−1	0	3	−1	−3	2	0	0	0	2	1	1	2	−1	−1
22 Girl #4	−1	0	−1	0	−1	0	−1	−1	−3	0	−2	0	−3	−3	2	0	1
23 Girl #5	−1	−1	−1	0	−1	−1	−1	−1	−2	−1	−1	0	−2	−2	2	−1	−1
24 Girl #6	−1	0	0	0	−1	3	0	0	−3	0	0	0	0	−3	2	3	2
25 Boy #19	−1	−3	−3	3	−1	−3	−3	−1	−3	−1	−3	−1	−2	−3	−3	−3	3
26 Girl #7	−1	1	−1	0	−1	1	2	0	−3	0	3	0	1	−3	3	1	3

(continued)

(continued)

Pupil	1	2	3	4	5	6	7	8	9	10	11	12	13	14	15	16	17
27	Girl #8	-2	-3	-1	0	-1	1	1	-2	0	0	-3	0	-2	-3	3	1	3
28	Boy #20	-1	-3	-2	3	0	-2	-2	-1	-3	0	-3	1	-2	-3	-1	-2	-2
29	Boy #21	0	2	3	2	0	0	3	2	1	1	3	1	1	1	3	-1	3
30	Girl #9	-1	-1	-2	0	-1	3	-3	-1	0	0	0	0	-1	-2	2	2	1
31	Boy #22	3	2	2	1	3	0	0	0	3	0	-1	3	-2	-1	2	0	2
32	Girl #10	-1	2	1	0	-1	2	2	1	0	0	3	0	-3	-2	2	1	3
33	Boy #23	1	3	3	1	-1	0	3	3	-3	-1	3	1	1	3	2	0	3
34	Girl #11	0	0	1	-2	0	2	-2	-1	0	0	2	0	-1	0	3	2	3

Pupil	18	19	20	21	22	23	24	25	26	27	28	29	30	31	32	33	34
1	Boy #1	0	3	2	2	0	-1	-1	0	-1	0	0	0	0	3	0	0	0
2	Boy #2	1	-1	0	1	0	1	0	-2	2	-1	0	2	0	1	2	3	2
3	Boy #3	0	0	1	-1	0	0	0	1	0	0	1	2	0	2	0	2	1
4	Boy #4	0	-2	1	2	0	-2	-1	3	-1	0	2	0	0	0	0	0	-2
5	Boy #5	1	3	2	-1	0	0	0	0	0	1	0	-2	0	3	0	0	1
6	Girl #1	2	-1	0	0	2	2	2	0	2	1	1	3	2	0	1	0	2
7	Boy #6	1	1	0	0	0	-1	0	0	2	1	1	1	0	-1	2	3	1
8	Boy #7	1	1	0	1	0	-1	0	2	1	1	0	2	0	2	0	3	2
9	Boy #8	1	3	1	3	0	1	0	-3	1	1	1	1	0	3	0	-2	1
10	Boy #9	0	1	1	0	0	0	-1	0	0	1	1	2	0	0	-1	1	1
11	Boy #10	1	-2	0	0	0	-1	0	-3	2	-1	0	1	0	-2	2	3	2
12	Boy #11	1	-1	3	0	0	-1	-2	1	-1	1	1	2	0	2	0	0	0
13	Boy #12	1	0	1	1	-1	0	-2	0	0	-2	0	2	0	-1	-2	3	2
14	Boy #13	0	-3	0	0	0	0	-2	0	0	0	0	0	0	0	-2	2	3
15	Boy #14	1	0	0	1	0	0	-1	-1	0	1	1	3	0	0	0	2	3
16	Girl #2	2	-1	0	0	2	-1	2	0	2	1	0	-1	2	0	2	0	2
17	Boy #15	1	0	-3	0	-1	-1	0	-3	0	0	-1	-2	-1	-3	0	0	3

(continued)

(continued)

Pupil	18	19	20	21	22	23	24	25	26	27	28	29	30	31	32	33	34
18	Girl #3	3	−1	0	0	0	2	2	0	2	3	0	1	1	2	3	−1	2
19	Boy #16	1	3	1	2	0	0	0	−3	0	1	1	2	0	3	1	−1	1
20	Boy #17	0	−2	3	−2	0	−1	−2	1	−1	1	−1	0	0	0	0	−1	−2
21	Boy #18	0	3	2	3	0	0	0	0	0	1	0	2	0	3	1	−1	0
22	Girl #4	1	−1	0	−2	3	2	1	0	2	1	0	0	1	0	1	−2	2
23	Girl #5	2	−1	0	−1	2	3	1	0	3	2	0	−2	1	0	1	−1	2
24	Girl #6	2	−1	0	0	2	2	3	0	2	2	0	−1	3	0	2	0	3
25	Boy #19	0	−3	1	−2	0	−3	−2	3	−3	−2	2	−3	−1	−1	−1	−1	−3
26	Girl #7	2	−1	0	0	2	3	1	0	3	2	0	2	1	0	1	0	2
27	Girl #8	3	−1	0	0	0	2	2	3	2	3	3	1	1	2	3	−1	1
28	Boy #20	0	−2	1	−2	0	−1	0	1	−1	0	1	−1	0	0	1	0	0
29	Boy #21	1	−1	1	1	0	1	0	1	1	2	0	3	0	2	1	2	2
30	Girl #9	2	−1	0	0	2	2	3	0	2	1	0	−1	3	0	1	0	2
31	Boy #22	1	3	2	3	0	0	0	1	0	1	1	2	0	3	1	0	0
32	Girl #10	2	−1	0	0	2	2	2	0	2	2	0	2	1	1	3	0	2
33	Boy #23	1	−1	2	0	0	1	0	0	1	2	1	2	0	−1	0	3	1
34	Girl #11	1	−1	0	0	2	2	2	−3	2	1	1	1	1	0	1	1	3

no value, a few students used lower values). All other values are without change, of course.

Our (and each) social network can be described by an oriented graph with weighted (or colored) edges and/or by the so-called *fuzzy relation*, which is nothing else than a function from $B \times B$ to the set of values (which is often ordered). In this cases it is appropriate to use some multi-valued version of the formal concept analysis. In our experiment, which is such case, we use the so-called *one-sided concept lattice*, which work with classical subsets of objects but the so-called *fuzzy subsets* of attributes.

Our relational table of data is simply an object-attribute model: The rows are the pupils, i.e. objects of our research, the columns represent relationship with schoolmates and they can be understood as attributes. It is easy to see our (psychologically useful) values from $\{-3, -2, -1, 0, 1, 2, 3\}$ can be transformed (using the function $x \mapsto \frac{x+3}{6}$ to values from $\left\{0, \frac{1}{6}, \frac{1}{3}, \frac{1}{2}, \frac{2}{3}, \frac{5}{6}, 1\right\}$ which is a subset of classical fuzzy interval $[0, 1]$. If we want to use formal concept analysis for these data, we have to think about some multi-valued version of FCA. The values in the row corresponding to the object can be understood as a function from the set of all attributes to $[0, 1]$, i.e. as a fuzzy subset of all attributes. Hence, we work with classical (crisp) subsets of all objects and with fuzzy subsets of all attributes. The application of our method had two phases:

- Firstly, we apply one-sided fuzzy formal concept analysis (from the Sect. 4). (As its interproduct we obtain about 25,000 s of concepts.)
- Secondly, we apply modified modified Rice-Siff algorithm (from the Sect. 5). In this way we reduce number of clusters to the following list. Numbers in brackets mean the step numbers. Clusters denoted by (0.*) are non-singletons which have arisen in the 0th step, they are closures of underlined objects. Singletons are omitted because they are not so interesting.

(0.1) girl #3, girl #8
(0.2) girl #6, girl #1
(0.3) girl #6, girl #9
(0.4) girl #7, girl #5
(0.5) boy #11, boy #17
(0.6) boy #20, boy #19, boy #4
(1) the same as (0.3)
(2) girl #6, girl #1, girl #9
(3) girl #6, girl #1, girl #2, girl #9
(4) boy #10, boy #6, boy #2
(5) boy #23, boy #7

(6) the same as (4)
(7) girl #3, girl #10
(8) boy #18, boy #5
(9) boy #21, boy #14
(10) the same as (0.4)
(11) boy #8, boy #16
(12) girl #7, girl #10, girl #5, girl #4
(13) boy #18, boy #5, boy #22
(14) girl #3, girl #10, girl #8
(15) boy #23, boy #7, boy #3
(16) boy #8, boy #16, boy #18, boy #5, boy #22
(17) boy #12, boy #23, boy #7, boy #3
(18) boy #12, boy #23, boy #7, boy #3, boy #13
(19) girl #6, girl #7, girl #1, girl #10, girl #2, girl #5, girl #4, girl #9
(20) boy #20, boy #4
(21) girl #11, boy #14, boy #21
(22) boy #10, boy #12, boy #23, boy #7, boy #3, boy #13, boy #6, boy #2
(23) boy #9, girl #11, boy #7, boy #14, boy #21
(24) the same as (0.6)
(25) boy #9, girl #11, boy #11, boy #7, boy #14, boy #21
(26) girl #3, girl #6, girl #7, girl #1, girl #10, girl #2, girl #8, girl #5, girl #4, girl #9
(27) boy #9, girl #11, boy #8, boy #11, girl #10, boy #16, boy #7, boy #18, boy #5, boy #22, boy #14, boy #21
(28) boy #9, girl #3, girl #11, girl #6, girl #7, boy #8, boy #11, girl #1, girl #10, girl #2, girl #8, girl #5, boy #23, boy #16, boy #7, boy #18, boy #5, girl #4, boy #22, boy #14, boy #21, girl #9
(29) boy #10, boy #1, boy #12, boy #23, boy #7, boy #3, boy #13, boy #6, boy #14, boy #2, boy #21
(30) boy #9, girl #3, girl #11, girl #6, girl #7, boy #8, boy #11, girl #1, girl #10, girl #2, girl #8, girl #5, boy #23, boy #16, boy #7, boy #18, boy #5, boy #17, girl #4, boy #22, boy #14, boy #21, girl #9
(31) boy #9, girl #10, boy #10, boy #1, boy #12, boy #23, boy #7, boy #3, boy #13, boy #18, boy #5, boy #6, boy #14, boy #2, boy #15, boy #21
(32) all pupils but boy #10, boy #1, boy #12, boy #13, boy #6, boy #15
(33) all pupils.

The following diagram express these clusters (except the last three). In the left side there are all objects, to right they are glued to bigger and bigger groups. The dark-gray color is used for boys and boy-groups, light-gray color is for girls and girl-groups. Mixed groups are white.

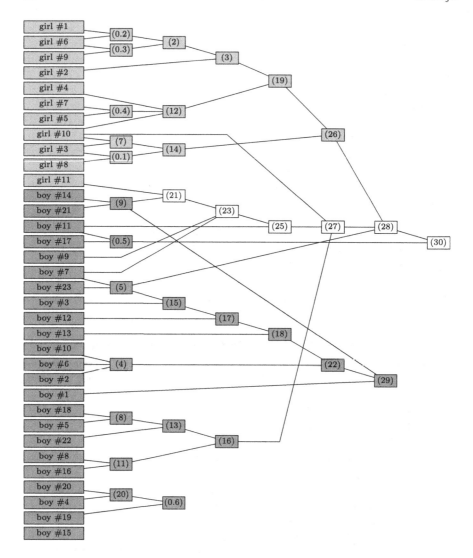

7 Interpretation

Our interpretation is based on the well-known observation expressed by the Slovak proverb "Vrana k vrane sadá." (lit. "A crow sits beside a crow."; i.e. "Birds of feather flock together."), i.e. that people with similar characteristics are friends (if they have enough time and opportunity to make friends, which is clearly true in our case). So we can deduce with a great degree of certainty that clusters based on evaluations provided by schoolmates who know them well, consist of persons with good relationships, the smaller group the stronger friendship. Hence we can valuate our result clusters from this point of view. Let us say a few notes:

- The most visible feature of our clusters is the rather strict division to boys and girls (there are only 6 mix-gender clusters, 3 of them are drawn in the diagram). This observation corresponds to a great extent to the commonly observed behavior of children aged 12. The only exception is girl #11, the girl with boyish interests like the karate.
- A very interesting group is (0.6), which consist of three boys who are underrated because they have two common features which are sensed by their schoolmates rather negatively, namely relatively worse social conditions in combination with not very good school progress.
- The cluster (31) (not shown in the diagram) is interesting too. Besides many (although not all) boys it contains the only girl which is treated as the beauty of the class. Moreover she is very friendly and, as we can see from the diagram, she is connected to two small groups of girls.
- boy #15 is the class exhibitionist. He often says that he wants to be the center of attention of schoolmates and teachers. This is probably the reason why he is not popular. He is in the last three clusters (which are not depicted) only.
- It is an old habit in schools in Slovakia (and in other central Europe countries too) that pupils sit at the tables in pairs. This year, the class teacher has given the students in question the freedom to make a pair with the person of their choice. Some of them were rather stable, some of them slightly changed in time: It is interesting that our research revealed the most of them, as we can see from the diagram:

  - girl #1 & girl #2
  - girl #6 & girl #9
  - girl #7 & girl #5
  - girl #3 & girl #8
  - girl #10 & girl #4
  - boy #14 & boy #21
  - boy #11 & boy #17
  - boy #7 & boy #23
  - boy #10 & boy #2
  - boy #4 & boy #19.

The interesting case is the following quadruple. Original pairs were boy #8 & boy #16 and boy #18 & boy #5 as it is visible in the diagram. But they were too noisy, we re-arranged them to the configuration boy #8 & boy #18 and boy #16 & boy #5.

The remaining pairs girl #11 & boy #15 (above-mentioned two pupils), boy #12 & boy #13, boy #22 & boy #1, boy #9 & boy #3 and boy #6 & boy #20 were not revealed.

All this result clusters were verify by a school teacher which can be understood as a domain expert in this environment.

Of course, the natural question arises whether there are some other interesting groups of these pupils. We do not know about any, but it does not mean that there are not some. Maybe they can be found by some another method.

8 Conclusion

In this contribution we offer (certain modification of) Formal Concept Analysis as a data-mining method which can be useful in a social network research. We have presented the experiment on a class of children of a certain Slovak secondary grammar school. After the evaluation of their mutual relationships we consider the social network of this class as the object-attribute model in order to apply the method of formal concept analysis, namely its one-sided fuzzy version to gain interesting group of pupils. Because the number of the (classical crisp and/or one-sided fuzzy) concepts was very large and therefore unusable in practice, to reduce of this amount we have tried to use modified Rice-Siff's method which makes hierarchy of some of concepts also based on their metric properties. We have obtained a reasonable number of concepts and it has shown surprisingly that the most of the clusters obtained were really interesting and relatively easy to interpret.

Of course, this method can be applied to an arbitrary relatively closed and small group of people where everyone knows everyone else. After our experiment we are convinced that such application of the formal concept analysis can help a manager of group (e.g. a class teacher) to understand better the structure of the managed social network.

Let us note that our interpretation deliberately uses a common speech of ordinary people which manage some group (like school teachers or firm managers). They usually are not academically educated and specialized psychologists or sociologists but they have a strong psychological and/or sociological intuition. We hope that presented method can help them discover interesting groups of managed people, without the need of naming them explicitly by some scientific terms. However, it will be interesting to express such observations by the scientific terminology of the psychology and/or sociology.

Acknowledgments This work was partly supported by grant VEGA 1/0832/12, by the Slovak Research and Development Agency under contract APVV-0035-10 "Algorithms, Automata, and Discrete Data Structures", and by the Agency of the Slovak Ministry of Education for Structural Funds of the EU under project ITMS: 26220120007.

References

1. Bělohlávek, R.: Fuzzy Galois connections. Math. Logic Quarterly 45(4), 497–504 (1999)
2. Bělohlávek, R.: Concept lattices and order in fuzzy logic. Ann. Pure Appl. Logic 128(1–3), 277–298 (2004)

3. Bělohlávek, R., Sklenář, V., Zacpal, J.: Crisply generated fuzzy concepts. In: Ganter, B., Godin, R. (eds.) ICFCA 2005, Lecture Notes in Computer Science, vol. 3403, pp. 268–283. Springer, Berlin (2005)
4. Bělohlávek, R. Vychodil, V.: Reducing the size of fuzzy concept lattices by hedges. In: FUZZ-IEEE 2005, The IEEE International Conference on Fuzzy Systems, May 22–25, 2005, Reno (Nevada, USA), pp. 663–668 (proceedings on CD), Abstract in Printed Proceedings, p. 44, ISBN 0-7803-9158-6
5. Yahia, S.B., Jaoua, A.: Discovering knowledge from fuzzy concept lattice. In: Kandel, A., Last, M., Bunke, H. (eds.) Data Mining and Computational Intelligence, pp. 169–190. Physica, Heidelberg (2001)
6. Ganter, B. Wille, R.: Formal Concept Analysis, Mathematical Foundation. Springer, Berlin (1999), ISBN 3-540-62771-5
7. Krajči, S.: Cluster based efficient generation of fuzzy concepts. Neural Netw. World **13**(5), 521–530 (2003)
8. Krajči, S., Krajčiová, J.: Social network and one-sided fuzzy concept lattices. In: FUZZ-IEEE 2007, The IEEE International Conference on Fuzzy Systems, July 23–26, 2007, London, pp. 222–227 (proceedings on CD), IEEE Catalog Number: 07CH37904C, ISBN: 1-4244-1210-2, ISSN: 1098-7584
9. Nettleton, D. F.: Data Mining of Social Networks Represented as Graphs, vol. 7, p. 1–34. Computer Science Review, Elsevier, (2013)
10. Pollandt, S.: Datenanalyse mit Fuzzy-Begriffen. In: Stumme, G., Wille, R. (eds.) Begriffliche Wissensverarbeitung. Methoden und Anwendungen, pp. 72–98. Springer, Heidelberg (2000)
11. Rice, M.D., Siff, M.: Clusters, concepts and pseudometrics. Electr. Notes Theor. Comput. Sci. **40**, 323–346 (2001)
12. Snášel, V., Horák, Z., Abraham, A: Understanding social networks using formal concept analysis. In: WI-IAT '08 Proceedings of the 2008 IEEE/WIC/ACM International Conference on Web Intelligence and Intelligent Agent Technology, vol. 03 pp. 390–393

Explaining Variation in State Involvement in Cyber Attacks: A Social Network Approach

Eric N. Fischer, Ciprianna M. Dudding, Tyler J. Engel,
Matthew A. Reynolds, Mark J. Wierman, John N. Mordeson
and Terry D. Clark

Abstract Cyber attacks pose an increasing threat to the international system. However, not all states are involved in cyber attacks to the same degree. This chapter asks how we might best explain variation in the involvement of states in cyber attacks across the international system. Conceiving the inter-state system as a social network, we hypothesize that as the global interconnectedness of a state increases so will its involvement in cyber attacks. Our regression models provide support for the contention that the more connected a state is to other states in the international system, the more likely it is to be involved in cyber attacks.

Keywords Network · Cyber attack · Connectivity · Social network analysis

1 Cyber Attacks

Cyber attacks pose a significant threat to an interconnected global system that is increasingly dependent on computer technology. Such attacks can alter systems to perform functions that interfere with their intended operations or they can alter data and algorithms [1]. Given the vulnerability of states and international business, it is not surprising that scholars of national security have devoted considerable attention to the phenomenon. Thus far, three key issues have emerged in this literature. The first revolves around the question of just what constitutes a cyber attack [1, 2]. An associated set of questions address whether it is possible to distinguish between cyber attacks and cyber war, whether a cyber attack must involve a state sponsor to be considered a national security threat, and whether a cyber attack must originate from another state. A second issue in the literature is

E. N. Fischer · C. M. Dudding · T. J. Engel · M. A. Reynolds · M. J. Wierman (✉) ·
J. N. Mordeson · T. D. Clark
Creighton University, Omaha, USA
e-mail: mwierman@creighton.edu

W. Pedrycz and S.-M. Chen (eds.), *Social Networks: A Framework*
of Computational Intelligence, Studies in Computational Intelligence 526,
DOI: 10.1007/978-3-319-02993-1_4, © Springer International Publishing Switzerland 2014

the nature and degree of the threat that cyber attacks pose to military command and control systems, e-mail servers, oil refineries, air traffic control systems, rail traffic, networks of financial systems, the power grid, and space infrastructure (see [2]). Finally, the third issue focuses on deterrence. A key debate here is whether cyber attacks *can* be deterred [3–7]. One of the major problems that plagues prospects for cyber deterrence is attribution. Botnets make it possible for cyber attacks to originate from anywhere in the world, including from inside the state subject to the attack. In the absence of contextual attribution, this has significantly complicated the use of computer forensics to trace a cyber attack to its source [6]. Another problem that complicates deterrence is how to mount a credible response to cyber attack.

We are not aware of any study as yet that has sought to explain the factors that make a state more likely to be involved in cyber attacks. It is clear that not all states are as likely to be the source or target of a cyber attack. Indeed, many states are only minimally involved. This chapter seeks to isolate the factors that explain the variation in the involvement of nation states in cyber attacks. Explaining the variation is essential to any international attempt to deter or defend against their occurrence.

We hypothesize that *an increase in the interconnectedness of a state in the international system will lead to an increase in the likelihood that the state will be involved in cyber attacks.* We develop a social network of the international system. Based on the network, we derive four measures of a state's interconnectedness. We test the association of these four measures with the frequency of a state's involvement in cyber attacks. Our results argue that the more strongly embedded a state is in the international system, the more likely it is that the state will be subject to or the source of cyber attacks. Thus, core states in the international system are more vulnerable to cyber attacks.

2 The Global Social Network

Scholars of national security or international relations have traditionally represented the international system in geographic terms or in reference to the structure of the distribution of power. We re-conceptualize global relations as defined by exchange relations between states. Five such relations stand out in terms of their utility for defining inter-state relations: ambassadorial exchange, direct air flights between states, the volume of telephone traffic between states, the presence of a submarine cable linking states, and the direct export of electricity from one state to another. Collectively, these five measures of exchange capture the strength of the ties between states.

Using these five factors of exchange between states, we develop a social network defining the international system. The social network consists of a series of nodes representing states and edges representing the strength of the ties between them. The edges constitute paths along which both goods and information flow internationally. Moreover, states can be identified in terms of their degree of interconnectedness within the international system of exchange as a whole.

The world network is a constructed from five 196 × 196 adjacency matrices of (1) diplomatic exchange [8] (2) direct air travel routes [9] (3) telephone minutes exchanged [10], (4) submarine cables [11], and (5) electricity exports [12–21]. The five adjacency matrices are directional, meaning that the movement of any exchange is from country i to country j. The tie may or may not be reciprocated based on the nature of the exchange. Moreover, the five adjacency matrices are binary, with a one representing a direct connection from country i to country j, and a zero representing no connection from country i to country j on each of the measures.

For the diplomatic exchange matrix, a one is assigned for countries that post ambassadors or high consulates in another country. The matrix is directed because not all ties are reciprocal. Thus if country i has an ambassador in country j, cell (i, j) is coded one. However, if country j has not reciprocated by posting an ambassador in country i, then cell (j, i) is coded zero. For the air traffic matrix, two countries are determined to have a connection if there is a direct flight between them. The matrix is bi-directional, meaning that two countries share a mutual edge that represents an exchange (a direct air route) which goes in both directions. Thus, the matrix is symmetric. That is, the score for cell (i, j) and that for cell (j, i) are identical. The telephone use matrix has a minimum threshold of 500-million telephone minutes between two countries. If two countries meet this threshold, a one is placed in the corresponding cell. The matrix is symmetrical. The electricity matrix is directed, and the adjacency matrix is non-symmetric. A one is marked in cell (i, j) if country i exports electricity directly to country j, otherwise the cell is coded zero. The submarine cable matrix is bi-directional. Cells (i, j) and (j, i) are coded one if countries i and j share a direct submarine cable.

We add the five binary matrices to create a single matrix in which the values in the cells of the resulting matrix range from zero to five. These values provide us with an ordinal measure of the degree of connectedness for each dyad. Using UCINet [22], we construct a global social network from this weighted matrix. The resulting network has 196 nodes (state actors) and 9,075 edges connecting them. The descriptive statistics for the network are in Table 1.

Density is a measure of the dyadic connections in a population. It is calculated by dividing the number of edges in a network by the total number of possible edges [23]. The density of the network is 0.237 which indicates that the network has 23.7 % of the possible edges. Distance is the length of the shortest path between nodes. The mean distance for the network is 1.796, while the standard deviation of the distance is 0.481. This indicates that nearly all of the nodes have distances between one and three. Diameter is the largest distance in path length between any two nodes in the network [23]. The diameter of the network is four. Fragmentation is the proportion of pairs of nodes that cannot reach each other by any path [23]. The fragmentation of the network is 0.005 which indicates that only 0.5 % of nodes cannot reach another given node. The clustering coefficient is 1.18. This measures the density of a node's local neighborhood, which consists of all

Table 1 Descriptive statistics of the global social network

Density	0.237
Average degree	46.306
Average distance	195
Standard deviation of the distance	0.481
Diameter	4.00
Fragmentation	0.005
Clustering coefficient	1.18
Transitivity	0.743
Reciprocity	0.7199

Source Calculated by the authors using UCINet 6.0

nodes directly connected to a given node. Since the clustering coefficient is greater than 1, this indicates that neighborhoods with weighted connections higher than 1 tend to be clustered more often than those connections at 1. Similarly, the transitivity of the network measures the number of triads connecting nodes (i, j). The transitivity of this network is 0.743, indicating that 74.3 % of nodes in the network complete a triad between any two given nodes in the network. The reciprocity of the network is 0. 7199, meaning that about 72 % of nodes share a reciprocated relationship. Based on its summary statistics the global social network is relatively well connected.

Using the statistical program R, we construct a dendrogram from the weighted matrix. The dendrogram, at Fig. 1, places countries into clusters using an algorithm that determines how structurally similar or dissimilar nodes are to one another. The clusters are based on structural equivalence. Two nodes are structurally equivalent if they have identical ties with themselves and with all other nodes. Using hamming distance, the number of structural differences between each node's connections is calculated. Clusters are formed based on the number of dissimilar node connections. For example, two nodes with all links exactly the same except for one would have a hamming distance of one. Using a hierarchical ordering, countries are placed into clusters in which they are structurally similar within the dendrogram. On the basis of the dendrogram we define six clusters of states. Each level down on the dendrogram indicates the level of connectedness that separates clusters. The most clearly defined and distinct sets of connected states are those marked in red brackets in Fig. 1. There are six such clusters of states.

The clusters with high centrality scores, such as cluster 1, are read from the lowest levels up. Thus, countries placing in the lowest positions in the cluster are the most central (connected) to the cluster. Conversely, clusters with low centrality measures, such as cluster 4, have the least connected countries placed lowest in the cluster.

Figure 2 represents the global relations in terms of the six clusters identified from the dendrogram. The clusters show the strength of relationship to each other by the thickness of the lines connecting them. The numbers along the lines between clusters indicate their shared density.

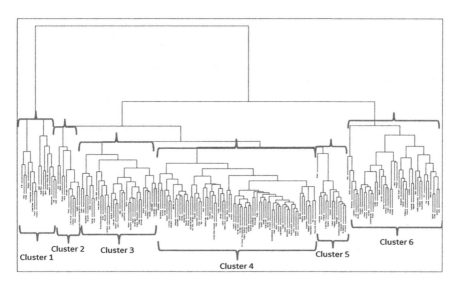

Fig. 1 Dendrogram of the global social network

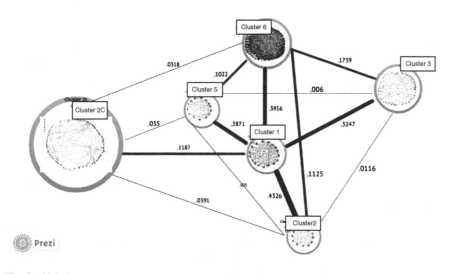

Fig. 2 Global map

Table 2 provides descriptive statistics for each of the clusters. Cluster 1 is the most densely connected cluster. It also has the strongest connections to all other clusters in the global network. The cluster has a density of 0.9779, with only one possible edge missing. Cluster 1 identifies seventeen states at the core of the international system, to include the United States, the United Kingdom, France, Russia, and China. The second most densely connected cluster is Cluster 6, with an

Table 2 Descriptive statistics of the six clusters

Cluster	# of nodes	Density	Example countries in the cluster
1	17	0.9779	China, France, Russia, UK, US
2	11	0.7818	Angola, Cote D'Ivoire, Ethiopia, Gabon, Nigeria
3	36	0.1452	Afghanistan, Belarus, Jordan, Lithuania, Sudan
4	74	0.0413	Chad, Fiji, Namibia, Rwanda, Tuvalu
5	14	0.7967	Costa Rica, Honduras, Nicaragua, Peru, Taiwan
6	44	0.7759	Australia, Poland, Saudi Arabia, South Korea, Venezuela

internal density of 0.7559. The 44 nodes in the cluster have their strongest connections with Cluster 1 (0.3956). Countries in this cluster represent the secondary core of the international system and include important regional actors such as South Korea, Saudi Arabia, and Australia.

The remaining clusters are less central to the global system. Moreover, their strongest external ties are with Cluster 1, the core of the international system. Cluster 2 is comprised of 11 of the best developed countries in the northern half of Africa. These countries are strongly connected with an internal density of 0.7818. The cluster's strongest external connections are with Cluster 1 (0.4326 density), and its next strongest set of connections is with Cluster 6. Its remaining connections are relatively sparse. Cluster 3 contains 36 countries in Eastern Europe and the Middle East. Its internal density is 0.1452, and its strongest connections are to Cluster 1 with a shared density of 0.3247. Cluster 4 is very sparse, with an internal density of 0.0413. The cluster contains 74 countries, to include most of the globe's island states as well as less connected countries from Asia and Africa. Cluster 4's strongest connections are with Cluster 1, but the density of these connections is rather weak (0.1187). Cluster 5 contains 14 countries from Latin America. It also includes Taiwan. Cluster 5 is one of the densest clusters at 0.7967, and the density of its shared connections with Cluster 1 is 0.3871.

3 Validity Testing

Before testing our paper's hypothesis, we wish to consider the validity of our global map. If the map represents the structure of the patterns of global exchange of goods and information, then it should be correlated with other measures of exchange, other than those from which it was constructed.

We begin by considering the degree to which our matrix of the global social network is associated with a matrix representing global arms trade. (Arms trade data are from the UN Register of Conventional Arms. The register records dyadic arms trade each year from 1992 to 2011.) Two procedures suggest themselves. The first, cross-validation, divides the dataset into separate units and then determines how well one part of the dataset predicts another. We use the second procedure, which is more widely employed in the social science literature, a Quadratic

Assignment Procedure (QAP). A QAP determines the probability that two states will have a tie in one matrix given that they have a tie in the other. It does so by permuting the matrices a set number of times, comparing the newly created matrices to the original, and determining the likelihood that the result is the same as that of the original matrix.

The QAP testing the association between our global matrix and the matrix of the global arms trade results in a Goodman–Kruskal Gamma of 0.904. The Goodman–Kruskal Gamma is the proportion of concordant/discordant pairs between the two matrices. A measure of 1 obtains when all pairs in the matrices are concordant, and 0 obtains when no pairs are concordant. Negative measures indicate that pairs are discordant. Our Goodman–Kruskal Gamma indicates that given a tie between two states in the global matrix, there is a 90.4 % chance of a reciprocated tie in arms trade. The result is statistically significant at less than the 0.001 level. Similar results obtain between our global social network and a matrix representing global conflict (Goodman–Kruskal Gamma of 0.910, statistically significant at less than 0.001). Conflict data are taken from the *2010 Conflict Barometer* published by the Heidelberg Institute for International Conflict Research. Conflicts are measured on an ordinal scale from 1 to 5, with 1–2 being non-violent and 3–5 being violent. If a dyad was involved in more than one conflict with each other, we recorded the higher of the two measures. The data are non-directional as any dispute inherently involves the same states. The minimum dyadic observation is 0, indicating no conflict. The maximum is 3, indicating the lowest violent score between states. The mean and median are 0 as most dyads are not engaged in dispute.

We next consider our global matrix a bit further by considering the degree to which measures of the embeddedness of individual states are correlated with states' involvement in trade, conflict, and IGO membership. Using UCINet, we obtain four measures of a state's embeddedness in the international system defined by exchange relationships: degree centrality, closeness centrality, betweenness centrality, and Eigenvector centrality. Degree centrality is a measure of how connected a state is in the system. It is the total number of incoming and outgoing ties a node has. Degree centrality is calculated using the expression

$$C_d(v) = deg(v) \tag{1}$$

where C_d is the degree centrality and deg is the degree of a node [23]. In basic terms the degree centrality of a node is identical to the degree of the node. The nodes in our global matrix have an average degree of 46.3, a minimum degree of 1.026 (Tuvalu), and a maximum degree of 92.82 (Belgium and the United States).

The second measure of embeddedness is closeness centrality, which calculates each node's average geodesic distance to all other nodes. Closeness centrality is calculated by the expression

$$C_C(v) = \sum_{t \in V/v} 2^{-d_G(v,t)} \tag{2}$$

where C_C is closeness centrality and all other variables remain the same as in other measures of centrality [23]. The average closeness is 1.76, with a range from 36.93 (Tuvalu) to 93.3 (Belgium and the United States).

Betweenness centrality is the ratio of paths that pass through a given node to the total number of paths in the network. The betweenness centrality of a node v is given by the expression

$$g(v) = \sum_{s \neq v \neq t} \frac{\sigma_{st(v)}}{\sigma_{st}} \tag{3}$$

where σ_{st} is the total number of shortest paths from node s to node t and $\sigma_{st(v)}$ is the number of those paths that pass through v [23]. The average betweenness of the weighted network is 0.004, with a minimum of 0 (several countries) and a maximum of 6.781 (United States).

The fourth centrality measure is Eigenvector centrality. Eigenvector centrality is a measure of the influence of a node. Eigenvector centrality is calculated using an adjacency matrix $a_{v,t}$. The expression to calculate eigenvector centrality from that matrix is

$$x_v = \frac{1}{\gamma} \sum_{t \in M(v)} x_t = \frac{1}{\gamma} \sum_{t \in G} a_{v,t} x_t \tag{4}$$

where M_v is a set of neighbors of v and γ is a constant [23]. States with high eigenvector centrality are connected with other nodes that are important [23]. Nodes have a higher Eigenvector centrality score as their number of connections with other well connected states increases. The average Eigenvector centrality score is 0.369, with a range from 0.11 (Tuvalu) to 20.69 (Belgium).

The centrality scores indicate that the world network is fairly sparse. The low betweenness measurement indicates that only a small minority of nodes act as key connectors to other nodes. The low average Eigenvector centrality score in the world matrix indicates that a small percentage of nodes in the network are central, powerful actors.

We use four maximum likelihood estimation models to consider the correlation between each measure of embeddedness and trade, conflict, and IGO membership. Trade data are for 2008 (International Monetary Fund). The data are a measure of the number of states with which the country trades. IGO membership is taken from *Country Memberships in Selected Intergovernmental Organizations and Accession to Selected Regional and Global Treaty Regimes: Global, Country-Year Format, 1995–2010.* The data record a state's membership in several international regional IGOs. The minimum observation for IGO membership is 0, and the max 16. The median is 9 and the mean 9.46. We create a vector by summing a state's total number of memberships in these IGOs. Since our data are relational, we cannot use OLS regression models, and therefore violate the assumption of independence of the observations. The procedure that we use obtains initial estimators by regressing each measure of embeddedness on trade, conflict, and IGO membership. It then

Table 3 Results of regression models estimating the effects of trade, conflict, and IGO membership on measures of embeddedness

Independent variable	Model 1	Model 2	Model 3	Model 4
Constant	3.215	48.376	1.871	−0.326
Trade	22.107**	9.346**	5.618**	0.761*
Conflict	545.4**	242.5**	129.8**	28.05**
IGO membership	1.521**	0.611**	0.459**	0.028
F-statistic	59.93**	56.12**	65.39**	27.67**
Adjusted r-squared	0.473	0.456	0..495	0.287

Source Calculated by the authors using UCINet 6.0
*—significant at <0.05
**—significant at <0.001

randomly permutes the values of the dependent variable and re-estimates the coefficients. This is done hundreds of times to estimate standard errors. Table 3 reports the results. Model 1 regresses degree centrality on trade, conflict, and IGO membership, Model 2 regresses closeness centrality on trade, conflict, and IGO membership, Model 3 regresses Eigenvector centrality on trade, conflict, and IGO membership, and Model 4 regresses betweenness centrality on trade, conflict, and IGO membership.

All four models are statistically significant at less than the 0.001 level, and the goodness of fit is relatively high. Moreover, with the exception of IGO membership in Model 4, state-level measures of trade, conflict, and IGO membership are significantly associated with our four measures of the degree to which a state is embedded in the international system. These results strongly suggest that our global matrix is a valid representation of global relations between states. We now turn to our paper's hypothesis.

4 Embeddedness and Involvement in Cyber Attacks

We hypothesize that as states become more central to the social network defining global relations, they are more likely to be involved in cyber attacks. We measure the dependent variable using data from Akamai [24]. Akamai reports the percentage of total global cyber attacks that a state has been involved in. These data are non-directional and do not indicate which states are the aggressors and which are the victims of cyber attacks. Moreover, given the difficulty of attribution associated with cyber attacks, we are uncomfortable with trying to distinguish between the two. In either case, the data permit us to consider a country's degree of involvement in cyber attacks, which is the question that we raise.

The mean involvement for states in cyber attacks is 0.52 %. The median involvement is 0 %, the maximum involvement is 11 % and the standard deviation is 0.15 %. This indicates that well over half of the states have no involvement in cyber attacks. Russia, China, and the United States each account for more than

Table 4 Results of regression models estimating the association of cyber attacks with measures of embeddedness

Variable	Model 1	Model 2	Model 3	Model 4
Constant	−0.000084	−0.002056	−0.040155	−0.003088
Betweenness centrality	0.000010*			
Degree centrality		0.000043		
Closeness centrality			0.000516*	
Eigenvector centrality				0.81253*
Internet penetration	0.004239	0.003399	0.003937	0.002297
Total internet users	0.0000*	0.0000**	0.0000	0.00000**
N	185	185	185	185
F score	0.00	0.00	0.00	0.00
R squared	0.611	0.594	0.600	0.605

Source Calculated by the authors using UCINet 6.0
*—significant at <0.05
**—significant at <0.001

10 % of the total global cyber attack activity. These states are significantly more than three standard deviations from the mean.

As before, we test four models using permutation analysis to arrive at maximum likelihood estimators of the association of cyber attack involvement on each of our four measures of embeddedness. We include two control variables in the analysis: the percentage of a state's population that has access to the internet and the total number of internet users in each state. The data are obtained from *Internet World Statistics* (2011). The mean population percent with access to the internet (internet penetration) is 36.92 %. Iceland has the highest internet penetration with 97.8 % of the population having access to the internet. The state with the lowest internet access is Myanmar with a penetration rate of 0.2 %. The mean number of internet users per state is 12,099,903. The state with the maximum number of internet users is China with 513,100,100 users. The state with the fewest Internet users at is Tuvalu (4,300).

The results of the regression models are reported in Table 4. Model 1 regresses cyber attacks on betweenness centrality, internet penetration, and total number of internet users. Model 2 regresses cyber attacks on degree centrality, internet penetration, and total number of internet users. Model 3 regresses cyber attacks on closeness centrality, internet penetration, and total number of internet users. Model 4 regresses cyber attacks on Eigenvector centrality, internet penetration, and total number of internet users.

The results support our hypothesis. Eigenvector, closeness and betweenness centrality all have a statistically significant association with the involvement of states in cyber attacks when controlling for internet penetration rate and total internet users. All three measures of a state's embeddedness in the international system are significant at the 0.05 level. Furthermore, all three models have r-squared values greater than 0.600, indicating the centrality measures, internet penetration rate, and total internet users explains 60 % of a state's involvement in cyber attacks. This indicates that as states become more embedded in the network they are more likely to be involved in cyber attacks, either as the initiator or the target. This makes intuitive sense in that as states become more central in the network they are also generally more important and powerful in the system. States on the periphery of the network are less likely to have significance in the system, thus making them less likely to be targeted for cyber attack. As mentioned above, Russia, China, and the United States cumulatively account for over 30 % of all states' involvement in cyber attacks. These three states are also part of cluster 1; the most central states in the network.

5 Conclusions

The models lend support to our argument that the degree to which a country is connected with other countries in the international system is associated with its likelihood of involvement in cyber attack. However, it is not merely a matter of connections. Model 1 found no statistical relationship between the number of connections that a state has and the likelihood of its involvement in cyber attacks. Rather, what seems to be important is its relative position in the global network connecting states. Just how important is a state in the overall global social network? States that lie along routes connecting other states (betweenness) as well as those that can reach other states quickly along the least number of paths (closeness) and those that are connected to states that are highly connected (Eigenvector) are more likely to be engaged in cyber attack.

This chapter is one of the first to examine how a state's position in relations to exchange relations between states in the international system affects it likelihood to be involved in cyber attacks. The results of the analysis leave significant questions to be explored by social scientists interested in studying cyber attacks.

References

1. Libicki, M.: Cyber Deterrence and Cyber War. Rand Publishing, Santa Monica (2009)
2. Clarke, R.: Cyber War: The Next Threat to National Security and What to Do about It. Haper Collins, New York (2010)
3. Alperovitch, D.: Towards the Establishment of Cyberspace Deterrence Strategy. CCD COE Publications, Tallinn (2011)
4. Arquilla, J.: From Blitzkrieg to Bitskrieg: the military encounters with computers. Commun. ACM **10**, 54 (2011)
5. Eijndhoven, D.: Cyber Deterrence: Methods and Effectiveness. Argent Consulting, Zeist (2010)
6. Kramer, F., Stuart, S., Wentz, L.: Cyberpower and National Security. Center for Technology and National Security. National Defense University Press, Washington, DC (2009)
7. Raduege, H.: Fighting Weapons of Mass Disruption: Why America Needs a Cyber Triad. The East West Institute, New York (2010)

8. Embassy Pages: Embassies and Consulates Around the World. www.embassypages.com (2012)
9. Commercial Airline Traffic Data (CATD): International Civil Aviation Organization. http://www.icaodata.com/Trial/Links.aspx (2010)
10. TeleGeography: Global Traffic Map 2010. http://www.telegeography.com/assets/website/images/maps/global-traffic-map-2010/global-traffic-map-2010-l.jpg (2010)
11. TeleGeography: Submarine Cable Map. http://www.submarinecablemap.com/ (2012)
12. Arnson, C.J., Fuentes, C., Aravena, F.R., Varat, J.: Energy and Development in South America: Conflict and Cooperation. Woodrow Wilson Center for International Studies (2008)
13. Central Intelligence Agency (CIA) Electricity—Consumption. https://www.cia.gov/library/publications/the-world-factbook/rankorder/2042rank.html (2010)
14. Central Intelligence Agency (CIA) Electricity—Exports. https://www.cia.gov/library/publications/the-world-factbook/fields/2044.html (2010)
15. Central Intelligence Agency (CIA) Electricity—Imports. https://www.cia.gov/library/publications/the-world-factbook/fields/2043.html (2010)
16. Lang, M., Mutschler, U.: BDEW: Germany Remains Electricity Exporter, But Imports Increase Significantly. German Energy Blog. http://www.germanenergyblog.de/?p=7218 (2011)
17. National Energy Grid Map Index: National Energy Grid. http://www.geni.org/globalenergy/library/national_energy_grid/index.shtml (2012)
18. News from Uzbekistan: UzNews.net. http://www.uznews.net/news_single.php?lng=en (2012)
19. Patel, T.: Germany Becomes Net Power Importer From France After Atomic Halt. Bloomberg. http://www.bloomberg.com/news/2011-05-30/germany-becomes-net-power-importer-from-france-after-atomic-halt.html (2011)
20. RGCE SG Network Models and Forecast Tools: Indicative values for Net Transfer Capacities (NTC) in Continental Europe Winter 2010/11, working day, peak hours (non binding values) (2010)
21. Tajikistan, Pakistan Discuss Electricity Exports: RadioFreeEurope/RadioLiberty. http://www.rferl.org/content/tajikistan_pakistan_discuss_electricity_exports/24441274.html (2012)
22. Borgatti, S.P., Everett, M.G., Freeman, L.C.: Ucinet for Windows: Software for Social Network Analysis. Analytic Technologies, Harvard (2002)
23. Jackson, M.: Social and Economic Networks. Princeton University Press, Princeton (2008)
24. Akamai: The State of the Internet. http://www.akamai.com/stateoftheinternet/ (2010)

Moblog-Based Social Networks

A. Abdul Rasheed and M. Mohamed Sathik

Abstract Social Network (SN) represents the relationship among the social entities, like friends, co-workers, co-authors. Online Social Network (OSN) sites attracted the people who are scattered all over the world. It is an application of web 2.0, which facilitates the users to interact among themselves, nevertheless of considering the geographical locations. Hence, these sites are having unprecedented growth. The members of these sites can establish networking by viewing the profiles of similar-interested persons. A blog is a content posted over a website or a web page, usually arranged in reverse chronological order. Blogs which are hosted by using specialized mobile devices like iPad, Personal Digital Assistants (PDA), are called mobile blogs (or shortly called *moblog*). A moblog facilitates habitual bloggers to post write-ups directly from their phones even when "on-the-move". There are different ways to establish the communication among the users of social networking sites and to form the social networking environment among them. One such way is through blog posts hosted on a website. The habitual users who responded to a blog post can be connected by links between them and this structure will grow as a network. Hence, the users (responders) will form the graph structure. An interesting research phenomenon from such an environment would be to extract a subgraph of users based on some common property, and such structure can be called as communities. Discovering communities by partitioning a graph into subgraph is an NP-hard problem. Hence, a machine learning method, clustering, is applied to discover communities from the graph of social network which is formed from the mobile bloggers.

A. Abdul Rasheed (✉)
Department of Computer Applications, Valliammai Engineering College,
Chennai, Tamil Nadu, India
e-mail: profaar@gmail.com

M. Mohamed Sathik
Sadakathullah Appa College, Palayamkottai, Tirunelveli, Tamil Nadu, India
e-mail: mmdsadiq@gmail.com

W. Pedrycz and S.-M. Chen (eds.), *Social Networks: A Framework*
of Computational Intelligence, Studies in Computational Intelligence 526,
DOI: 10.1007/978-3-319-02993-1_5, © Springer International Publishing Switzerland 2014

Keywords Social networks · Moblog · Computational intelligence · Graph clustering · Community discovery

1 Introduction

1.1 Social Networking

The concept of social networking is as older than four decades as of now. It started with an Online Bulletin Board System (BBS) which is used by the people of an organization. They used the system to share news, stories, and entertainments among each other. These systems gain its popularity for a decade and dedicated to a special interest. The BBS was used like an Intranet application a small group of people of any organization. The primary purpose of using such systems was to share the content one who like with the others inside the organization. The people also used the system to share information through individual postings, and hence the virtual community began to emerge. The system evolved over a period of time and the members were used the system to inform the group's other members of their group about meetings and some other important announcements without making phone calls. The capabilities of the system were transformed from "*stand-alone*" to "*multi-user*" environment.

When the telecommunication facilities conceived, the communication among the individuals was enhanced, nevertheless of considering the geographical location all over the world. This facility was also attracted the people to use it to communicate among themselves, as they are dispersed throughout the world.

A German sociologist named Georg Simmel, who first talked about social networks. He defined the society as a "*higher unity*". Eric Thomas, who is the creator of LISTSERV, played a crucial role in the development of later social networks. The users could send email to an entire list of people very easily by using one email address by this system. The founders—Jonathan Abrams and Peter Chin—of the well-known social networking site Friendster, dubbed the '*grandaddy*' of social networks. Tim O'Reilly, the founder of Web 2.0 described the usage of it as building social relationship and making technology more interactive for people. Andrew Weinrich found *sixdegrees.com*, an early social network that would give ways to more modern types of online networking like Friendster and MySpace with the advantage of *user profiling*. The idea is based on the "*six degrees of separation*".

Social networking sites have some classifications. Group of people can choose a topic and they can discuss among them over the topic of their interest. This type of social networking is called as *Forum*. For example, online social networking site *Facebook* offers this type of social networking. People can host their interest, feelings, opinions, suggestion over a website or a web page, called as *blogs*. The well-known online social networking site *LinkedIn* provides this type of service. Some individuals can maintain their own postings about what is going on in an individual's life or can be information the individual wants to share with the others and get feedback like things from the public. This category is called as *Microblogging*. This service is provided by *Twitter*, and is such blog is so called *tweets*. Apart from content sharing, people can use social networks to share photos over the website and it can be viewed by the visitors all over the world. The social networking sites can also be used to *share photos*. This type of service is offered by *Flickr*. Another important and vital usage of social networking site is *bookmarking*. *Del.icio.us* is a social bookmarking web service for storing, sharing, and discovering web bookmarks. The user of a social networking site can bookmark a webpage or content of a website and this can be shared by all the others in a network. Blogging and bookmarking are the two categories which are predominantly used in social networks.

Online Social Networking (OSN) sites are used by various user categories. Friends can use these sites to find friends. Such services are called Friend-of-A-Friend (FoAF). OSN like *Friendster* and *hi5* provides these types of services to connect friends over the world. Many professionals want to keep in touch with people belonging to their chosen profession or industry and would usually like to make that network separate from their network of friends. LinkedIn and *Jobster* are the OSN sites which provide this sort of services to professionals. The familiar OSN sites *YouTube* and *Flickr* are offering services for online media sharing such as video contents.

Social networking attracted all age groups from 18 to 65+. The main objective of using these sites are for sharing links, photos, videos, news and status updates with a growing network of contacts among the group of people. The following Fig. 1 shows the usage pattern in different age groups.

Social networking made popular due to its ease of use, the simple way to know the people who have common interest and making the network easier, without considering the geographical location over the globe. Due to this easiness, it has been adopted in many application domains. Now-a-days, almost all the products and services are having their product information in any one of the social networking sites, to promote their business by reaching the prospective customers in each every nook and corners of the world, wherever they launched their business. This is not only used to provide the information to the customers, but also can know the demand in the market.

Government agencies are utilizing social networking to get the opinion of the public and to keep the public updated on their activity. Some government agencies are using social networking for disease vaccinations and disease control. Business professionals heavily depend on social networking for various purposes like interaction among their own colleagues wherever they are, to know the trend of the product, market research and sales force automation. Social networking provides advantages to these professionals like reducing the cost for travelling, save time, and it also offers interaction at reduced cost. Students and teachers can use social networking for online learning, which connects the educators of inter-disciplines

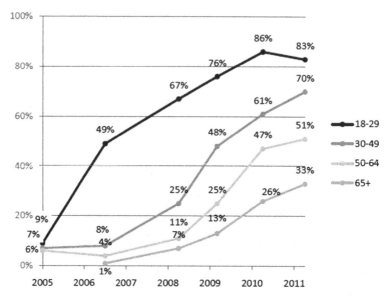

Fig. 1 Social networking site use by age group. *Source* http://pewinternet.org/Commentary/
2012/March/Pew-Internet-Social-Networking-full-detail.aspx

throughout the world. Healthcare professionals are using these sites to manage
institutional knowledge, disseminate peer to peer knowledge and to highlight
individual physicians and institutions. Social networking is widely used for
entertainment applications such as music, and videos.

Social networks can be represented in two ways: (i) adjacency matrix and (ii)
graph structure. In the adjacency matrix representation, the cell entries are filled
with one when there exist ties between two actors, and zero otherwise. In graph
structure, the *actors* (or person) in the network are represented by nodes (vertices)
and the *ties* (or relationship that exists among the actors) are represented by links
(edges). This structure represents how a person is connected with another person or
a group of persons. This representation helps the researchers to carry out further
research.

A *graph* G is represented as G = (V, E), consists of a set of objects $V = \{v_1, v_2, v_3, \ldots., v_n.\}$ called *vertices* (also called *nodes*) and other set $E = \{e_1, e_2, e_3, \ldots., e_n\}$ whose elements are called *edges* (also called *arcs*). The set V(G) is called the
vertex set of G and E(G) is the edge set. Let G be a graph and {u, v} an edge of G.
It is often more convenient to represent this edge by uv, alternatively as (u, v). If
e = uv is an edge of a graph G, then u and v are adjacent in G and that e joins u
and v. Figure 2 represents the multi-relationship among the people, as a simple
social networking graph (so called sociogram).

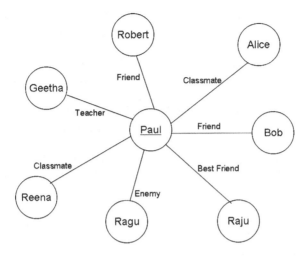

Fig. 2 A Simple Sociogram

1.2 Computational Intelligence

As defined by Wikipedia, Computational Intelligence (CI) is a set of nature-inspired computational methodologies and approaches to address complex real-world problems to which traditional approaches are ineffective or infeasible. An important aspect of Computational Intelligence is adaptivity which is covered by the fields such as machine learning. Though there is no specific, well-defined and universally accepted definition of computational intelligence, Computational Intelligence is considered as umbrella techniques which include cognitive information, data mining, machine learning, intelligence agents, pattern recognition, soft computing and graphical methods. IEEE Computational Intelligence Society defines its subjects of interest as neural networks, fuzzy systems and evolutionary computation, including swarm intelligence [1]. Machine learning usually focuses on prediction, based on known properties learned from the training data and Data mining focuses on the discovery of (previously) unknown properties on the data. Very importantly, computational intelligence is considered as part of or sub-branch of Artificial Intelligence (AI).

A computational Intelligence (CI) technique contains Artificial Neural Networks (ANN), Artificial Immune Systems (AIS), Fuzzy Systems, meta-heuristics, Evolutionary Algorithms (EA) such as Genetic Algorithms (GA), Bayesian Networks, Particle Swarm Optimization (PSO) and Ant Colony Optimization (ACO). In the above list, there is no predefined application for each and every technique.

ANN imitates the human brain structures and its internal neurons. It just resembles the mathematical model inspired by biological neural networks. It consists of an interconnected group of artificial neurons which processes information using a connectionists approach to computation. A neural network is an adaptive system changing its structure during a learning phase and is used for modeling complex relationships between inputs and outputs. They are used to find patterns in data. Artificial neural networks have been applied to problems ranging from speech recognition to prediction of protein secondary structure and classification of cancers. There are six major components in a general ANN such as: a set of inputs, a set of weights, a threshold, a weight training function, an activation function, and a single neural output. There are two topologies can be used in ANN. *Feed-forward networks*, in which the data from the input to output units is strictly feed forward. The classical example of feed-forward neural network is the Perceptron. As the other structure, so called as *recurrent neural networks*, contain feedback connections. Though there are numerous advantages by applying ANN to variety of applications or problems, it has its own disadvantages:

(i) It requires training to operate.
(ii) An emulator is required so as to imitate the neural structure of human being.
(iii) It will take high processing time, if the network is large.

Artificial Immune System (AIS) is the study of combined areas such as immunology and engineering. It uses immune system metaphors for the creation of novel solutions to problems. AIS is moving into an area of true interdisciplinary and one of genuine interaction between immunology, mathematics and engineering [58]. These algorithms typically exploit the immune system's characteristics of learning and memory to solve a problem.

An optimization problem is the problem of finding the best solution from all feasible solutions. These are the special classes of algorithms, known as NP-hard, as the solution to the given problem falls into the category of non polynomial. The running time and the analysis of such algorithms would be so tricky and hard. These types of problems (problem space) find the best solution (solution space) among the possible solutions for the given problem. There are two optimization techniques fall into computational intelligence approach: (i) Particle Swarm Optimization (PSO) and (ii) Ant Colony Optimization (ACO).

Particle Swarm Optimization (PSO) is a computational method that optimizes a problem by iteratively trying to improve a candidate solution with regard to a given measure of quality. PSO optimizes a problem by having a population of candidate solutions and moving these particles around in the search-space according to simple mathematical formulae over the particle's position and velocity (as defined in Wikipedia). PSO is a population-based, bio-inspired optimization method. It is a metaheuristic as it makes few or no assumptions about the problem being optimized and can search very large spaces of candidate solutions. PSO can be used on optimization problems that are irregular, noisy, change over time, etc.

The Ant Colony Optimization Algorithm (ACO) is a probabilistic technique for solving computational problems which can be reduced to finding good paths through graphs. It constitutes some metaheuristic optimizations. The concept of this optimization came into existence from the foraging behavior of real ant colonies. These algorithms are part of swarm intelligence which is made up of simple individuals that cooperate through self-organization as they don't have any form of central control over the swarm members. The essential trait of ACO algorithms is the combination of a priori information about the structure of a promising solution with a posteriori information about the structure of previously obtained good solutions. The technique of ACO can be applied to variety of optimization problems that includes scheduling, telecommunication networks and vehicle routing problem.

There are two types of machine learning (i) supervised learning and (ii) unsupervised learning. In supervised learning, the class label by which the objects can be classified is known. Classification technique falls into the category of supervised learning. When the class label is unknown then such learning is called as unsupervised learning. Clustering technique falls into the category of unsupervised learning. A cluster is a collection of objects which are "similar" between them and are "dissimilar" to the objects belonging to other clusters. The goal of clustering is to determine the intrinsic grouping in a set of unlabeled data. This research concentrated on applying graph clustering, as part of computational intelligence, over the moblog dataset which was created by this research.

1.3 Graph Clustering

Graph clustering and graph partitioning are the two classical terms which are used to subdivide the graph into number of partitions. Though these two terms are synonymously used, the domain and its applicability may differ. The generic meaning of clustering itself is in principle exactly that of partition. The semantics of clustering in a graph varies with respect to the other term called overlapping, which means the same nodes may appear in two or more number of clusters. A partition is strictly defined as a division of some set S into subsets which satisfies (a) all pairs of subsets are disjoint and (b) the union of all subsets yields S. Graph clustering works on the basis of finding the difference between objects. For this purpose, clustering process assume that entities are represented by vectors, describing how each entity scores on a set of characteristics or features. This makes convenient to find the dissimilarity, and hence to cluster the objects.

Graph clustering is the task of grouping the vertices of the graph into clusters taking into consideration the edge structure of the graph in such a way that there should be many edges within each cluster (intra-cluster) and relatively few between the clusters (inter-cluster). Partitioning the graph into subgraph can be done on various criteria depends on the type of the application or domain, usually grouped according to the fact the nodes present a relatively high number of connections. For a given data set, the goal of clustering is to divide the data set into clusters such that the elements assigned to a particular cluster are similar or connected in some predefined sense.

The graph clustering problem consists on dividing G into k disjoint partitions and the partitioned subgraph is called a cluster. Due to the complexity in partitioning the graph into k disjoint subgraphs, the graph clustering problem is considered as NP-hard problem. The goal is minimize the number of cuts in the edges of the partition in the entire graph.

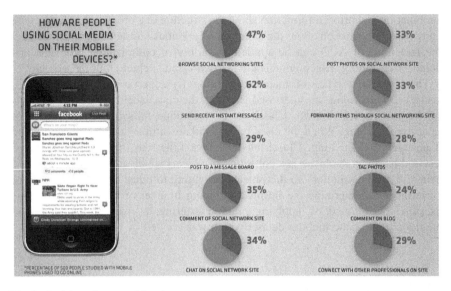

Fig. 3 Social media on mobile phones

Real–world complex systems are represented by network structures which consist of nodes to represent entities and edges to represent connectivity. The typical size of these networks grows rapidly in billions over a period of time. In the network representation, identifying the sub-structures is an important phenomenon. Such compartments appear as sets of nodes with a high density of internal links, whereas links between compartments have a comparatively lower density. These subgraphs are called communities. Communities can be discovered by clustering the graph.

Mobile technology is ruling the world in today's scenario. It is almost become as part of every human life. It is used to establish communication among the people to keep-in-touch with their friends, neighbours, relatives and co-workers when on-the-move. More than that, it is being used as a play station and media player. The following are some interesting facts provided by the leading mobile usage forecasting website, *mobile youth around the world*:

- There are 1.8 billion mobile owning youth in the world.
- 60 % of youth sleep with their mobile phone.
- The mobile industry has made $1 trillion from SMS over the last decade.
- Youth spend $350 billion annually on mobile services.
- Mobile internet is now the *defacto access* point for youth in most of the countries.

The following Fig. 3 shows how social media sites are accessed on mobile devices [56].

A blog (or web log) is a type of website or part of a website usually maintained by an individual with regular entries of some content like events, graphics or video, which are displayed in reverse chronological order. Most blogs are interactive, allowing visitors to leave comments and messages, as an application of Web 2.0. A typical blog combines text, images and links to other blogs, web pages, and other media related to its topic. The visitor or user who hosts a blog post is called as a blogger. In today's scenario, blogs are used for collective wisdom. The bloggers used to share their content, views and reviews on any selected subject for discussion.

The term "moblog" is the combination of two terms "mobile" and "Web log". A Web log (also called a blog) is a Web page that serves as a publicly accessible personal journal for an individual, a moblog is a blog which has been posted to the Internet from a mobile device such as a mobile phone, Personal Digital Assistance (PDA), or Blackberry. Mobile blogging has emerged as a further convenience for bloggers who want to update their readers while "on the go". Any mobile blogger with an Internet-capable cell phone can utilize this method of updating their blogs. A mobile blog photo can be taken with a camera phone and then uploaded to the user's blog via their mobile browser. As cell phones continue to develop technologically, the use of mobile blogging will expand and enjoy continuous growth for bloggers "on the go".

Mobile blogging architecture may have the following components for the successful receiving of messages from the mobile devices. Though the application domain may vary depends upon the needs and the type of users and blogging, the following are the components included in the generic framework.

- *Blog server*: generates pages and allows authors to upload new blog entries. The users can upload the blog entries into this server.
- *Blog server Application Program Interface* (API): As blogging has grown in popularity, alternate blog clients have emerged. This is in the form of clients in an HTML form in a browser.
- *XML editor:* This component is useful to accept blog posts sent by the bloggers and to process the request as XML documents.

This research focused on discovering communities from a moblog website, as an application of computational intelligence.

The area of community discovery has been started from the origin of Physics by power law distribution. Then, it is being applied to the other domains like social science, biological science and computer science. The research on community discovery in social network has been studied in various dimensions by researchers in the recent past.

1.4 Related Works

A general idea about social network, the type of data and how it differs from other data was introduced by [2]. With the advent of Internet technologies, online social networking already had a profound impact on identifying like-minded people. The interesting properties, the methodologies used to study them, and the challenges faced by researchers in measuring OSNs are enumerated in [3]. A general criticism about social networking can be found in [4]. Web mining techniques are also used for mining the web content from a web site, through which the social network can also be established. The authors of [5] studied the issues around using web mining techniques for analysis of on-line social networks. They also set out a process to use web mining for on-line social networks analysis. An online social networking site Slashdot was analyzed by [6]. A common methodology for the research of the e-Government Research (EGR) field in social network and citation analysis terms was proposed by [7]. It also introduced and described a data set that allows for such analysis and outlines the prospects for further research. Social network can be applied to many application domains. Marketing area is no way exempted from social networks as an application area. A case study on eBay by applying SNA has been studied by [8]. The amount of benefit that the multiple industries can have by applying social network is numerous.

In today's scenario, blogs are used for collective wisdom. The blog posts are of different type and which are analyzed in various ways. The people who host the blog posts over a website, review blogs, and the response for an already hosted blog posts are the examples of social networks. Blog-oriented social network analysis was carried out by various researchers in the past decade. References [9–12] were conducted different type of analysis by using e-mail as a source of information.

Graph clustering has become ubiquitous in the study of relational data sets like social networks. A survey on various graph clustering methods can be found in [13]. The work summarized the various methods that evolved over a period of time. A new novel algorithm called structural clustering algorithm for networks was proposed by [14], which detects clusters, hubs and outliers in networks. It clusters vertices based on a structural similarity measure. A graph partitioning method based on link distribution was proposed by [15]. It approximates a graph affinity matrix under the corresponding distortion measure. Clustering is an example of unsupervised machine learning. A brief introduction to machine learning can be found at [16]. A graph clustering coefficient called global clustering coefficient was used as a measure to cluster the graph was discussed in [17].

Bayesian models are used for different kinds of applications. The Block-Clustering model can be described in a full Bayesian framework. The posterior distribution, of all the parameters and latent variables, can be approximated efficiently applying Variational Bayes (VB) was described in [18]. A new criterion that overcomes the limitation of cluster overlapping problem by combining internal density with external sparsity in a natural way was introduced by [19]. The authors of [20] formalized the problem of discovering a priori unspecified number of clusters in the context of joint cluster analysis of attribute and relationship data.

Discovering a community in a social networking environment is a clustering problem, which partition the entire graph into subdivisions, or sub-graphs. The clustered or partitioned division of nodes (people) in the network is called as a community. Simply, a cluster in a graph is called as a community. Community discovery methods are evolved over a period of time by the researchers. This section reviews such works.

A comprehensive survey on community detection algorithms can be found in [21]. A framework for identifying communities in social networks that change over time was proposed by [22]. Problem decomposition to discover communities was applied by [23]. In this approach, the network is decomposed into manageable sub networks using a multilevel graph partitioning procedure.

Biologically inspired algorithms are applied for wide variety of problems. It is applied for discovering communities using genetic algorithmic approach, which is described by [24]. The algorithm uses a fitness function able to identify groups of nodes in the network having dense intra—connections, and sparse inter—connections. Community detection using modularity was proposed by [25]. It is an agglomerative hierarchical clustering method. The basic idea of the algorithm was modularity. Its subsequent work, called eigenvector of matrices for community discovery was introduced by [26]. An extremal optimization method for community discovery was proposed by [27] which is a divisive algorithm for graph partitioning. It optimizes the modularity using a heuristic search based on the Extremal Optimization (EO) algorithm which improves the work based on the modularity algorithm.

Applying statistical mechanics to community detection was described by [28]. By this method, the community detection was interpreted as finding the ground state of an infinite range spin glasses. It also discussed about overlapping in the community structure. The statistical properties to discover communities were studied by [29]. The basic concept that lies behind this method was community profile plot, which characterizes the "best" possible community in the various type of network in different domains taken for study.

A new proposal for discovering communities and a measure was introduced by [30]. Though the blogs can also be used to identify similar-interested persons, it can also be used to identify hate groups. This situation was studied by [31]. This study was conducted over a blog hosting site *Xanga*. The process of target group selection by adding the social elements derived from the behaviors of people in marketing and advertising domain was discussed in [32]. This work also coined a new term called "*human filtering*". A simple label propagation algorithm that uses the network structure alone as its guide and requires neither optimization of a predefined objective function nor prior information about the communities was introduced by [33]. The authors of [34] compare and evaluated different metrics for community structure in networks.

A recent survey of community discovery [54] provides a manual for the problem. The authors organized the main categories of community discovery based on their own definition of community. The work is designed to provide a set of approaches that researchers could focus on. The authors of [35] made an attempt to map between community detection with AIS. They proposed a novel community detection algorithm based on the immune clonal selection principle to govern the system and the authors claimed that their algorithm obtained more scalable and accurate solution with a lower computational cost.

The online social networking site, The *Facebook*, has been studied in various dimensions in the recent past, both in positive and negative sense. One such recent work is reported by [36]. The author adopted two well-known clustering algorithms and the communities were extracted from Facebook users. The communities may evolve overtime in dynamic networks than static ones. This property has been studied by a novel multiobjective immune algorithm by [37]. The authors conducted experiments over both synthetic datasets and real-world datasets. The authors of [38] addressed the problem of discovering topically meaningful communities from a social network. It discovers both community interests and user interests based on the information and linked associations. They also demonstrated the effectiveness of our model on two real word datasets and show that it performs better than existing community discovery models.

A genetic algorithmic approach to discover communities has been proposed by [39] as a novel multiobjective evolutionary algorithm. This proposed method found results by applying synthetic data sets in comparison with modularity method. Though there are numerous methods for discovering communities, a new concept of community seed, vector and relation matrix was proposed by [40]. This method that the proposed algorithm is highly effective at detecting community structures in both computer-generated and real-world networks.

Optimization techniques are part of computational intelligence. These techniques are well applied to varieties of problems. One such work to discover communities was demonstrated by [55]. This algorithm does not require any prior knowledge about the community structure. It is a quiet common scenario in community discovery that a node (or vertex) may present in more than one community, as the clustering has been done. This phenomenon is called as overlapping community structure. In simple terms, a node belongs to several communities is called as overlapping community structure. An algorithm to find some seed and to expand this seeding information to discover communities in a complex network structure has been studied by [41].

Particle Swarm Optimization technique to discover communities and to effectively conduct personalized recommendation in a scientific paper management system was studied by [42]. An ant colony optimization algorithm to discover communities in network was presented by [43]. This algorithm uses artificial ants to travel on a logical digraph to construct solutions of community detection. Rather than using social network analysis techniques to discover communities, [44] proposed a new algorithm using ant colony optimization. The ACO techniques are used to find cliques in the network and assign these cliques as nodes in a reduced graph to use with SNA algorithms.

Fig. 4 A group of three
communities and the
interaction among the
members

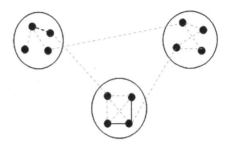

2 Problem Description

Social networking is the grouping of individuals into specific groups, or a
neighborhood subdivision. Although it is possible to network socially in person the
establishment of networking is tend to be different now-a-days. Online social
networking sites gains its popularity due to its ease of use and the way it simplified
the visitors to identify the like-minded individuals, as friends. The primary purpose
of these kinds of network sites is to facilitate the visitors with large volume of
users who have similar profile. The study of networks is an active area of research
due to its capability of modeling many real-world complex systems. This real-
world system includes millions of web pages, co-authorship network, Friend-of-a-
Friend (FoAF). Social networks can be represented as a graph structure. The
typical size of this type of networks is now counted in millions, even in billions, of
nodes and these scales demand new methods to retrieve comprehensive information from their structure (Fig. 4).

The division of network nodes into groups is called as community. Among
them the network connections are dense, but between which they are sparser.
Community structure is closely related to the ideas of graph partitioning in graph
theory and computer science. Communities can be discovered by clustering the
graph. They provide a natural division of graph nodes into densely connected
subgroups. Communities are formed by subsets of nodes in a graph, which are
closely related.

When the network is represented by graph structure, in which the people are
represented by vertices and the relationship is represented by edges, then the
interesting pattern in the network can be done by clustering the entire graph using
some clustering technique. Such a type of interesting pattern over a commonality
of group is called as community.

Graph clustering is the task of grouping the vertices of the graph into clusters
taking into consideration the edge structure of the graph in such a way that there
should be many edges within each cluster (*intra-cluster*) and relatively few
between the clusters (*inter-cluster*). Partitioning the graph into subgraph can be
done on various criteria depends on the type of the application or domain, usually
grouped according to the fact the nodes present a relatively high number of
connections. For a given data set, the goal of clustering is to divide the data set into

Fig. 5 A graph with three
partitions

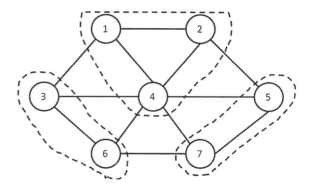

clusters such that the elements assigned to a particular cluster are similar or connected in some predefined sense (Fig. 5).

The graph clustering problem consists on dividing G into k disjoint partitions and the partitioned subgraph is called a cluster. Due to the complexity in partitioning the graph into k disjoint subgraphs, the graph clustering problem is considered as NP-hard problem. The goal is minimize the number of cuts in the edges of the partition in the entire graph.

Online social networking sites gains its popularity due to its ease of use and the way it simplified the visitors to identify the like-minded individuals, as friends. The primary purpose of these kinds of network sites is to facilitate the visitors with large volume of users who have similar profile. It allows individuals to connect with the others over the Internet worldwide, nevertheless of considering the geographical location over the globe. The profile sharing is the main aspect of these kinds of web sites. The number of users participating in these networks is very large, say in hundred millions, and growing day by day. Users of these sites form a social network, which provides a powerful means of sharing, organizing, and finding content and contacts. Due to the ease of use with mobile phones for getting updates and to host the posts of their choice and to browse their own friend's networks on social networking websites, people prefer to use mobile devices now-a-days. This change in using mobile devices than desktop users reported the usage from 91 % (mobile users) to 79 % (desk top computer users), according to leading World Wide Web and social networking site reporter [57]. This phenomenon says that the people prefer to approach online social networking sites than by using traditional way of meeting people in-person and to establish and grow their network, like friendship.

When the graph G has V vertices and E edges the graph partitioning splits the entire graph into number of subgraphs. This graph partitioning problem arises in many real-time environments, especially in computer science. Generally, finding an exact solution to a partitioning task of this kind is believed to be an NP-hard problem, due to its complexity in structure of the real-world systems, as the size of the graph is too large. The number of vertices in a graph is called as size of the graph. Community discovery falls under the classification of combinatorial optimization problem. The graph partitioning problem consists of dividing G into k disjoint partitions.

Community discovery in online social networking environment provides many advantages which includes, Identifying the like-minded individuals, Identifying the like-minded groups, Identifying the organizations involved in similar type of projects, People who shared the similar content over a web site, People who commented a response over a blog post, People who identified similar profiles to establish new network, and People who established communication by means of exchanging messages.

3 Methodology

Discovering communities in large real–world complex systems can be done by partitioning the graph into subgraphs. This partitioning can be done by various approaches. When the graph G has V vertices and E edges the graph partitioning splits the entire graph into number of subgraphs. Finding an exact solution to a partitioning task is believed to be an NP-hard problem, due to its complexity in structure of the real-world systems, as the size of the graph is too large.

Many methods have been developed during the past decade by various researchers. This research focused on discovering such communities from a moblog social network by applying the concept of mutual accessibility. The objective of the method is to cluster the graph dataset into clusters so that the members in it know among themselves each other. Social network is represented by a graph structure in which the individual or person is represented by nodes and the relationship among the person is represented by edges.

Consider a directed graph $G = (V, E)$ where V is the set of vertices and E is the set of edges. A *path* from node v_0 to v_k in G is an alternating sequence $(v_0, (v_0, v_1), v_1, (v_1, v_2), \ldots, v_{k-1}, (v_{k-1}, v_k), v_k)$ of nodes and edges that belong to V and E, respectively. Two nodes v and w in G are *path equivalent* if there is a path from v to w and path from w to u. Path equivalence partitions V into maximal disjoint sets of path equivalent nodes. This path equivalence property of the graph is utilized to discover communities so that when the graph is partitioned into disjoint subgraphs, the nodes in individual subgraphs are accessible by each other. With reference to social networking terminology, the members in individual communities have known among themselves each other.

Social network represents relationship data. Reachability is an equivalence relation ($u \# v$), due to the following properties:

i. It is *reflexive*: there is a trivial path of length zero from any vertex to itself. That is $u \# u$.

ii. It is *symmetric*: if there is a path from u to v, then there should be a path from v to u. That is, if $u \# v$, then $v \# u$.

iii. It is *transitive*: If there is a path from u to v and a path from v to w, the two paths may be concatenated together to form a path from u to w. That is, $u \# v$ and $v \# w$, then $u \# w$.

Fig. 6 A framework to
extract communities from
moblog website

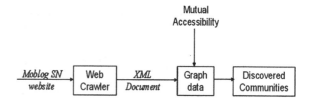

For any equivalence relation $u \# v$, equivalence classes can be defined by:

$$[u] = \{v|u\#v\}$$

Mutual accessibility can be achieved by the following four step process:

1. Apply DFS (G) and compute fn[u] for each vertex v.
2. Compute G^T = (V, E^T).
3. Apply DFS (G^T):
   visited[1..n] = 0
   for each vertex u in decreasing order of fn[u] do
   if visited[u] = 0 then call dfs(u)
4. Output the vertices of each tree in the depth first order.

In the above steps, DFS(G) represents the Depth First Search of the digraph. G^T represents the transpose graph of the digraph, which just reverses the edges of the digraph. For example, when there is an edge from u to v its transpose edge is represented from v to u.

Let x and y be two arbitrary vertices which are in the same subgraph (component) of G. A serial number $vn(v)$ is assigned to each vertex v the first time that it is visited. Assume, without loss of generality, the $vn(x) < vn(y)$. By definition, there is a path p in G from x to y. Since there is a subtree which with its cross edges and backward edges contains p, let r be the root of the minimal such subtree. Because of the minimality of the subtree, if p does not pass through r, there would have edges from r to an ancestor of each of x and y. It implies that p would contain at least one cross edge. Thus p must passes through r. It implies that r is in the same subgraph as x and y. It can be concluded if S is a subgraph of G, then the vertices of S define a tree which is a subgraph of the DFS forest. By this proposed method, mutually accessible nodes form individual communities (Fig. 6).

A framework was proposed to extract the blog posts from a moblog website. It accepts the posts which are hosted by using specialized mobile devices like iPad. The following are the components of the proposed framework:

A moblog web site or a Universal Resource Link (URL) is given as an input to the web crawler. This website contains the blog posts done by the bloggers (the persons who posted blogs over the moblog website. The crawler's output is generated in the form of Extensible Markup Language (XML). It contains the details like blogger's name, blog post and the time stamp information. This XML data are converted into graph data in node list format. While doing like so, the names of the blogger is converted into numbers, to preserve the privacy of the

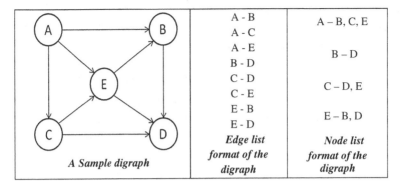

	A - B	A – B, C, E
	A - C	
	A - E	B – D
	B - D	
	C - D	C – D, E
	C - E	
	E - B	E – B, D
	E - D	
A Sample digraph	*Edge list format of the digraph*	*Node list format of the digraph*

Fig. 7 A digraph and its edge list and node list formats

bloggers. The outcome of this conversion yields moblog dataset. There are two methods to represent graph data. (i) Edge list format and (ii) Node list format. The edge list format represents how one node in the graph is being associated with the other node of the graph. The second representation (node list format) represents how one node is associated with what are all the other nodes of the graph. This graph data in edge list format will be used by the proposed method to discover communities over the moblog dataset (Fig. 7).

A moblog dataset which contains 63592 vertices and 813098 edges has been created by crawling a moblog website. It contains blog posts and its responses. This large moblog dataset is utilized as input to the proposed method. It is able to discover 142 communities of variant sizes.

The proposed method is also tested with other two well-known methods called (i) Modularity and (ii) Extremal Optimization, to evaluate the scalability and effectiveness. It is being apparent that the proposed method outperforms well for variant sizes of dataset. The results outperform to fit into any size of the graph. The results are discussed in the following section.

4 Results

When two persons know among themselves in the network (graph), then they are "mutually known themselves". Hence the method of mutual accessibility was applied to discover the communities over the created dataset. These two members are partitioned through the proposed method to form the subgraph from the existing graph. The implemented method has its uniqueness in such a way that the members of the individual communities (subgraph) known among themselves. This also provided an enhancement that the interaction among the individual communities (intra-cluster) is strengthen than the members in other community (inter-cluster). Hence, the proposed method satisfied the property of the community by graph partitioning method. The proposed method decomposed the moblog dataset as a graph into subgraph of various sizes, called as clusters. There are 142

Table 1 Summary of moblog dataset and discovered communities

Property	Value
Number of vertices	63592
Number of edges	813098
Type of graph	Connected
Number of communities discovered	142

Table 2 Comparative analysis of results obtained by various methods

Network structure	No. of nodes	No. of edges	No. of communities discovered		
			Modularity	Extremal optimization	Mutual accessibility
Zachary karate club [45]	34	77	2	4	1
American college football team [46]	115	613	6	–	1
Jazz musicians network [47]	198	2742	4	5	1
E–mail network [48]	1133	5452	13	15	1
Online social network [49]	1899	13838	–	–	4
Synthetic mobile network [50]	10000	86619	–	–	1
PGP network [51]	10680	24316	80	965	1
Co-authorship network [52]	40421	175693	–	–	954
Moblog social network [53]	63592	813098	–	–	142

clusters have been discovered by the proposed method for the moblog dataset. The summary of the dataset and the number of communities discovered are given as in the Table 1:

The effectiveness of the proposed method was compared with two existing methods called Modularity and Extremal Optimization (EO). The results obtained by these two methods using different datasets were also compared with the proposed method.

The modularity method falls into the category of agglomerative hierarchical clustering method. It is based on the concept of modularity Q. The extremal optimization method is a new divisive algorithm that optimizes the modularity Q using a heuristic search. It operates optimizing a global variable by improving extremal local variables that involve co-evolutionary avalanches. Table 2 represents the datasets compared and the results obtained by the three methods:

There are two benchmarking datasets tested by the proposed method. They are:

(i) Zachary Karate Club dataset. This dataset is made a remarkable study in the field of social networking. Hence, it is vital to compare the results with the existing dataset with the proposed dataset (i.e.) the moblog dataset. It is also essential to compare the results obtained by the proposed method with the existing method.

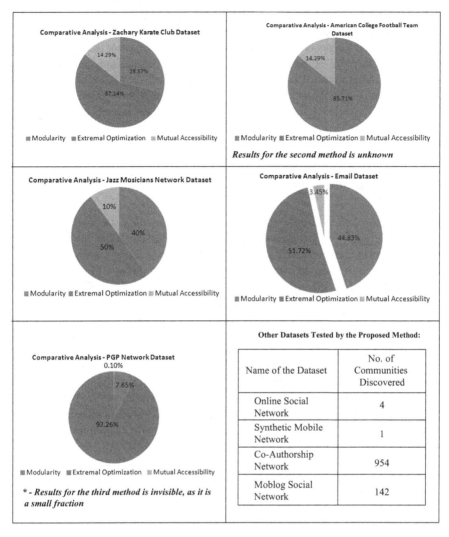

Fig. 8 Comparative analysis for different datasets

(ii) Co-Authorship Network. Since inception, this dataset is studied in various dimensions. The members can have ties with many members. Hence, it is also essential to compare the results obtained with the proposed dataset with this dataset also.

In the above table, some of the datasets are exclusively tested by the proposed method. For some datasets, the results may neither discussed by the particular method nor unknown for the corresponding dataset. Hence, its results are just filled with hyphen.

As it is known from the above table, the data ranges from 34 to 63592, the number of edges ranges between 77 and 813098, the results obtained by the different methods ranges between 0 and 965. These data may not fit into a common scale. Hence, they are normalized by the method of "normalization with division by average". After normalization, the scale fit ranges between 0 and 10. Again, for simplicity, the results obtained by the various datasets that corresponds to the three methods are converted into percentage level, to prepare bar chart. The results obtained after these conversions, are as shown in Fig. 8:

5 Conclusion and Discussion

Blog-based research is a new phenomenon. It is used for collective wisdom. It attracted many diversified areas of research. Mobile communication facilities shrink the world and it became as part of human life in today's scenario.

This research focused on establishing social network through blog posts which are hosted by using specialized mobile devices such as iPad. The network is formed from the blog posts (bloggers) and their responses from a website, which is specialized in mobile blogs (moblog). The users' interactions are converted into numbers, rather than using the names of bloggers, to preserve their privacy.

Discovering communities in a social networking environment is an active area of research. Though there are several methods to discover communities are proposed over a period of time by researchers, this research introduced a method known as "mutual accessibility". Strong connectedness can be used to discover the clustering of vertices in the entire graph, since the strong connectivity is an equivalence relation. This property was used to discover the communities over the moblog dataset. A framework also utilized to serve as a basis for the creation of moblog dataset. It enables the members of the intra-cluster community have more interaction than the inter-clusters, known among themselves each other.

This research is able to create a moblog dataset, by crawling a moblog website, which accepts the blog posts hosted by specialized mobile users. There are 63592 vertices, 81308 edges in the graph and the introduced method discovered 142 communities. It is able to cluster the entire graph into 142 disjoint subgraphs. Since, graph clustering is NP-hard problem, the new method is able to cluster the graph of various sizes. The proposed method decomposed the moblog dataset as a graph into subgraph of various sizes. The scalability is also tested with different datasets of various sizes. The result shows that the proposed method outperforms well and scaled well to all the datasets. Another advantage of moblog is the blog posts hosted by the users are registered users.

The proposed method has the following advantages:

(i) The members in the individual communities known among themselves each other. Hence, the members in the intra-cluster communities are tightly-coupled when compared with members in the inter-cluster communities.

(ii) Another advantages by the research is that the bloggers are all registered users. Hence, there is no possibility of having spam-blogs and the anonymous bloggers.

References

1. Duch, W.: What is computational intelligence and what could it become? http://cogprints.org/5358/1/06-CIdef.pdf
2. Hanneman, R.A., Mark, R.: Introduction to Social Network Methods. University of California, Riverside. http://faculty.ucr.edu/∼hanneman/ (2005)
3. Nazir, A., Raza, S., Chuah. C-N.: Unveiling facebook: a measurement study of social network based applications. In: Proceedings of the 8th ACM SIGCOMM Conference on Internet Measurement, pp. 43–56, (2008)
4. Goth, G.: Are social networking sites growing up? IEEE Distrib. Syst. Online **9**(2), 3 (2008). doi: 0802-mds2008020003
5. Duch, J., Arenas, A.: Community detection in complex networks using extremal optimization. Phys. Rev. **E72**, 027104 (2005)
6. Gomez, V., Kaltenbrunner, A., Lopez, V.: Statistical analysis of the social network and discussion threads in slashdot. In: Proceedings of International Conference on World Wide Web, ACM, (2008)
7. Erman, N.: Citation analysis for e-government research. In: Proceedings of the 10th Annual International Conference on Digital Government Research: Social Networks: Making Connections between Citizens, Data and Government, pp. 244–253, (2009)
8. Kumar, P., Zhang, K.: Social network analysis of online marketplaces. In: IEEE International Conference on e-Business Engineering, pp. 363–367, (2007)
9. He, B., Macdonald, C., He, J., Ounis, L.: An effective statistical approach to blog post opinion retrieval. In: Proceedings of International Conference, ACM, (2008)
10. Shen, D., Sun, JT., Yang, Q., Chen, Z.: Latent friend mining from blog data. In: Proceedings of IEEE International Conference on Data Mining, (2006)
11. Zhu, L., Sun, A., Choi, B.: Online spam-blog detection through blog search. In: Proceedings of International Conference, ACM, (2008)
12. Sekiguchi, Y., Kawashima, H., Okuda, H., Oku, M.: Topic detection from blog documents using users' interests. In: Proceedings of the 7th International Conference on Mobile data management, p. 108, (2006)
13. Schaeffer, SE.: Graph clustering. Comput. Sci. Rev. **1**(1), 27–64 (2007)
14. Xu, X., Yuruk, N., Feng, Z., Schweiger, T.A.J.: SCAN: a structural clustering algorithm for networks. In: Proceedings of the 13th ACM SIGKDD International Conference on Knowledge Discovery and Data Mining (KDD-07)
15. Long, B., Zhang, Z., Yu, PS.: Graph partitioning based on link distributions. Proceedings of the 22nd AAAI Conference on Artificial Intelligence, pp. 578–583, (2007)
16. Ratsch, G.: A brief introduction to machine learning. http://www.tuebingen.mpg.de/∼raetsch
17. Opsahl, T., Panzarasa, P.: Clustering in weighted networks. Soc. Netw. **31**(2), 155–163 (2009)
18. Latouche, P., Birmelé, E., Ambroise, C.: Bayesian methods for graph clustering. advances in data analysis, data handling and business intelligence. In: Studies in classification, data analysis, and knowledge organization, Part 3, pp. 229–239. Springer, (2010)
19. Mishra, N., Schreiber, R., Stanton, I., Tarjan, R.E.: Clustering social networks. Lect. Notes Comput. Sci. **4863**, 56–67 (2007)
20. Moser, F., Ge, R., Ester, M.: Joint cluster analysis of attribute and relationship data without a-priori specification of the number of clusters. In: Proceedings of the 13th ACM SIGKDD

International Conference On Knowledge Discovery and Data Mining (KDD'07), pp. 510–519, (2007)

21. Lancichinetti, A., Fortunato, S.: Community detection algorithms: a comparative analysis. arxiv:0908.1062v1 [physics.soc-ph], (2009)

22. Tantipathananandh, C., Berger-Wolf, T., Kempe, D.: A Framework for community identification in dynamic social networks. In: Proceedings of the 13th ACM SIGKDD International Conference on Knowledge Discovery and Data Mining (KDD-07) (2007)

23. Narasimhamurthy, A., Greene, D., Hurley, N., Cunningham, P.: Community finding in large social networks through problem decomposition. In: 19th Irish conference on Artificial Intelligence and Cognitive Science (AICS'08), (2008)

24. Pizzuti, C.: Community detection in social networks with genetic algorithms. In: Proceedings of the 10th annual conference on Genetic and evolutionary computation, pp. 1137–1138, (2008)

25. Newman, M.E.J.: Fast algorithm for detecting community structure in networks. Phys. Rev. **E69**, 066133 (2004)

26. Newman, M.E.J.: Finding community structure using the eigenvectors of matrices. Phys. Rev. E **74**, 036104 (2006)

27. Duch, J., Arenas, A.: Community detection in complex networks using extremal optimization. Phys. Rev. **E72**, 027104 (2005)

28. Reichardt, J., Bornholdt, S.: Statistical mechanics of community detection. Phys. Rev. **E74**, 016110 (2006)

29. Leskovec, J., Lang, K. J., Dasgupta, A., Mahoney, M. W.: Statistical properties of community structure in large social and information networks. In: Proceedings of the 17th International World Wide Web Conference (WWW2008) ACM, pp. 695–704, (2008)

30. Newman M. E. J., Girvan M.: Finding and evaluating community structure in networks. Phys. Rev. **69**(2) (2004)

31. Michael C, Xu, J.: Mining communities and their relationships in blogs: a study of online hate groups. Int. J. Hum. Comput. Stud. **65**, 57–70 (2007)

32. Kazienko, P., Musiał, K.: On utilizing social networks to discover representatives of human communities. Int. J. Intell. Inf. Database Syst. **1**(3/4), 293–310 (2007)

33. Raghavan, U.N., Albert, R., Kumara, S.: Near linear time algorithm to detect community structures in large-scale networks. Phys Rev E **76**, 036106 (2007)

34. Steinhaeuser, K., Chawla, N.V.: Identifying and evaluating community structure in complex networks. J. Pattern Recogn. Lett. **31**(5), 413–421 (2010)

35. Li, X., Gao, C.: A novel community detection algorithm based on clonal selection. J. Comput. Inf. Syst. **9**(5), 1899–1906 (2013)

36. Ferrara, E.: Community structure discovery in facebook. Int. J. Social Network Mining **1**(1), 67–90 (2012)

37. Gong, M.-G., Zhang, L.-J., Ma, J.-J., Jiao, L.-C.: Community detection in dynamic social networks based on multiobjective immune algorithm. Int. J. Social Netw. Min. **27**(3), 455–467 (2012)

38. Sachan, M., Contractor, D., Faruquie, T. A., Subramaniam, LV.: Using content and interactions for discovering communities in social networks. In: Proceedings of the 21st International Conference on World Wide Web WWW'12, pp. 331–340, (2012)

39. Chen, G., Wang, Y., Wei, J.: A new multiobjective evolutionary algorithm for community detection in dynamic complex networks. Mathematical problems in engineering, Volume 2013, Article ID 161670, (2013). http://dx.doi.org/10.1155/2013/161670

40. Ma, R., Deng, G., Wang, X.: A cooperative and heuristic community detecting algorithm. J. Comput. **7**(1), 135 (2012)

41. Wang, C., Li, Y.: Algorithm for detecting overlapping community in complex networks. J. Inf. Comput. Sci. **10**(9), 2625–2632 (2013)

42. Ma, R., Deng, G., Wang, X., Ailinna.: The Application of the PSO Based Community Discovery Algorithm in Scientific Paper Management SNS Platform. Software Engineering

and Knowledge Engineering: Theory and Practice. Adv. Intell. Soft Comput. **162**, 661–667 (2012)

43. Chen, B., Chen, L., Chen, Y.: Detecting community structure in networks based on ant colony optimization. http://elrond.informatik.tu-freiberg.de/papers/WorldComp2012/IKE2561.pdf

44. Sadi, S., Etaner-Uyar, S., Gündüz-Öğüdücü, S.: Community detection using ant colony optimization techniques. http://web.itu.edu.tr/etaner/mendel09_2.pdf

45. Zachary, W.W.: An information flow model for conflict and fission in small groups. J. Anthropol. Res. **33**, 452–473 (1977)

46. Girvan, M., Newman, M.E.J.: Community structure in social and biological networks. Proc. Natl. Acad. Sci. **99**(12), 7821–7826 (2002)

47. Gleiser, P., Danon, L.:Community structure in Jazz. Adv. Complex Syst. **6**(4), 565–573 (2003)

48. Guimera, R., Danon, L., diaz-Guilera, A, Giralf, F., Arenas, A.: Self-similar community structure in a network of human interactions. Phys. Rev. E **68**, 06503 (2003)

49. Opsahl, T., Panzarasa, P.: Clustering in weighted networks. Soc. Netw. **31**(2), 155–163 (2009)

50. Narasimhamurthy, A., Greene, D., Hurley, N., Cunningham, P.: Scaling community finding algorithms to work for large networks through problem decomposition. In: 19th Irish Conference on Artificial Intelligence and Cognitive Science (AICS'08), Cork, Ireland (2008)

51. Guardiola, X., Guimera, R., Arenas, A., diaz-Guilera, A., Streib, D., Amaral, L. A. N.: Macro- and micro-structure of trust networks. arXiv: cond-mat/0206240v1 (2002)

52. Newman M. E. J.: The structure of scientific collaboration networks. In: Proceedings of the National Academy of Sciences, pp. 404–409 (2001)

53. Mohamed Sathik, M., Abdul Rasheed, A.: Community discovery in mobile blog. J. Glob. Res. Comp. Sci. **2**(2), 12–16 (2011)

54. Coscia, M., Giannotti, F., Pedreschi, D.: A classification for community discovery methods in complex networks. 15 June 2012 arXiv:1206.3552v1 [cs.SI]

55. Sobolevsky, S., Campari, R., Belyi, A., Ratti, C.: A general optimization technique for high quality community detection in complex networks. 15 Aug 2013 arXiv:1308.3508v1 [cs.SI]

Web References

56. http://www.flowtown.com/blog/how-are-mobile-phones-changing-social-media?display=wide

57. http://www.web-strategist.com

58. http://www.artificial-immune-systems.org/people-new.shtml

Uncertainty-Preserving Trust Prediction in Social Networks

Anna Stachowiak

Abstract The trust metric has became an increasingly important element of social networks. While collecting, processing and sharing information is becoming easier and easier, the problem of the quality and reliability of that information remains a significant one. The existence of a trust network and methods of predicting trust between users who do not know each other are intended to help in forming opinions about how much to trust information from distant sources. In this chapter we discuss some basic concepts like trust modeling, trust propagation and trust aggregation. We briefly recall recent developments in the area and present a new approach that focuses on the uncertainty aspect of trust value. To this end we utilize a theory of incompletely known fuzzy sets and we introduce a new, uncertainty-preserving trust prediction operator based on group opinion and on relative scalar cardinality of incompletely known fuzzy sets. Motivated by the need for proper uncertainty processing, we have constructed a new method of calculating relative scalar cardinality of incompletely known fuzzy sets that ensures the monotonicity of uncertainty. We outline the problem of uncertainty propagation, and we illustrate by examples that the proposed operator provides most of the desirable properties of trust and uncertainty propagation and aggregation.

Keywords Trust propagation · Trust aggregation · Uncertainty · Incompletely known fuzzy sets · Relative cardinality

A. Stachowiak (✉)
Faculty of Mathematics and Computer Science, Adam Mickiewicz University,
Umultowska 87 61-614 Poznań, Poland
e-mail: aniap@amu.edu.pl

W. Pedrycz and S.-M. Chen (eds.), *Social Networks: A Framework*
of Computational Intelligence, Studies in Computational Intelligence 526,
DOI: 10.1007/978-3-319-02993-1_6, © Springer International Publishing Switzerland 2014

1 Introduction

Widespread access to the Internet and the emergence of the new generation of web applications referred to as Web 2.0 enables collaboration, interaction and knowledge sharing between an unprecedented number of users. In such a large community the question of whom and what to trust has become increasingly important. On the Internet, as in other areas of life, trust information can help a user to sort and filter information, make proper decisions, evaluate recommendations, act under uncertainty. Therefore, along with social networks' growth, virtual networks of trust are built by their members by explicitly indicating users who are trustworthy, with the intent to help the community to use web content safely. When there are many alternative sources available, the existence of a trust network allows users to choose the most reliable one.

A thorough study of a notion of trust in computer science was carried out by Artz and Gil in [1]. The authors quote the definition from Mui et al. [2] saying that trust is "a subjective expectation an agent has about another's future behavior based on the history of their encounters." They recognize four major areas of trust research. The first is policy-based trust, where a trusted third party serves as an authority for issuing and verifying credentials assigned to a specific party. Next is reputation-based trust, which uses past actions of a user to assess its future behavior. In the absence of first-hand knowledge, trust is computed over social networks or along paths of trust. The third area covers general models of trust, working on trust modeling and defining, helping to better understand subtle and complex aspects of composing, capturing, and using trust in computational settings. The last category refers to trust in information resources, including Web resources and web sites' reliability.

In the present chapter we refer to the category of reputation-based trust. We consider trust networks that offer mechanisms to verify the source of the information. The idea is different from the centralized certification system, which is not always possible and is not personalized. In a trust network, trust relations are dynamically created and constantly modified by the users themselves, by making subjective judgments based on their knowledge about other users and past personal experience.

Trust networks have already become a central component of many intelligent web applications and have proved to contribute to the improvement of the systems' work. Well-known examples of such applications include peer-to-peer (P2P) networks [3, 4], wikis (like Wikipedia, an online encyclopedia created by users, with the possibility of rating moderators), social networking sites and communities (like Facebook, Stack Overflow, CouchSurfing), e-commerce and recommender systems (like FilmTrust or Epinions, which is a consumer review site where users can trust or not trust each other based on their ratings and review of products), social news (Slashdot), email anti-spam techniques, question answering systems [5] and others. In particular, a lot of research effort has been devoted to the role of trust in recommender systems [6–12]. It has been proven that trust-enhanced recommender systems can generate more personalized, more reliable and more relevant recommendations, as data are tailored to the individual user, and mediated through the sources trusted by the user. Also, incorporating trust into recommendations makes them significantly more accurate for users who disagree with the average rating for a specific item. Another benefit from maintaining a trust system is that it encourages individuals to act in a trustworthy manner, having a positive impact on social network evolution.

In the following chapters we discuss in details the problem of trust modeling and processing with particular reference to the uncertainty (incompleteness) factor. Such incompleteness may occur during the process of trust construction, when the positive and negative evidence being collected do not necessarily complement each other. Possibly, a user will never be certain about how much to trust another user. As a modeling tool we use the theory of I-fuzzy sets (incompletely know fuzzy sets) which allows for imprecise statements and leaves room for uncertainty. We recall some basic definitions and facts of this theory. At the same time we argue that the notion of relative cardinality of I-fuzzy sets has not been satisfactorily discussed in the literature and there is still a need for a thorough study, especially in the context of uncertainty preserving. Consequently, we propose a new method for calculating relative cardinality of I-fuzzy sets and we place it at the heart of a trust prediction operator that is the main topic of this chapter.

In the final section we introduce and study a new cardinality-based operator for trust prediction that incorporates both trust propagation and aggregation, paying special attention to uncertainty preservation during the process of propagation and aggregation of trust. To complete, we give examples of trust prediction and study properties of this operator.

2 An Overview of Trust Modeling and Processing

Much work has been done towards formalizing trust as a computational concept. One possible method is represented by a probabilistic approach, studied for example in [3, 5, 13], which deals with trust in a binary way, meaning that a user can be either trusted or not. Next the probability that a user can be trusted is computed, often based on the number of positive and negative transactions. An example system is EigenTrust [3] for P2P networks, which computes global trust values for peers, based on their previous behavior and has been shown to be highly resistant to attack.

On the other hand, the need to express a more imprecise concept like "to trust somebody *very much*" or "*rather* do not trust somebody" led to a gradual model of trust. This approach fits better with many applications where the outcome of some action is not just positive or negative, but may be positive to some extent. A common example of gradual trust can be found in [6, 14, 15]. A well-suited mathematical model to represent such vague terms is fuzzy logic.

Recently, the potential of adding a distrust degree is being discovered. Experience with real trust systems such as Epinions and eBay suggests that distrust is at least as important as trust. Considering a pair (t, d) of trust and distrust gives a dual perspective that is valuable for modeling incomplete and inconsistent information about trust. In a probabilistic context this was done by Jøsang in [13], and examples of dual gradual models can be found in [15, 16]. The most comprehensive solution was proposed by Victor et al. in [8], where trust and distrust are drawn from a bilattice [17]. This approach enables one to distinguish the situation of lack of trust $(0, 1)$ from that of lack of knowledge $(0, 0)$, and to model inconsistency $(1, 1)$. Thereby the provenance of information is preserved, since additional knowledge about the source's degree of certainty about trust is passed on.

Besides trust modeling, the problems that motivated much of the existing work include trust propagation and aggregation. Clearly, in a large network not all the users know each other. To establish a trust score between two users we may need to transfer trust information from a third party, and how to transfer it is a problem of *trust propagation*. Usually we assume that trust is transitive-if a user a trusts (to some degree) user b and user b trusts c, then we can somehow estimate the trust score of a in c. In general, using propagation we are able to calculate a trust score of a in some z if there is a path connecting users a and z in a trust network.

It is not so clear, however, how to propagate distrust. Several scenarios are possible: "the enemy of my enemy is my friend" or "don't respect someone not respected by someone you don't respect". In [15] the authors consider and compare two different models for the propagation of distrust, one-step distrust propagation and propagated distrust, and show the favorable influence of this mechanism for predicting trust score.

When trust recommendations come from many different sources, combining them to synthesize a new trust score is another commonly addressed problem called *trust aggregation*. The final trust estimation is an effect of both trust propagation and aggregation. Some authors offer separate propagation and aggregation operators. In [8] four propagation operators are defined, each designed to serve different purposes:

$$Prop_1((t_1, d_1), (t_2, d_2)) = (T(t_1, t_2), T(t_1, d_2)),$$
$$Prop_2((t_1, d_1), (t_2, d_2)) = (T(t_1, t_2), T(N(d_1), d_2)),$$
$$Prop_3((t_1, d_1), (t_2, d_2)) = (S(T(t_1, t_2), T(d_1, d_2)), S(T(t_1, d_2), T(d_1, t_2))),$$
$$Prop_4((t_1, d_1), (t_2, d_2)) = (T(t_1, t_2), S(T(t_1, d_2), T(d_1, t_2))),$$

where T, S, N denote, respectively, triangular norm, conorm and negation, as described in the next section. Next, propagated trust values are aggregated into one final trust score by means of the OWA-based operator (for more details, see [18]). Another trust metric, Advogato [19], computes trust relative to a "seed" of trusted accounts. TidalTrust, proposed by Golbeck et al. in [20], calculates a trust-based weighted mean. Other examples include Appleseed [21], Sunny [22] and Mole-Trust [23]. For an empirical comparison of TidalTrust with MoleTrust and the

O'Donovan trust metric [9] on an Epinions.com data set, we refer to [10]. Although the experiment did not yield any clear winner, it did show that each of the considered algorithms has its own merits.

3 Trust Networks with Uncertainty—Theoretical Basis

3.1 Modeling Trust with I-Fuzzy Sets

The trust model that we consider here makes it possible to deal with both imprecision and incompleteness of trust information. Imprecision appears in the ways in which humans express their trust in others (rarely in a black-and-white manner, more often in a variety of shades). There could be several sources of uncertainty that preclude a precise definition of trust. The most evident one is lack of knowledge—consider for example a newcomer who at the beginning is not acquainted with anyone in a social network; then he forms his opinion about other's reliability continuously, gathering pieces of information, probably never reaching a stage where he is completely free of doubts.

The imprecision of information can be successfully handled by the fuzzy set theory proposed by Zadeh in [24]. Let A be a fuzzy set in the universe X, i.e. a set characterized by a membership function $A : X \rightarrow [0, 1]$. Then $A(x)$ describes the degree to which an element $x \in X$ belongs to a fuzzy set A; this value can be interpreted for example as the degree of fulfillment of some property, as a satisfaction degree, and in particular as a trust degree. However, this model appears to be insufficient in the presence of incomplete (uncertain) information, when the exact degree of membership cannot be specified.

To be able to capture the incompleteness of information we consider a model in which each piece of information is described from a double-positive and negative-perspective. In terms of the set theory this means that an incompletely known set S in the universe X is given by a pair of sets (S^+, S^-) such that S^+ contains those $x \in X$ that we know belong to S, and S^- contains those $x \in X$ that we know do not belong to S. Thus (S^+, S^-) represents incomplete knowledge about a set S. If we assume S^+ and S^- to be disjoint then $S^+ \subseteq S \subseteq (S^-)^c$ and we may define an area $S^? = (S^+ \cup_S S^-)^c$ that consists of those $x \in X$ whose status is unknown (uncertain), but we do not allow inconsistency. A more general contradictory model, without this restriction, was considered for example in [25]. In the context of trust networks it was thoroughly studied among others in [8].

It is clear that a fuzzy set carries imprecise but complete information, since

$$A = A^+ = (A^-)^c.$$

In 1975, Zadeh [26] introduced interval-valued fuzzy sets (IVFS); later in 1983 Atanassov proposed the concept of intuitionistic fuzzy sets (AIFS) [27, 28]. Both of these theories are able to model incompletely known fuzzy sets as described above: by relaxing the equality $A^+ = (A^-)^c$ from fuzzy set theory and allowing for an uncertainty area by providing lower and upper bounds of a "true" membership value. Despite the semantic difference, these two models—AIFS and IVFS—are mathematically equivalent, as is shown for example in [29–31]. Therefore all of the presented results might be expressed equivalently by using either AIFSs or IVFSs. For convenience we adopt the interval-derived notation and use a common name-I-fuzzy set (incompletely-known fuzzy set).

An I-fuzzy set $\tilde{A}$ is a pair $(\underline{A}, \overline{A})$ of fuzzy sets $\underline{A}, \overline{A} : X \to [0, 1]$ that are interpreted, respectively, as a lower and upper bound of an unknown fuzzy set A such that

$$\underline{A} \subseteq A \subseteq \overline{A}.$$

Clearly, $\underline{A} = A^+$ and $\overline{A} = A^+ \cup_S A^? = (A^-)^c$. We denote the set of all fuzzy sets by $\mathcal{F}$ and the set of all I-fuzzy sets as $\mathcal{I}$, $\mathcal{F} \subset \mathcal{I}$.

Let us recall some basic facts of fuzzy set and I-fuzzy set theories. The standard relations and operations on fuzzy sets A and B are defined in the following way:

$$
\begin{aligned}
A = B &\Leftrightarrow \forall_x A(x) = B(x), && (equality) \\
A \subseteq B &\Leftrightarrow \forall_x A(x) \leq B(x), && (inclusion) \\
(A \cap_T B)(x) &= T(A(x), B(x)), && (intersection) \\
(A \cup_S B)(x) &= S(A(x), B(x)), && (union) \\
A^c(x) &= N(A(x)). && (complement)
\end{aligned}
\tag{1}
$$

where T denotes an arbitrary t-norm, i.e. an increasing, commutative and associative $[0, 1]^2 \to [0, 1]$ mapping satisfying $T(1, x) = x$ for all $x \in [0, 1]$ and S denotes an arbitrary t-conorm, i.e. an increasing, commutative and associative $[0, 1]^2 \to [0, 1]$ mapping satisfying $S(0, x) = x$ for all $x \in [0, 1]$. The most commonly used t-norms are:

- minimum t-norm: $T_\wedge(x, y) = x \wedge y$,
- algebraic t-norm: $T_a(x, y) = x * y$,
- Òukasiewicz t-norm: $T_L(x, y) = 0 \vee (x + y - 1)$.

By N we denote a negation, i.e. a decreasing $[0, 1] \to [0, 1]$ mapping satisfying $N(0) = 1$ and $N(1) = 0$; here we consider only the most popular Òukasiewicz negation: $N(a) = 1 - a$.

For I-fuzzy sets $\tilde{A} = (\underline{A}, \overline{A})$ and $\tilde{B} = (\underline{B}, \overline{B})$ in X we define the following relations and operations:

$$
\begin{aligned}
\tilde{A} = \tilde{B} &\Leftrightarrow \underline{A} = \underline{B} \ \& \ \overline{A} = \overline{B}, && (equality) \\
\tilde{A} \subseteq \tilde{B} &\Leftrightarrow \underline{A} \subseteq \underline{B} \ \& \ \overline{B} \subseteq \overline{A}, && (inclusion) \\
\tilde{A} \cap_T \tilde{B} &= [\underline{A} \cap_T \underline{B}, \overline{A} \cap_T \overline{B}], && (intersection) \\
\tilde{A} \cup_S \tilde{B} &= [\underline{A} \cup_S \underline{B}, \overline{A} \cup_S \overline{B}], && (union) \\
\tilde{A}^c &= [(\overline{A})^c, (\underline{A})^c]. && (complement)
\end{aligned}
\tag{2}
$$

Moreover, a scalar cardinality of a fuzzy set, the so-called sigma-count, was defined in [32] as:

$$\sigma_f(A) = \sum_{x \in X} f(A(x)) \tag{3}$$

with f being a non-decreasing function $f : [0, 1] \rightarrow [0, 1]$ such that $f(0) = 0$, $f(1) = 1$ called a cardinality pattern. This was next extended to the case of relative cardinality in the following way:

$$\sigma_f(A|B) = \frac{\sigma_f(A \cap_T B)}{\sigma_f(B)}. \tag{4}$$

The cardinality of an I-fuzzy set has been defined [33] as an interval whose length corresponds to the size of uncertainty of $\tilde{A}$:

$$\tilde{\sigma}_f(\tilde{A}) = [\sigma_f(\underline{A}), \sigma_f(\overline{A})]. \tag{5}$$

The relative cardinality of I-fuzzy sets will be discussed in details in the next section.

It appears that the I-fuzzy set framework can serve as a pre-eminent tool for modeling trust in trust networks, as it can be used to conveniently model both imprecision and incompleteness of trust values. The following definition formalizes the idea.

Definition 1 A trust network is a triplet (U, E, R) such that (U, E) is a directed graph with a set of nodes (users) U and edges (directed trust links) $E \subseteq U \times U$. R is an I-fuzzy mapping $E \rightarrow [0, 1]^2$. For each edge $(a, b) \in E$:

$$R(a, b) = [\underline{t}, \bar{t}]$$

where

- $R(a, b)$ is called the trust score of a in b;
- $\underline{t} \leq \bar{t}$ is called the necessary trust degree of a in b;
- $\bar{t}$ is called the possible trust degree of a in b;
- $u = \bar{t} - \underline{t}$ is an uncertainty degree (the amount of information deficiency).

The degree $\bar{t}$, "possible trust", refers to one of the possible interpretation of distrust: as the upper limit of trust. Referring to the previously used notation (t, d) we may establish the equality: $(t, d) = [\underline{t}, 1 - \bar{t}]$ for $t + d \leq 1$ and $(t, d) = [1 - \bar{t}, \underline{t}]$ for $t + d > 1$. In particular, $(0, 1) = [0, 0]$, $(0, 0) = [0, 1]$ and $(1, 1) = [0, 1]$. The length of the interval $[\underline{t}, \bar{t}]$ reflects the uncertainty about what the "true" trust degree t should be. This uncertainty is equally high for both cases $(0, 0)$ and $(1, 1)$ (although the amount of information is different). It may also be interesting to consider the case $[0.5, 0.5]$, which is different from $[0, 0]$ and $[1, 1]$ though assigned the same zero uncertainty. This problem was noticed and resolved in [34] where the general t-norm dependent form of the uncertainty factor was proposed:

$$u = T(N(t), N(d)). \tag{6}$$

Then for [0.5, 0.5] and for example for the algebraic t-norm we get $u = 0.25 > 0$.

Introducing uncertainty to trust modeling is conceptually simpler then dealing with distrust factor (that is difficult to interpret and operate on) but still it is more powerful then just a single trust value. We equip the user with the possibility to express his hesitation. As revealed by experiments conducted in [35], people tend to define their preferences in a incomplete way. Notice that the formulation "I rather trust y" may indicate that, under certain circumstances, the user allows for the full trust in y, but with some doubts; then we found convenient to model his opinion not just with 1 or 0.8 but with the interval [0.8, 1], to highlight the spectrum of possibilities that he has in mind saying "rather trust". On the other hand, statement "I trust y no more then a little" can be modeled by [0, 0.4], and it is hard to express the intention of this user without using intervals.

Information about the amount of uncertainty is essential for a user to make a well-informed decision. One of our main objectives when operating on trust scores would thus be preserving information not only about the trust value, but also about a scale of uncertainty. The main challenge would be how to propagate and aggregate uncertainty together with trust without loosing any vital information. Next sections would be devoted to this subject.

3.2 Relative Cardinality of I-Fuzzy Sets

After setting theoretical basis for trust modeling, this section provides an introduction to the trust prediction operator described in Sect. 4. Its main intention is to measure the opinion of a group of users and to this end we use the notion of relative cardinality.

The proper (the meaning of the word "proper" will be explained below) definition of the operator of relative cardinality of I-fuzzy sets requires the introduction of the notion of a *representation set* of an I-fuzzy set $\tilde{A}$. The representation set $Rep(\tilde{A})$ is defined as:

$$Rep(\tilde{A}) = \left\{ A^* \in \mathcal{F} \mid \underline{A}(x) \leq A^*(x) \leq \overline{A}(x), \text{ for each } x \in X \right\}. \tag{7}$$

We claim that each I-fuzzy set $\tilde{A}$ can be represented by an infinite set of fuzzy sets that possibly correspond to a true form of A "hidden" behind $\tilde{A}$. In other words, $A \in Rep(\tilde{A})$. In particular, $\underline{A} \in Rep(\tilde{A})$ and $\overline{A} \in Rep(\tilde{A})$.

For practical reasons, we distinguish $\hat{Rep} \subseteq Rep$ defined as:

$$Re\hat{p}(\tilde{A}) = \{A^* \in \mathcal{F} \mid A^*(x) = \underline{A}(x) \text{ or } A^*(x) = \overline{A}(x), \text{ for each } x \in X\}$$

which we call a *boundary representation set*. Obviously, $Re\hat{p}(\tilde{A}) = Rep(\tilde{A}) = \tilde{A}$ only when the I-fuzzy set $\tilde{A}$ is a fuzzy set. From the point of view of applications it is important that the cardinality of $Re\hat{p}(\tilde{A})$ for a finite universe X is finite, and is equal to $2^{|prop(\tilde{A})|}$, where $prop(\tilde{A}) = \{x \in X \mid \underline{A}(x) \neq \overline{A}(x)\}$. A similar idea was developed in [36].

The practical meaning of the representation set is twofold. Firstly it directs our attention to fuzzy representations other than $\underline{A}$ and $\overline{A}$, but equally legitimate, for I-fuzzy sets $\tilde{A}$. We use a representation set as a means to capture and emphasize the uncertainty resulting from deficiency of knowledge about which piece of information is true. Secondly, it suggests a way of operating on I-fuzzy sets-not directly but by applying relevant operations on fuzzy sets. We utilize these ideas to obtain a relative cardinality of I-fuzzy sets. We postulate that the relative cardinality operator should give the smallest interval from $[0, 1]$ that covers all possible relative cardinality values $\sigma_f(A^*|B^*)$ of fuzzy sets $A^* \in Rep(\tilde{A})$ and $B^* \in Rep(\tilde{B})$ and should be monotonic. This leads to the following general and intuitive formula for $\tilde{\sigma}_f(\tilde{A}|\tilde{B})$.

Definition 1 *The relative cardinality of two I-fuzzy sets $\tilde{A} = (\underline{A}, \overline{A})$ and $\tilde{B} = (\underline{B}, \overline{B})$ is given by the interval:*

$$\tilde{\sigma}_f(\tilde{A}|\tilde{B}) = \left[\min_{\substack{A^* \in Rep(\tilde{A}) \\ B^* \in Rep(\tilde{B})}} \sigma_f(A^*|B^*), \max_{\substack{A^* \in Rep(\tilde{A}) \\ B^* \in Rep(\tilde{B})}} \sigma_f(A^*|B^*) \right]. \quad (8)$$

The Definition 1 does not produce a specific formula for $\tilde{\sigma}_f$ for I-fuzzy sets, and instead is based on the well known relative cardinality operation for fuzzy sets σ_f. The following theorem states that $\tilde{\sigma}_f(\tilde{A}|\tilde{B})$ defined by (8) is monotonic with respect to uncertainty inclusion $\sqsubseteq$. This important property guarantees that reduction of the uncertainty of I-fuzzy sets $\tilde{A}$ and $\tilde{B}$ (increase of knowledge) implies reduction of the uncertainty about the value of $\tilde{\sigma}_f(\tilde{A}|\tilde{B})$ (down to zero for complete $\tilde{A}$ and $\tilde{B}$). We have found the monotonicity property essential when applying the operation $\tilde{\sigma}_f(\tilde{A}|\tilde{B})$ in practice, and so we pay a special attention to it. Absence of this property makes it difficult to draw meaningful conclusions from analysis of results, and to compare them.

Theorem 1 (Monotonicity). *Let $\tilde{A}$, $\tilde{B}$, $\tilde{A}_1$, $\tilde{A}_2$, $\tilde{B}_1$, $\tilde{B}_2$ be I-fuzzy sets over X, $\sqsubseteq$ be the uncertainty-based inclusion defined as:*

$$\tilde{A} \sqsubseteq \tilde{B} \Rightarrow Rep(\tilde{A}) \subseteq Rep(\tilde{B}), \quad (9)$$

and $\subseteq$ be the standard inclusion relation of intervals; then the following holds:

$$\tilde{A}_1 \sqsubseteq \tilde{A}_2 \;\&\; \tilde{B}_1 \sqsubseteq \tilde{B}_2 \Rightarrow \tilde{\sigma}_f(\tilde{A}_1|\tilde{B}_1) \subseteq \tilde{\sigma}_f(\tilde{A}_2|\tilde{B}_2).$$

Proof From the definition

$$\tilde{\sigma}_f(\tilde{A}_1|\tilde{B}_1) = [\alpha_1, \beta_1],$$
$$\tilde{\sigma}_f(\tilde{A}_2|\tilde{B}_2) = [\alpha_2, \beta_2],$$

where

$$\alpha_1 = min\{\sigma_f(A_1^*|B_1^*) \mid A_1^* \in Rep(\tilde{A}_1), B_1^* \in Rep(\tilde{B}_1)\},$$
$$\beta_1 = max\{\sigma_f(A_1^*|B_1^*) \mid A_1^* \in Rep(\tilde{A}_1), B_1^* \in Rep(\tilde{B}_1)\},$$

and

$$\alpha_2 = min\{\sigma_f(A_2^*|B_2^*) \mid A_2^* \in Rep(\tilde{A}_2), B_2^* \in Rep(\tilde{B}_2)\},$$
$$\beta_2 = max\{\sigma_f(A_2^*|B_2^*) \mid A_2^* \in Rep(\tilde{A}_2), B_2^* \in Rep(\tilde{B}_2)\}.$$

It follows from the assumptions and from definition of $\sqsubseteq$ that

$$Rep(\tilde{A}_1) \subseteq Rep(\tilde{A}_2) \;\&\; Rep(\tilde{B}_1) \subseteq Rep(\tilde{B}_2) \Rightarrow \alpha_2 \leq \alpha_1,$$

and, analogously,

$$Rep(\tilde{A}_1) \subseteq Rep(\tilde{A}_2) \;\&\; Rep(\tilde{B}_1) \subseteq Rep(\tilde{B}_2) \Rightarrow \beta_2 \geq \beta_1.$$

Thus

$$\tilde{\sigma}_f(\tilde{A}_1|\tilde{B}_1) \subseteq \tilde{\sigma}_f(\tilde{A}_2|\tilde{B}_2).$$

Obviously, as the number of $A^* \in Rep(\tilde{A})$ and $B^* \in Rep(\tilde{B})$ is infinite, the formula (8) is not applicable in practice, and we need a finite, easy to compute method. It can easily be seen that (8) can be simplified as:

$$\tilde{\sigma}_f(\tilde{A}|\tilde{B}) = \left[\min_{B^* \in Rep(\tilde{B})} \sigma_f(\underline{A}|B^*), \max_{B^* \in Rep(\tilde{B})} \sigma_f(\overline{A}|B^*)\right] \tag{10}$$

due to the monotonicity property of any t-norm. In general, however, it is a difficult problem which B^* to choose to minimize (maximize) the lower (upper) bounds of the interval (10). Let us recall briefly some finite approaches proposed in the literature which approximate (8) using only $\underline{B}$ and $\overline{B}$.

The first, interval-based approach, was mentioned in [37–39], and it is based on standard interval calculus. Let $\tilde{A}, \tilde{B} \subset \mathcal{I}$. Then:

$$\tilde{\sigma}_f^1(\tilde{A}|\tilde{B}) = \left[\frac{\sigma_f(\underline{A} \cap_T \underline{B})}{\sigma_f(\overline{B})}, \frac{\sigma_f(\overline{A} \cap_T \overline{B})}{\sigma_f(\underline{B})}\right].$$

According to the lower–upper approach proposed in [40] in the context of conditional probability, relative cardinality of $\tilde{A}$ and $\tilde{B}$ equals:

$$\tilde{\sigma}_f^2(\tilde{A}|\tilde{B}) = \left[\sigma_f(\underline{A}|\underline{B}), \sigma_f(\overline{A}|\overline{B})\right].$$

Another, necessary-possible, approach considered in [41] assumes that:

$$\tilde{\sigma}_f^3(\tilde{A}|\tilde{B}) = \left[\sigma_f(\underline{A}|\overline{B}), \sigma_f(\overline{A}|\underline{B})\right].$$

It can be verified that $\tilde{\sigma}_f^1(\tilde{A}|\tilde{B})$ overestimates (8) (in fact, the resulting interval may go beyond [0, 1], which is hard to interpret). On the other hand $\tilde{\sigma}_f^2(\tilde{A}|\tilde{B})$ and $\tilde{\sigma}_f^3(\tilde{A}|\tilde{B})$ underestimate (8) and are not monotonic in the sense of Theorem 1. It is thus of questionable merit to use these methods in practical applications, due to the risk of error accumulation and problems with interpreting the results.

Therefore, in the following we propose novel approximations of (8). We must consider separate formulas for distinct t-norms. For our further purposes, the most appropriate t-norms out of the popular ones would be the algebraic t-norm and Òukasiewicz t-norm. The presented approximations provide much better accuracy than $\tilde{\sigma}_f^1$, $\tilde{\sigma}_f^2$ and $\tilde{\sigma}_f^3$ and are easy to apply. The better accuracy comes from the fact that the presented approximations benefit from the representation set approach and consider $B^* \in Rep(\tilde{B})$ other than $\underline{B}$ and $\overline{B}$ only. That is why we call this approach a representation-based approach.

Proposition 1 *The relative cardinality of I-fuzzy sets $\tilde{A} = (\underline{A}, \overline{A})$ and $\tilde{B} = (\underline{B}, \overline{B})$ with algebraic t-norm T_a can be approximated by the interval:*

$$\tilde{\sigma}_f^4(\tilde{A}|\tilde{B}) = \left[\sigma_f(\underline{A}|B_L^*), \sigma_f(\overline{A}|B_R^*)\right] \tag{11}$$

where

$$B_L^*(x_i) := \begin{cases} \overline{B}(x_i) & \text{if} \quad \underline{A}(x_i) \in min_x\{\underline{A}(x)\} \\ \underline{B}(x_i) & \text{otherwise,} \end{cases}$$

$$B_R^*(x_i) := \begin{cases} \underline{B}(x_i) & \text{if} \quad \overline{A}(x_i) \in max_x\{\overline{A}(x)\}, \\ \overline{B}(x_i) & \text{otherwise.} \end{cases}$$

Proposition 2 *The relative cardinality of I-fuzzy sets $\tilde{A} = (\underline{A}, \overline{A})$ and $\tilde{B} = (\underline{B}, \overline{B})$ with Òukasiewicz t-norm $T_{\dot{O}}$ can be approximated by the interval:*

$$\tilde{\sigma}_f^5(\tilde{A}|\tilde{B}) = \left[\sigma_f(\underline{A}|B_L^*), \sigma_f(\overline{A}|B_R^*)\right] \tag{12}$$

Table 1 Relative
cardinality: $\tilde{\sigma}_f^1 - \tilde{\sigma}_f^5$ versus
def

Method	Algebraic t-norm	Łukasiewicz t-norm
$\tilde{\sigma}_f^1$	[0.2, 1.11]	[0.07, 0.86]
$\tilde{\sigma}_f^2$	[0.4, 0.56]	[0.14, 0.43]
$\tilde{\sigma}_f^3$	[0.33, 0.7]	[0.21, 0.57]
$\tilde{\sigma}_f^4$	[0.32, 0.7]	–
$\tilde{\sigma}_f^5$	–	[0.08, 0.67]
def	[0.32, 0.7]	[0.08, 0.67]

where

$$B_L^*(x_i) := \begin{cases} \underline{B}(x_i) & if \quad \underline{A}(x_i) \geq 1 - \underline{B}(x_i), \\ \overline{B}(x_i) & if \quad \underline{A}(x_i) \leq 1 - \overline{B}(x_i), \\ 1 - \underline{A}(x_i) & otherwise. \end{cases}$$

$$B_R^*(x_i) := \begin{cases} \overline{B}(x_i) & if \quad \overline{A}(x_i) \geq 1 - \frac{1}{2}(\underline{B}(x_i) + \overline{B}(x_i)) \\ \underline{B}(x_i) & otherwise. \end{cases}$$

Example 1 Let us consider two I-fuzzy sets $\tilde{A}$ and $\tilde{B}$ in the universe $X = \{x_1, x_2\}$:

$$\tilde{A} = \{[0.2, 0.3]/x_1, [0.4, 0.7]/x_2\},$$
$$\tilde{B} = \{[0, 0.5]/x_1, [0.7, 0.9]/x_2\}.$$

Then the relative cardinality calculated according to $\tilde{\sigma}_f^1, \tilde{\sigma}_f^2, \tilde{\sigma}_f^3, \tilde{\sigma}_f^4$ and $\tilde{\sigma}_f^5$ for $f = id$ and two t-norms T_a and $T_\text{Ò}$ is as given in Table 1. The approximate results given by $\tilde{\sigma}_f^1 - \tilde{\sigma}_f^5$ are confronted with the exact ones according to definition formula (8).

4 Cardinality-Based Operator for Predicting Trust Score with Uncertainty

The process of predicting the trust score of a user in a social network can be done both:

- locally; then the predicted trust score is a subjective level of trust from the point of view of a single user;
- globally; then the predicted trust score represents the trust of the whole community of users; it is then often called "reputation".

On the other hand, when predicting trust scores two mechanisms are applied:

- trust propagation—if a trust score is to be determined based on the opinion of distant, not directly related, users;
- trust aggregation—if there is a need to combine trust scores provided by several users.

The basic postulate for a trust prediction operator is that it should be sensitive to the amount of input trust and the amount of uncertainty about the input trust. Obviously, the higher the input trust/uncertainty degree, the higher output trust/uncertainty we may predict. Also, reduction of the input trust/uncertainty implies a reduction of output trust/uncertainty. Hence a trust prediction operator should be monotonic with respect to two relation: $\subseteq$ and $\sqsubseteq$.

When considering the propagation aspect of the operator, we adopt the wildly accepted assumption that the length of a propagation path influences the certainty about the final trust score (the effect is similar to the passing of rumors; the longer the path, the less reliable is the outcome). As for aggregation, it should be idempotent (aggregation of any number of equal trust scores gives the same trust score) and commutative (the order of trust scores is not important). Importantly, the propagation mechanism is neither idempotent nor commutative. More detailed properties will be discussed together with examples in the later part of the chapter.

4.1 Operator UTrust$_\sigma$: Construction

In the following we propose and discuss an uncertainty-preserving trust prediction operator $UTrust_\sigma$ that combines both trust propagation and aggregation functionality and can be used for prediction of a local trust score with uncertainty. We will also show that it has the properties postulated for a trust propagation and aggregation operator.

The operator has been developed from a simple observation: if most of our friends trust somebody, we would also be willing to trust that person. Thus, the truth degree of the following statement:

"Most of the users trusted by x trust y"(s)

may be used as a predicted, local trust score of user x in y.

Aggregating opinions from many users is a natural approach however, it involves a risk of attack—it is easy for a coalition of users to express a large number of biased opinions. Instead, in the presented approach a user bases the conclusion about trust on the opinions of trusted acquaintances only, not those of strangers. Next, a trusted acquaintance also trusts the beliefs of his friends, propagating trust (with appropriate discounting) through the relationship network.

(S) is a linguistically quantified sentence in the sense of Zadeh [42] with the fuzzy quantifier "most". Fuzzy quantifiers are a commonly accepted tool for describing uncertain facts and quantitative relations, and are used to implement data aggregation and fusion [43, 44]. The process of truth evaluation of such a sentence refers to the notion of cardinality and relative cardinality of a set. Notice that a cardinality-based approach is justified in trust prediction since it reflects a way in which trust between users may be constituted—by counting a proportion of

Fig. 1 A simplified model of trust network

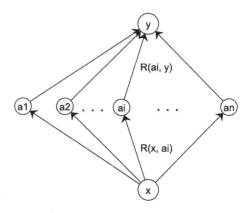

good and bad past experiences. In the case where a user x cannot directly evaluate his trust in y, he has the possibility of taking into account the opinions of a group of his friends, by employing (S).

The main issue considered here is the problem of counting elements in the presence of uncertainty. Since trust scores between users may not be complete, it is desirable and reasonable to preserve information about the amount of uncertainty during trust propagation. To this end we use the new method for determining relative cardinality of I-fuzzy sets introduced in the previous section, and we take advantage of the uncertainty-based monotonicity property of this operator.

To evaluate sentence (S) two I-fuzzy sets need to be constructed: "users trusted by x" and "users that trust y". Using notation from Fig. 1 we define those sets as:

- $U_x : \{R(x, a_i)| i = 1, \ldots, n\}$—an I-fuzzy set of users $a_1, \ldots, a_n$ trusted by x to a degree $R(x, a_1), \ldots, R(x, a_n)$;
- $U_y : \{R(a_i, y)| i = 1, \ldots, n\}$—an I-fuzzy set of users $a_1, \ldots, a_n$ that trust y to a degree $R(a_1, y), \ldots, R(a_n, y)$.

We formalize the statement (S) as:

$$Q\ U_x\ (x, a) \in E \text{ are } U_y \tag{13}$$

where Q is a linguistic quantifier of "most" type; the simplest example of such a quantifier is an identity function. Finally the uncertainty-preserving trust prediction operator $UTrust_\sigma : I \times I \to [0, 1]^2$ which gives a predicted trust of x in y is defined as the truth degree of (S):

$$UTrust_\sigma(x, y) = Q(\tilde{\sigma}_f(U_y | U_x)). \tag{14}$$

In the case where users x and y are connected via one and only one intermediate user, trust prediction is reduced to atomic trust propagation and we obtain $R(x, y)$ immediately by applying a simplified version of $UTrust_\sigma$:

$$UTrust_\sigma(x, y) = [Q(\sigma_f(\underline{U_y}|\overline{U_x})), Q(\sigma_f(\overline{U_y}|\underline{U_x}))]. \tag{15}$$

Usually, however, the network of connection between users is much more complex. Not only is the path between users x and y longer than 2, but there exist many paths of different length. We assume that in the initial phase all cycles in the trust network are removed. Then the procedure of predicting the trust of x in y proceeds recursively: x requests its direct neighbors for their opinion about y and then aggregates their answers with $UTrust_\sigma$. If one of those neighbors does not have an opinion about y then he proceeds via the same procedure by asking his friends, and so on. Notice that a single operation only requires knowledge about the direct neighbor of a given node, not about the whole network.

$UTrust_\sigma$ is in fact a family of operators. To use it, it is necessary to set a number of parameters: the definition of Q, the type of t-norm T, and the type of cardinality pattern f. Using $T = T_a$, $Q = id$ and $f = id$, and considering only complete trust degrees (I-fuzzy sets reduced to fuzzy sets), we obtain the well-known trust-based weighted mean formula that is at the heart of the Golbeck trust metric TidalTrust. The presented operator gives more flexibility: first of all by making it possible to operate on incomplete (uncertain) data, and secondly by providing the possibility of choosing different Q, t-norms and cardinality patterns that lead to different properties of the operator. For example, Golbeck observed that taking account of only higher trust values yields better trust prediction. Thus by setting an adequate cardinality pattern f we can eliminate all users who are not sufficiently trusted. On the other hand, by considering nilpotent t-norms like the Òukasiewicz t-norm, we tend to be more rigorous, requiring the aggregated values to cross a threshold to return a positive value. It seems that the specific properties of a t-norm minimum makes it inappropriate in the context of trust propagation and aggregation, and so we will not consider this t-norm. Instead, interesting properties can be obtained for the algebraic and Òukasiewicz t-norms, as we will demonstrate in the next section.

4.2 Operator UTrust_σ—Properties and Examples

We now discuss in detail some of the basic properties of the proposed operator. Again, at the center of our interest is the uncertainty factor and how it propagates and aggregates. For simplicity, the following discussion and examples refer to the "most" quantifier $Q = id$; we consider two t-norms: algebraic and Òukasiewicz, and two cardinality patterns: $f = id$ and $f = f_{p=0.2}$ where:

$$f_p(x) = \begin{cases} \frac{x-p}{1-p}, & \text{if } x > p, \\ 0, & \text{otherwise.} \end{cases}$$

In the following we justify the choice of these parameters as the most appropriate one. Whenever it is convenient, we will use the notation $UTrust_\sigma(x, y) = UTrust_\sigma([\underline{t}_x, \overline{t}_x], [\underline{t}_y, \overline{t}_y])$ for atomic propagation (propagation with only one intermediate user).

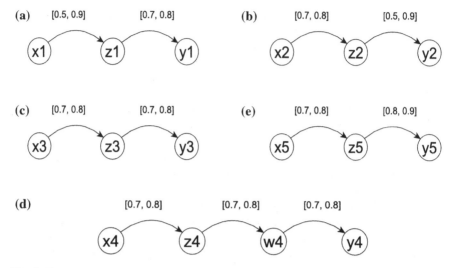

Fig. 2 Fragments of trust networks illustrating trust propagation

We will consider separately the properties connected with the propagation aspect of the $UTrust_\sigma$ operator and the properties connected with the aggregation aspect of that operator.

4.2.1 Propagation Properties

(P1) Propagation—not commutative. The first important feature of the proposed trust prediction operator is non commutativity when propagating trust. This property reflects the difference in meaning of the sets U_x and U_y. The opinion of user who directly know y is of greater importance when estimating final trust, while the opinions of U_x serve as a kind of weight. Thus, for example, the final trust scores for the subnetworks in Fig. 2a, b would be different. First we need to construct for each case two I-fuzzy sets: U_x of users trusted by x, and U_y of users that trust y:

- $U_{x1} = \{[0.5, 0.9]/z1\}$, $U_{y1} = \{[0.7, 0.8]/z1\}$
- $U_{x2} = \{[0.7, 0.8]/z2\}$, $U_{y2} = \{[0.5, 0.9]/z2\}$.

Then we obtain the following results:

(a) $T = T_{\partial}$ and $f = id$: $UTrust_\sigma(x1, y1) = [0.4, 0.78]$ with $u = 0.38$;
    $T = T_a$ and $f = f_{0.2}$: $UTrust_\sigma(x1, y1) = [0.61, 0.74]$ with $u = 0.13$;
(b) $T = T_{\partial}$ and $f = id$: $UTrust_\sigma(x2, y2) = [0.29, 0.87]$ with $u = 0.59$;
    $T = T_a$ and $f = f_{0.2}$: $UTrust_\sigma(x2, y2) = [0.33, 0.87]$ with $u = 0.53$.

(P2) Propagation—not idempotent. An operator O is idempotent if for each argument a it holds that $aOa = a$. Notice that such a property would imply the possibility of propagating information along an infinite path without any information discount. When propagating trust it seems to be more realistic that $aOa < a$—that is, that lengthening a connection path entails, on one hand, a reduction in the strength of the propagated trust, and on the other hand, an increase in the uncertainty margin. At some point, the trust value reduces to zero, and uncertainty attains the highest value-this means that the distance between x and y has become too great to make reliable inferences about trust. Note that $UTrust_\sigma$ with the minimum t-norm does not have this property. For this reason we propose to use the Òukasiewicz t-norm with cardinality pattern $f = id$ and algebraic t-norm with $f = f_{0.2}$ for which it is true that $UTrust_\sigma([\underline{t}_x, \overline{t}_x], [\underline{t}_y, \overline{t}_y]) < [\underline{t}_y, \overline{t}]$ if $[\underline{t}_x, \overline{t}_x], [\underline{t}_y, \overline{t}_y] \in [0, 1)$. It is easy to see that for Òukasiewicz t-norm, $Q = id$ and $f = id$, according to (15), we can express a left bound of $UTrust_\sigma([\underline{t}_x, \overline{t}_x], [\underline{t}_y, \overline{t}_y])$ as $\frac{\underline{t}_y \overline{t}\overline{t}_x}{\overline{t}_x} = 0 \vee \frac{\underline{t}_y + \overline{t}_x - 1}{\overline{t}_x} < \underline{t}_y$. We proceed similarly with the right bound. Also, for the algebraic t-norm, $Q = id$ and $f = f_{0.2}$, after a basic arithmetic operation we obtain a left bound of $UTrust_\sigma([\underline{t}_x, \overline{t}_x], [\underline{t}_y, \overline{t}_y])$ equal to $\frac{\sigma_f(\underline{t}_y \cdot \overline{t}_x)}{\sigma_f(\overline{t}_x)} = 0 \vee \frac{\underline{t}_y \cdot \overline{t}_x - p}{\overline{t}_x - p} < \underline{t}_y$; analogously, the right bound is less than $\overline{t}_y$.

Let us illustrate this property with the following example. For the small subnetwork depicted in Fig. 2c we construct two I-fuzzy sets: U_{x3} of users trusted by $x3$ and U_{y3} of users who trust $y3$:

- $U_{x3} = \{[0.7, 0.8]/z3\}$
- $U_{y3} = \{[0.7, 0.8]/z3\}$.

The predicted trust degree of $x3$ in $y3$ is equal to:

(a) $T = T_\text{Ò}$ and $f = id$: $UTrust_\sigma(x3, y3) = [0.57, 0.75]$ with $u = 0.18$
(b) $T = T_a$ and $f = f_{0.2}$: $UTrust_\sigma(x3, y3) = [0.6, 0.73]$ with $u = 0.13$.

Notice that, first of all, the estimated trust values are smaller than the input trust values of U_y. Moreover, the final uncertainty is greater than the initial uncertainty, which was equal to 0.1. If we further lengthen the path by introducing an additional user, as shown in Fig. 2d we obtain an even smaller trust value and greater uncertainty:

(c) $T = T_\text{Ò}$ and $f = id$: $UTrust_\sigma(x4, y4) = [0.39, 0.69]$ with $u = 0.3$
(d) $T = T_a$ and $f = f_{0.2}$: $UTrust_\sigma(x4, y4) = [0.47, 0.64]$ with $u = 0.18$.

(P3) Propagation—trust monotonicity. The trust monotonicity property ensures that a positive change in trust opinions—an increase in the trust of x in his friends or an increase in the trust of some user in y—does not bring about a decrease in the predicted trust of x in y (the predicted trust is greater or does not change in certain circumstances). Once again we consider the case in Fig. 2c. A user $z3$ acquires some new information about $y3$ that allows him to update his opinion about $y3$ to a more favorable one. The updated network is depicted in Fig. 2e. The new trust score is equal to:

(a) $T = T_{\dot{\partial}}$ and $f = id$: $UTrust_\sigma(x5, y5) = [0.71, 0.87]$ with u $= 0.16$
(b) $T = T_a$ and $f = f_{0.2}$: $UTrust_\sigma(x5, y5) = [0.73, 0.87]$ with u $= 0.13$.

(P4) Propagation—uncertainty monotonicity. The uncertainty monotonicity of the operator $UTrust_\sigma$ was shown in Theorem 1. This important property ensures that the reduction in uncertainty about the trust scores of involved users will bring about reduction in uncertainty about the output trust (and vice versa). Let us say that a user $x1$ from Fig. 2a reduces his uncertainty about how much to trust $y1$ from [0.5, 0.9] to [0.7, 0.9]; then
    (a) $T = T_{\dot{\partial}}$ and $f = id$: $UTrust_\sigma(x1', y1) = [0.57, 0.78]$ with u $= 0.21$
    (b) $T = T_a$ and $f = f_{0.2}$: $UTrust_\sigma(x1', y1) = [0.61, 0.74]$ with u $= 0.13$.

(P5) [1, 1]—neutral element for propagation. For a simple case, when

- $U_x = \{[1, 1]/z\}$
- $U_y = \{[0.7, 0.8]/z\}$.

  we obtain:

$$UTrust_\sigma(x, y) = [0.7, 0.8]$$

In fact, for all $\underline{t}, \bar{t} \in [0, 1]^2$ and for all t-norms T and $f = id$ it holds that:

$$UTrust_\sigma([1, 1], [\underline{t}, \bar{t}]) = [\underline{t}, \bar{t}].$$

Thus, the propagation operator $UTrust_\sigma$ copies information from a fully trusted source. This basic property also holds for all of the operators $Prop_1$, $Prop_2$, $Prop_3$ and $Prop_4$ considered in [8].

Notice that, since $UTrust_\sigma$ is not commutative, it is in general not true, that $UTrust_\sigma([\underline{t}, \bar{t}], [1, 1]) = [\underline{t}, \bar{t}]$. In fact, it is equal to one if we do not set a threshold for an acceptable trust degree, as will be discussed later. However, we should add here that according to our idea of a trust network with uncertainty, the value [1] would be achieved only in a hypothetical, infinitely distant past, as the process of constituting trust assumes that the trust value is always accompanied by some (even if very small) uncertainty.

4.2.2 Aggregation Properties

(P5) Aggregation—commutative. Unlike for propagation, commutativity is a desirable feature for the aggregation aspect of the $UTrust_\sigma$ operator. It means that

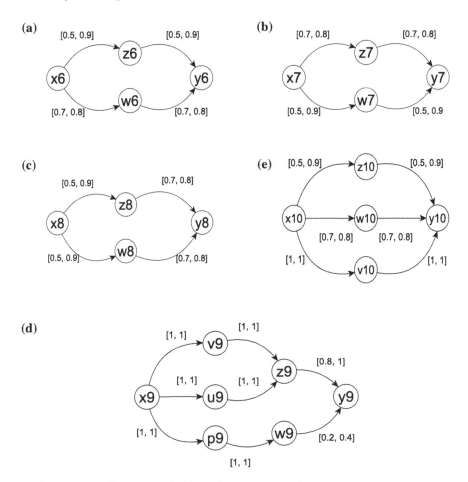

Fig. 3 Fragments of trust networks illustrating trust aggregation

the order of aggregated paths (where the order can be established based on, for
example, the time when two users became friends) does not influence the final trust
degree. Compare the two networks from Fig. 3a, b—in both cases the final trust
score equals:

(a) $T = T_{\eth}$ and $f = id$: $UTrust_\sigma(x6, y6) = UTrust_\sigma(x7, y7) = [0.33, 0.82]$ with
   $u = 0.5$
(b) $T = T_a$ and $f = f_{0.2}$: $UTrust_\sigma(x6, y6) = UTrust_\sigma(x7, y7) = [0.45, 0.81]$ with
   $u = 0.36$.

(P6) Aggregation—idempotent. Being idempotent is another preferred property
of the aggregation operator. On the introduction of an additional path between
users x and y with identical trust degrees as the existing ones, the final estimation
of $UTrust_\sigma(x, y)$ should not change—compare Figs. 2a and 3c.

However, consider the example in Fig. 3d. If we duplicate only a path connecting user $x9$ and $z9$, then the opinion of user $z9$ gains greater importance and consequently has a greater influence on the final result, which will change from:

(a) $T = T_{\partial}$ and $f = id : UTrust_{\sigma}(x9, y9) = [0.5, 0.7]$ with u = 0.2
(b) $T = T_a$ and $f = f_{0.2} : UTrust_{\sigma}(x9, y9) = [0.375, 0.625]$ with u = 0.25.

(if there was only one path connecting $x9$ and $z9$) to the higher trust values:

(a) $T = T_{\partial}$ and $f = id : UTrust_{\sigma}(x9, y9) = [0.6, 0.8]$ with u = 0.2
(b) $T = T_a$ and $f = f_{0.2} : UTrust_{\sigma}(x9, y9) = [0.49, 0.69]$ with u = 0.2.

(if there were two paths, through $v9$ and $u9$).

(P7) Aggregation—trust monotonicity. Regarding the aggregation aspect, trust monotonicity guarantees that adding a more trustful connection between x and y results in an increase in the final trust $UTrust_{\sigma}(x, y)$. An analogous property is observed on adding a less certain/less trustful path.

(P8) Aggregation—uncertainty monotonicity. Similarly as for trust, adding a new, more certain path between x and y results in a decrease in the uncertainty of $UTrust_{\sigma}(x, y)$; analogously, if a new path carries less certain information, the resulting trust score will be less certain. For example, on adding a new path to the subnetwork in Fig. 3a we obtain the network depicted in Fig. 3e and the nfollowing trust prediction:

(a) $T = T_{\partial}$ and $f = id : UTrust_{\sigma}(x10, y10) = [0.64, 0.89]$ with u = 0.25
(b) $T = T_a$ and $f = f_{0.2} : UTrust_{\sigma}(x10, y10) = [0.67, 0.88]$ with u = 0.21.

(P9) [0, 0]—neutral element for aggregation. It is easy to observe that an opinion of a totally untrusted user—a user trusted by x to a degree [0, 0]—does not contribute to the final outcome. In the extreme case, when none of the users is trusted enough we may obtain $\frac{0}{0}$ then we assume 0 as the result.

We may also consider setting a threshold on trust degrees. As mentioned earlier, a better results of trust prediction can be obtained by taking into account only highly trusted sources. By setting an appropriate parameter p for a cardinality pattern f_p, we can zero all trust degrees that are not high enough. A slightly different effect can be achieved when using the Òukasiewicz t-norm-the aggregated value is reduced to zero, if both operators are not sufficiently high. For example,

(a) $T = T_{\partial}$ and $f = id : UTrust_{\sigma}([0.4, 0.7][0.4, 0.7]) = [0, 0.57]$,
(b) $T = T_a$ and $f = f_{0.2} : UTrust_{\sigma}([0.4, 0.7][0.4, 0.7]) = [0.16, 0.58]$ with u = 0.17
but when $f = f_{0.4}$:
(c) $T = T_a$ and $f = f_{0.4} : UTrust_{\sigma}([0.4, 0.7][0.4, 0.7]) = [0, 0.3]$ with u = 0.17.

(P10) Contradiction. Let us consider the case of contradictory opinions. If two fully trusted acquaintances of x express opposite trust opinions about y then the final trust score for $f = id$ will always be symmetric: $UTrust_\sigma(x, y) = [s, 1 - s]$, $s \in (0, 0.5]$. Thus, for example, for:

- $U_x = \{[1, 1]/z1, [1, 1]/z2\}$
- $U_y = \{[0.8, 1]/z1, [0, 0.2]/z2\}$

we obtain:

$$UTrust_\sigma(x, y) = [0.4, 0.6].$$

The discussed $UTrust_\sigma$ operator incorporates the uncertainty factor in the process of trust modeling and predicting, starting from a natural definition based on group opinion, up to the intuitive properties, particularly concerning uncertainty monotonicity. Although the idea of counting elements is very simple, it seems that it gives a powerful tool, enabling a user to set parameters-quantifier, t-norm and cardinality pattern-to adopt behavior of the operator to particular problems. Since articulating beliefs in a way that is imprecise and incomplete, is a human nature, the main effort was directed to the proper handling and processing of imprecision and uncertainty so as not to loose any valuable information intended by a user during the process of trust prediction.

5 Conclusions

Computer modeling of human traits, emotions and relations-such as trust-still presents significant challenges for researchers in view of the complexity and ambiguity involved. Methods based on fuzzy sets theory and I-fuzzy sets theory appear to be an effective tool for describing those types of human reality that are imprecise and incomplete.

The problem of trust in computer science is considered from the point of view of many different aspects; in this chapter we have considered just one of them-reputation-based trust-in the context of social networks. A social network enriched with the possibility of predicting trust allows users to evaluate others' reliability and the quality of the information obtained. Various proposals have included both probabilistic and gradual models of trust, with and without a distrust degree. In the present chapter these approaches have been mentioned and a new one proposed, in which trust is modeled as a lower and upper approximation, and so carries uncertainty. An uncertainty degree gives us additional information about the quality of the opinion about trust—whether it is certain or may be weakened by incomplete knowledge—and makes it possible to reflect the impact of the propagation path length on the final trust score. With a new method of calculating the relative cardinality of I-fuzzy sets, it was possible to construct an uncertainty-preserving trust propagation and aggregation operator $UTrust_\sigma$ which satisfies the uncertainty monotonicity condition as well as other desirable properties. The idea of the $UTrust_\sigma$ operator is to build on group opinion, which makes this operator attack-resistant—it follows the opinions of trusted users only. An additional merit of this operator is the possibility of setting parameters—the linguistic quantifier Q, the t-norm and the cardinality pattern—to adapt its behavior to particular applications. The aggregation and propagation operation depends only on the information acquired from adjacent neighbors, which makes it scalable.

There are still numerous open challenges in the area of trust modeling and processing. One question is how to deal with the trust-based "cold start" problem. Newcomers should be guided wisely through the early process of making connections with users in the network. Another direction of research involves investigation of the potential of distrust. There is still no general consensus on how to propagate and use a distrust degree. Such research is made even more difficult by the lack of publicly available data sets containing suitable gradual and incomplete trust data.

References

1. Artz, D., Gil, Y.: A survey of trust in computer science and the semantic web. J. Web Semant. **5**, 58–71 (2007)
2. Mui, L., Mohtashemi, M., Halberstadt, A.: A computational model of trust and reputation. In: Proceedings of the 35th International Conference on System Science, pp. 280–287 (2002)
3. Kamvar, S., Schlosser, M., Garcia-Molin, H.: The eigentrust algorithm for reputation management in p2p networks. In: Proceedings of the 12th International World Wide Web Conference, pp. 640–651 (2003)
4. Cornelli, F., Damiani, E., Capitani, S.D., Paraboschi, S., Samarati, P.: Choosing reputable servents in a p2p network. In: Proceedings of the 11th World Wide Web Conference, pp. 376–386 (2002)
5. Zaihrayeu, I., Silva, P.P.D., Mcguinness, D.L., Zaihrayeu, I., Pinheiro, P., Deborah, S., Mcguinness, L.: Iwtrust: Improving user trust in answers from the web. In: Proceedings of 3rd International Conference on Trust Management, pp. 384–392. Springer (2005)
6. Massa, P., Avesani, P.: Trust-aware collaborative filtering for recommender systems. Lect. Notes Comput. Sci. **3290**, 492–508 (2004)
7. Golbeck, J.: Generating predictive movie recommendations from trust in social networks. In: Proceedings of the Fourth International Conference on Trust Management, Springer (2006)
8. Victor, P., Cornelis, C., Cock, M.D., da Silva, P.P.: Gradual trust and distrust in recommender systems. Fuzzy Sets Syst. **160**(10), 1367–1382 (2009)
9. O'Donovan, J., Smyth, B.: Trust in recommender systems. In: IUI, pp. 167–174 (2005)
10. Ricci, F., Rokach, L., Shapira, B., Kantor, P.B. (eds.): Recommender Systems Handbook. Springer, Berlin (2011)
11. Sinha, R., Sinha, R., Swearingen, K.: Comparing recommendations made by online systems and friends. In: Proceedings of the DELOS-NSF Workshop on Personalization and Recommender Systems in Digital Libraries (2001)
12. Dyczkowski, K., Stachowiak, A.: A recommender system with uncertainty on the example of political elections. In: Advances in Computational Intelligence; Communications in Computer and Information Science, vol. 298, pp. 441–449. Springer, Berlin (2012)
13. Josang, A., Knapskog, S.J.: A metric for trusted systems. In: Proceedings of NIST-NCSC1998, pp. 16–29 (1998)

14. Abdul-Rahman, A., Hailes, S.: Supporting trust in virtual communities. In: Proceedings of the 33rd Hawaii International Conference on System Sciences (HICSS'00), vol. 6, pp. 1769–1777. IEEE Computer Society, Orlando (2000)
15. Guha, R., Kumar, R., Raghavan, P., Tomkins, A.: Propagation of trust and distrust. In: Proceedings of the World Wide Web Conference, pp. 403–412. ACM Press, New York (2004)
16. Cock, M.D., da Silva, P.P.: A many valued representation and propagation of trust and distrust. In: WILF. Lecture Notes in Computer Science, vol. 3849, pp. 114–120. Springer, Berlin (2005)
17. Arieli, O., Deschrijver, G., Kerre, E.: Uncertainty modeling by bilattice-based squares and triangles. IEEE Trans. Fuzzy Syst. **15**, 161–175 (2007)
18. Victor, P., Cornelis, C., De Cock, M., Herrera-Viedma, E.: Bilattice-based aggregation operators for gradual trust and distrust. In: World Scientific Proceedings Series on Computer Engineering and Information Science, pp. 505–510. World Scientific, Singapore (2010)
19. Levien, R., Aiken, A.: Attack resistant trust metrics for public key certification. In: 7th USENIX Security Symposium, pp. 229–242. (1998)
20. Golbeck, J.A.: Computing and applying trust in web-based social networks (2005)
21. Ziegler, C.N., Lausen, G.: Spreading activation models for trust propagation. In: Proceedings of the IEEE International Conference on e-Technology, e-Commerce, and e-Service, IEEE Computer Society Press, Taipei (2004)
22. Kuter, U., Golbeck, J.: Using probabilistic confidence models for trust inference in web-based social networks. ACM Trans. Internet Technol. **10**(2), 1–23 (2006)
23. Massa, P., Avesani, P.: Trust metrics on controversial users: balancing between tyranny of the majority and echo chambers. Int. J. Semant. Web Inf. Syst. **3**, 39–64 (2007)
24. Zadeh, L.: Fuzzy sets. Inf. Control **8**, 338–353 (1965)
25. Ginsberg, M.L.: Multi-valued logics: a uniform approach to reasoning in artificial intelligence. Comput. Intell. **4**, 256–316 (1988)
26. Zadeh, L.: The concept of a linguistic variable and its applications to approximate reasoning i. Inf. Sci. **8**, 199–249 (1975)
27. Atanassov, K., Stoeva, S.: Intuitionistic fuzzy sets. In: Proceedings Polish Symposium on Interval and Fuzzy Mathematics, pp. 23–26. Poznan (1983)
28. Atanassov, K.: Intuitionistic Fuzzy Sets: theory and Applications. Physica-Verlag, Heidelberg (1999)
29. Atanassov, K., Gargov, G.: Interval valued intuitionistic fuzzy sets. Fuzzy Sets Syst. **31**, 343–349 (1989)
30. Cornelis, C., Atanassov, K., Kerre, E.: Intuitionistic fuzzy sets and interval-valued fuzzy sets: a critical comparison. In: Proceedings of Third European Conference on Fuzzy Logic and Technology (EUSFLAT'03), pp. 159–163. Zittau (2003)
31. Deschrijver, G., Kerre, E.: On the relationship between some extensions of fuzzy set theory. Fuzzy Sets Syst. **133**, 227–235 (2003)
32. Luca, A.D., Termini, S.: A definition of non probabilistic entropy in the setting of fuzzy sets theory. Inf. Comput. **20**, 301–312 (1972)
33. Szmidt, E., Kacprzyk, J.: Entropy for intuitionistic fuzzy sets. Fuzzy Sets Syst. **118**, 467–477 (2001)
34. Pankowska, A., Wygralak, M.: General if-sets with triangular norms and their applications to group decision making. Inf. Sci. **176**, 2713–2754 (2006)
35. Danan, E., Ziegelmeyer, A.: Are preferences incomplete? an experimental study using flexible choices. Papers on strategic interaction, Max Planck Institute of Economics, Strategic Interaction Group (2004)
36. Mendel, J.M., John, R.I.: Type-2 fuzzy sets made simple, IEEE Trans. Fuzzy Syst. **10**, 117–127 (2002)
37. Cornelis, C., Kerre, E.: Inclusion measures in intuitionistic fuzzy set theory. In: Nielsen, T., Zhang, N. (eds.) Symbolic and Quantitative Approaches to Reasoning with Uncertainty, vol. 2711, pp. 345–356. Lecture Notes in Computer ScienceSpringer, Berlin (2003)

38. Niewiadomski, A., Ochelska, J., Szczepaniak, P.S.: Interval-valued linguistic summaries of databases. Control Cybern. **35**, 415–443 (2006)
39. Stachowiak, A.: Propagating and aggregating trust with uncertainty measure. In Jedrzejowicz, P., Nguyen, N.T., Hoang, K. (eds) ICCCI (1). Lecture Notes in Computer Science, vol. 6922, pp. 285–293. Springer (2011)
40. Grzegorzewski, P.: Conditional probability and independence of intuitionistic fuzzy events. Notes Intuitionistic Fuzzy Sets **6**, 7–14 (2000)
41. Grzegorzewski, P.: On possible and necessary inclusion of intuitionistic fuzzy sets. Inf. Sci. **181**, 342–350 (2011)
42. Zadeh, L.: A computational approach to fuzzy quantifiers in natural languages. Comput. Math. Appl. **9**, 149–184 (1983)
43. Yager, R.R.: Interpreting linguistically quantified propositions. Int. J. Intell. Syst. **9**, 541–569 (1994)
44. Kacprzyk, J.: Group decision making with a fuzzy linguistic majority. Fuzzy Sets Syst. **18**, 105–118 (1986)

Impact of Social Network Structure on Social Welfare and Inequality

Zhuoshu Li and Zengchang Qin

Abstract In this chapter, how the structure of a network can affect the social welfare and inequality (measured by the Gini coefficient) are investigated based on a graphical game model which is referred to as the Networked Resource Game (NRG). For the network structure, the Erdos–Renyi model, the preferential attachment model, and several other network structure models are implemented and compared to study how these models can effect the game dynamics. We also propose an algorithm for finding the bilateral coalition-proof equilibria because Nash equilibria do not lead to reasonable outcomes in this case. In economics, increasing inequalities and poverty can be sometimes interpreted as a circular cumulative causations, such positive feedback is also considered by us and a modified version of the NRG by considering the positive feedback (p-NRG) is proposed. The influence of network structures in this new model is also discussed at the end of this chapter.

Keywords Networked resource game · P-NRG network formation · Graphical games · Nash equilibrium

1 Introduction

Modern game theory began with the idea regarding the existence of mixed-strategy equilibria in two-person zero-sum games and its proof is given by John von Neumann. His book *Theory of Games and Economic Behavior* [27] published in

Z. Li (✉) · Z. Qin
Intelligent Computing and Machine Learning Lab, School of Automation Science and Electrical Engineering, Beihang University, Beijing 100191, China
e-mail: zslibuaa@gmail.com

Z. Qin
e-mail: zcqin@buaa.edu.cn

W. Pedrycz and S.-M. Chen (eds.), *Social Networks: A Framework of Computational Intelligence*, Studies in Computational Intelligence 526, DOI: 10.1007/978-3-319-02993-1_7, © Springer International Publishing Switzerland 2014

1944 considered cooperative games of several players. The second edition of this book provided an axiomatic theory of expected utility, which allowed mathematical statisticians and economists to treat decision-making under uncertainty. His later paper published in 1952 introduced a polynomial algorithm, it is the first work on algorithm even before the appearance of digital computer. The modern game-theoretic concept of *Nash Equilibrium* is defined in terms of mixed strategies, where players choose a probability distribution over possible actions. The contribution of John Forbes Nash in his 1951 paper *Non-Cooperative Games* was to define a mixed strategy Nash Equilibrium for any game with a finite set of actions and prove that at least one (mixed strategy) Nash Equilibrium must exist in such a game. How to develop efficient algorithms for calculating the Nash Equilibrium in a given system is a focus of modern *Algorithmic Game Theory* [21].

In recent years, graphical games have attracted much attentions for modeling social phenomena. This emerging research provides new approaches to investigate problems such as group consensus making, networked bargaining and trading strategy designs. In this chapter, we introduce the *Networked Resource Game* (NRG) to investigate the interactions of a society where actions are resource-bounded, i.e., agents have limits on how they are able to act across their network. In this model, agents have a finite number of resources and their network structure may affect how those resources can be coupled with others' resources in order to produce social rewards. One example of this is in professional networks where agents need to form partnerships and the payoffs of the partnerships are determined by a function of their capabilities.

Few work has been reported to study the network structure and its dynamics affect social welfare and inequality, measured by the Gini coefficient [7], of the resulting equilibria. For the network structure, we utilize the *Erdos–Renyi* (ER) model [23], the *Preferential Attachment* (PA) model [1], and several other structure models. We propose an algorithm to find bilateral coalition-proof equilibria because Nash equilibria do not lead to reasonable outcomes in this case. In previous research [18], preliminary results have been obtained to show the impact of network structures on game dynamics. In this chapter, more comprehensive results are presented .

In the NRG model, we only consider the cooperations between agents through bilateral resource consumption to obtain the reward. However, it is not a very good assumption considering the real-world cases where cooperations and competitions are co-existing. In this chapter, we present a new NRG model with positive feedback (p-NRG) to simulate both cooperation and competition scenarios in a game. For both the NRG and p-NRG models, we study the impact of network structures on the game dynamics in terms of resource allocation, social welfare and inequality.

The remaining of this chapter is structured as the following. Section 2 gives a full review on related works from game theory to graphical game models. Section 3 introduces the networked resource game in details. In Sect. 4, we introduce a set of network structure models for empirical evaluations presented in Sect. 5. In Sect. 6, we introduce the NRG model with positive feedback and empirical evaluations are also given. Finally, the conclusions and future work are given in the last section.

2 Related Works

2.1 Game Theory and Intelligent Decision Making

Game theory has influenced many research fields including economics (historically its initial focus), political science, biology, and many other fields. In recent years, its presence in computer science has become impossible to ignore. It features routinely in the leading conferences and journals of artificial intelligence (AI) theory, certainly electronic commerce, as well as in networking and other areas of computer science. According to Shoham [25]:

> One reason for such binding is application pull: the Internet calls for analysis and design of systems that span multiple entities, each with its own information and interests. Game theory, for all its limitations, is by far the most developed theory of such interactions. Another reason is technology push: the mathematics and scientific mind-set of game theory are similar to those that characterize many computer scientists.

Increasing requirements of e-commerce initiate the development of this interdisciplinary research field. Especially in the field like market mechanism designs, AI theory such as machine learning [29] and evolutionary computing [28] had played important roles. Game theory is a framework to explore the interaction among self-interested players. It can be explained as the study of mathematical models of conflict and cooperation between intelligent rational decision-makers. An alternative term suggested it as an interactive decision theory. Such decision making is inseparable to so called multiple interacting intelligent agents within an environment.

An *Intelligent Agent* (IA) is generally regarded as an autonomous entity which observes through sensors and acts upon an environment using actuators. Intelligent agents may also learn or use knowledge to achieve their goals. Though an intelligent agent may have a physical structure (e.g., an autonomous robot), it is generally a software entity that carries out some set of operations on behalf of a user or another program with some degree of independence, and in so doing, employ some knowledge or representation of the user's goals or desires. In game theory, agents can be used for experimental studies. Agents are modeled to have bounded rationality for some decision making. In this research, homogeneous agents are used in graphical games.

2.2 Graphical Games

In social and economic interactions—including public goods provision, job search, political alliances, trade, partnership, and information collection—an agent's well being depends on his or her own actions as well as on the actions taken by his or her neighbors. Such neighboring relations can form a network whose structure decides the direct interaction. The literature identifying the effects of agents' neighborhood patterns (i.e., their social network) on behavior and outcomes has grown over the past several decades. The emerging empirical evidence motivates the theoretical study of network effects. We would like to understand how the pattern of social connections shape the choices that individuals make and the payoffs they can hope to earn.

In this research, we mainly focus on the network structure in order to understand how the changes in network structure will reshape the game. In recent years, the games played on networks have been studied [14, 15]. A general framework for the study of games in such an incomplete-information setup has been developed [13]. Graphical games [16] are a representation of multiplayer games meant to capture and exploit locality or sparsity of direct influences. They are most appropriate for large population games in which the payoffs of each player are determined by the actions of only a small subpopulation. As such, they form a natural counterpart to earlier parametric models. Whereas congestion games[1] and related models implicitly assume a large number of weak influences on each player, graphical games are suitable when there is a small number of strong influences.

Generally, a graphical game can be described at the first level by an undirected graph G in which players are identified with vertices. The semantics of the graph is that a player or vertex i has payoffs that are entirely specified by the actions of i and those of its neighbor set in G. Thus G alone may already specify strong qualitative constraints or structure over the direct strategic influences in the game. To fully describe a graphical game, we must additionally specify the numerical payoff functions to each player but now the payoff to player i is a function only of the actions of i and its neighbors, rather than the actions of the entire population. In the many natural settings where such local neighborhoods are much smaller than the overall population size, the benefits of this parametric specification over the normal form are already considerable.

2.3 Nash Equilibria for Graphical Games

It is known that finding Nash equilibria for graphical games is difficult even for restricted structures [5]. Local heuristic techniques are commonly employed [4, 10]. A seminal work in using agent-based simulation to study human interaction was Axelrod's tournament for Prisoner's Dilemma [2]. Prisoner's Dilemma has also been studied in a graphical setting with simulated agents [20]. Dynamic networked games based on the Ultimatum Game have also been investigated [17] Research on identification and development of networks includes analyzing event-driven growth [24] and inferring social situations by interaction geometry [9].

Some other works have described algorithmic methods to discover temporal patterns in networked interaction data [11]. Researchers have formulated efficient

---

[1] http://en.wikipedia.org/wiki/Congestion_game

solution methods for games with special structures, such as limited degree of interactions between players linked in a network, or limited influence of their action choices on overall payoffs for all players [16, 22, 26]. Another line of research focuses on the design of agents that must maximize their payoffs in a multi-player setting. If self-play convergence to Nash equilibrium is a desiderata for agent policies, In [3], the authors show the convergence of certain kinds of policies in small repeated matrix games. If correlated Nash equilibrium is our goal, it has also been shown that using another set of adaptive rules will result in convergence to a correlated equilibrium [8]. Other work has taken a different approach, and does not presume the equilibrium is a goal; rather profit maximization is the only metric.

Part of the original motivation for graphical games came from earlier models familiar to the machine learning, AI and statistics communities collectively known as graphical models for probabilistic inference, which include Bayesian networks, Markov networks, and their variants. Both graphical models for inference and graphical games represent complex interactions between a large number of variables (random variables in one case, the actions of players in a game in the other) by a graph combined with numerical specification of the interaction details. In probabilistic inference the interactions are stochastic, whereas in graphical games they are strategic (best response). The connections to probabilistic inference have led to a number of algorithmic and representational benefits for graphical games.

Graphical games adopt a simple graph-theoretic model. An n-player game is given by an undirected graph on n vertices and a set of n matrices. The interpretation is that the payoff to player i is determined entirely by the actions of player i and his neighbors in the graph, and thus the payoff matrix for player i is indexed only by these players. We thus view the global n-player game as being composed of interacting local games, each involving (perhaps many) fewer players. Each player's action may have global impact, but it occurs through the propagation of local influences. Formally, a graphical game model is a tuple

$$[I, \{A_i\}, \{J_i\}, \{u_i(\cdot)\}]$$

where I and A_i are as before, J_i is a collection of players connected to i, and $u_i(a_i, a_{J_i})$ is the payoff to player i playing $a_i \in A_i$ when the players J_i jointly play a_{J_i}. The notation $a_{J_i} \subset a_{-i}$ means that each $j \in J_i$ plays its assigned strategy from a_{-i}. We define A_{J_i} to be the set of joint strategies of players in i's neighborhood. This game structure is captured by a graph

$$G = [V, E] \tag{1}$$

in which $V = I$ (i.e., each node corresponds to a player) and there is an edge

$$e = (i, j) \in E \quad if \quad j \in J_i$$

The graph topology might model the physical distribution and interactions of agents: each sales-person is viewed as being involved in a local competition (game) with the salespeople in geographically neighboring regions. The graph may be used to represent organizational structure: low-level employees are engaged in a game with their immediate supervisors, who in turn are engaged in a game involving their direct reports and their own managers, and so on up to the CEO. The graph may coincide with the topology of a computer network, with each machine negotiating with its neighbors (to balance load, for instance).

Graphical games also provide a powerful framework in which to examine the relationships between the network structure and strategic outcomes. Of particular interest is whether and when the local interactions specified by the graph G alone (i.e., the topology of G, regardless of the numerical specifications of the payoffs) imply nontrivial structural properties of equilibria. It can be shown that different stochastic models of network formation can result in radically different price equilibrium properties. In [21], the authors give an example that considers the simplified setting in which the graph G is a bipartite graph between two types of parties, buyers and sellers. Buyers have an endowment of 1 unit of an abstract good called cash, but have utility only for wheat; sellers have an endowment of 1 unit of wheat but utility only for cash. Thus the only source of asymmetry in the economy is the structure of G. If G is a random bipartite graph (i.e., generated via a bipartite generalization of the classical Erdos–Renyi model), then as n becomes large there will be essentially no price variation at equilibrium (as measured, for instance, by the ratio of the highest to lowest prices for wheat over the entire graph). Thus random graphs behave "almost" like the fully connected case. In contrast, if G is generated according to a stochastic process such as preferential attachment. the price variation at equilibrium is unbounded, growing as a root of the economy size n.

In previous work, there has been tremendous interest in agent strategies in different games and auctions. Many have shown to optimize the social welfare, reduce the inequality and reach equilibrium in the social network [19], or how to get stable in a network, where the participants have special preference [12]. Gini coefficient is commonly used as a measure of inequality of income or wealth. Its value depends on both income inequality and other factors such as the network structure [7, 30].

3 Networked Resource Game Model

Similar to a standard graphical game, the NRG is defined by a set of N players $\{p_i\}_{i=1}^{N}$, a card distribution C, a graph G and a reward function R. At each round, the games are played based on a matrix M where each element indicating a link exists between the two players. In the NRG, the actions are based on available resources, which we will informally call *cards*. Each player p_i has a set of cards:

$$C_i = \left\{ c_{i,1}, \ldots, c_{i, N_i^C} \right\}$$

where N_i^C is the number of cards for the player p_i. The cards represent a skill or resource that the player can play on a link. The graph specifies the links over which players may play their cards. Here, we include the restriction that a player may play at most one card on a link. Thus, the number of cards indicate a players ability to have multiple simultaneous partnerships. It is possible that a player has more links than cards and also more cards than links.

Each card has a type which comes from a predetermined type set T, i.e., $c_{i,j} \in T$. For simplicity, given a discrete type set, we can think of the type as a color and that each card has a particular color. The graph $G = \{e_{ij}\}$ which has undirected edges between the players, where $e(i, j)$ denotes the link between p_i and p_j. It is possible that some players have no links between them. Each card can only be used once per round. Each card is one of T types, and the matrix game can be defined by a $(|T| + 1) \times (|T| + 1)$ payoff matrix where each row/column is one of the $|T|$ card types or the null action, and where the value pairs in the cells are the rewards for the two players.

As usual, in this chapter, we assume that M is fixed, is symmetric, and is the same for every link. Based on what cards are played on a link, each player gets a reward specified by the function $R(a, \bar{a})$ and $a, \bar{a} \in A$ where A is the action space. The payoff matrix R (we use the same R for the reward function as well) is defined by:

$$R(a, \bar{a}) \qquad a, \bar{a} \in A$$

For example in Fig. 1, the payoff matrix is shown:

$$R(green, green) = (7, 7)$$

$$R(green, red) = (9, 7)$$

and so on. However, it implicitly contains the situation of null actions like the following:

$$R(null, red) = (0, 0)$$

So that the actual action space for player p_i on link e_{ij} is

$$A_{ij} = C_i \cup 0 \tag{2}$$

where 0 indicates that the player chose not to play one of their cards on that link. Similarly, for the player j its action space is:

$$A_{ji} = C_j \cup 0 \tag{3}$$

The reward function R has $(|T| + 1)^2$ inputs representing every combination of actions, i.e., all card types and not playing a card, for each player. The NRG is similar to a standard graphical game, however, the action space has restrictions over multiple links whereas in standard graphical games, actions on link are independent. Here, we have the restriction that $\cup_j a_{ij} \subset C_i$ where a_{ij} is player p_i's action on link e_{ij}. This states that a player cannot play more cards than they have, which introduces a coupling over links.

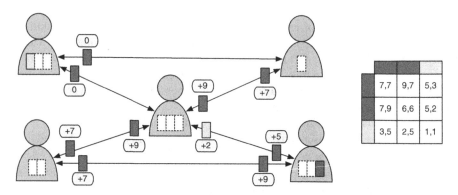

Fig. 1 An example of the NRG. The matrix on the *right-hand side* gives the payoff matrix by playing different type of cards. Each agent is initialized with different type and number of cards. For example, the player at the center of the network has 3 cards and they are 2 *greens* and one *yellow*. By playing these cards with 3 of his 4 linked neighbors he received rewards of $9 + 9 + 2 = 20$ that is calculated based on the given payoff matrix. For the agent on the *upper left* corner, he has received 0 payoff because there is no card-playing between him and his two linked neighbors

An illustration of the game is shown in Fig. 1. It shows a game involving three card types (*green*, *red* and *yellow*). One can imagine that these cards represent assets of value in an economy that yield different outcomes to each contributor in partnerships. For example, a green card could represent capital, red could represent skilled labor and yellow could represent unskilled labor. Different combination of these resources may result in different rewards. For example, capital plus skilled labor may yield much more rewards than capital plus unskilled labor for both sides.

4 Network Formation and Finding Equilibria

4.1 Network Structure Models

In this section we describe a few models we use to create social network graphs and how to find the equilibria for a given graph is discussed. Network formation is determined by various growth processes that describe how a link is added to an existing graph. In this chapter, we use the following four network structure models:

4.1.1 Erdos–Renyi

The Erdos–Renyi model is either of two closely related models for generating random graphs. In the $ER(n, M)$ model, where n is the number of nodes in a graph. A graph is chosen uniformly at random from the collection of all graphs which

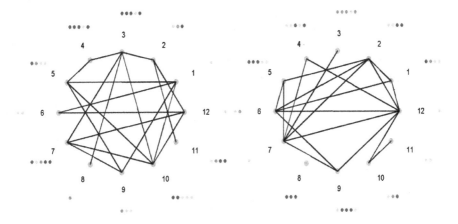

Fig. 2 An example of the ER model (*left*) and the PA model (*right*), where the *red, yellow,* and *green* dots represent the randomly distributed cards for each of the 12 players

have n nodes and M edges. For example, in the $G(3, 2)$ model, each of the three possible graphs on three vertices and two edges are included with probability $1/3$.

In the $ER(n, p)$ model, a graph is constructed by connecting nodes randomly with probability p independent from every other edge. This model is named after Paul Erdo and Alfred Renyi who published *On the Evolution of Random Graphs* in 1960 [6]. This is a baseline process where we add a link chosen uniformly from those links that do not already exist in the graph. In this model, we connect each pair of nodes with some given probability (See Fig. 2).

4.1.2 Preferential Attachment

If the input graph has zero or one link, we use the ER process. Thus, the network is seeded with two random links. After this, in order to add a link, we choose a node randomly and consider the links it could add to the graph, i.e., the set of links connected to the chosen node that are not already in the graph. Each such link is given a weight equal to the degree of the target node it connects to, and a link is chosen in proportion to these weights. Preferential attachment models have been proposed as a model that reflects how social networks are formed, particularly online.

4.1.3 Most Free Cards

Each node is given an *MFC score*: the number of cards it has minus the number of links it has, i.e., a measure of the number of free cards for that player. The process selects a node uniformly from those that have the highest MFC score. This node

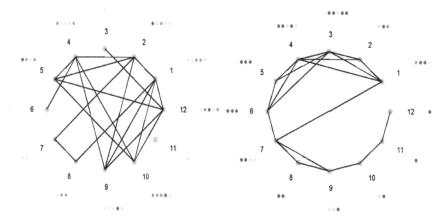

Fig. 3 An example of the MFC model (*left*) and the PRC model (*right*), where the *red*, *yellow*, and *green dots* represent the randomly distributed cards for each of the 12 players

then chooses a link uniformly from other nodes that have the highest MFC score. When the MFC scores are all zero, the algorithm becomes ER.

4.1.4 Poor-to-Rich Chain

We first associate each player with a wealth calculated as the sum of the value of their cards, where the value of each card is the maximum reward obtainable from applying that card (Fig. 3):

$$w_i = \sum_{c \in C_i} \max_{a \in T \cup 0} R(c, a) \tag{4}$$

We first create a chain, where agents are ordered by wealth with ties broken randomly. Then, a player is chosen uniformly from those with the highest MFC score adds a link. The target node is the closest node in the chain with a free card, i.e., an MFC score greater than zero.

The various processes described above capture various degrees of control that players may have over the network on which they play. In the ER and PA models, players have no control over links. One may consider PA as player driven, but the game properties (card, rewards) do not affect the formation of the links so the processes are not strategic. The MFC model is a decentralized strategic model where agents have partial information about the state of the world, namely the number of cards and links for each player. The PRC model is a centralized model that takes game parameters into account when making the graph and incorporates a social structure into the world where people with similar wealth are more likely to be connected to each other.

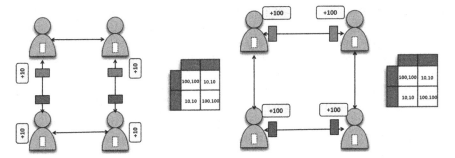

Fig. 4 Nash equilibria may not yield the best rewards for the Networked Resource Game. Considering the case that 4 players are connected sequentially. In the *left-hand side* figure, the game is with the Nash equilibria as none of the players has motivation to make unilateral change that will result in losing 10 point rewards though they may obtain much more points (+100) by making bilateral change as shown in the *right-hand side* figure

4.2 Finding Equilibria

Given a game structure (cards, rewards, and a graph), we would like to determine an appropriate outcome. Nash equilibria are often considered as a solution for graphical games, however, it has some issues for the Networked Resource Game. Consider a simple example of four players in a sequentially connected graph where each player has only one card. Two players have a single red card and two players have a single green card. Let the rewards for having two cards with same color on a link be 100 points of reward for each player, two cards with different colors on the same link be 10 points, and all links with one or zero links be worth nothing (see Fig. 4).

Consider the situation where we have two red-green links and each player receives 10 points of reward. For that case, each player has no incentive to deviate, i.e., move their one card to another link, because that would cause a loss of 10 points, even though each player has a link to a player with the same color card. Thus, in the Networked Resource Game, Nash equilibria lead to artificially poorer results than one would expect if one was playing this game assuming players could communicate over the links that they have. Thus, we consider equilibria where players can make bilateral deviations. An equilibrium in this context is a state where no player would choose to make a unilateral deviation and no two players would choose to make a bilateral deviation. We use the procedure below to discover such equilibria for a given game structure.

Each player first assigns cards randomly to available links. We then perform action updates in a series of rounds. In each round, we order the set of links. For each link, the players iterate back and forth on card choices for the link. On the first iteration, the first player assumes that the other player plays one of their cards, chosen from all cards that player has, i.e., not necessarily the card being played on the link currently. The first player then plays their best response on all links given the cards that are played on all the links that they have. In the second iteration and all following iterations, the acting player chooses their best response to the cards that are being played on their link. This procedure continues until an equilibrium is reached for that link or we reach a preset limit of interactions. We continue this procedure for all links in each round. The procedure terminates, when at the end of a round, the joint actions are the same as the joint actions in the previous round. The procedure continues for a preset number of rounds. Finding equilibria in graphical games is a challenging problem. The algorithm presented at below is sound in that if it terminates before reaching the preset number of rounds, we know that the resulting joint action is an equilibrium for the game, however, we may not find all equilibria.

Pseudo Code of FINDING-EQUILIBRIA for computing bilateral coalition-proof equilibria

```
Algorithm FINDING-EQUILIBRIA
Inputs: one game structure(cards,rewards and a graph)
Outputs: bilateral coalition-proof equilibria
  Each player first assigns cards randomly to available links
  equilibria ← 0
  for round ← 1 to n1
    do order the set of links
    num ← 1
    repeat
      (1)one player Pi assumes that the other player Pj
         which links with Pi plays one of its cards
      (2)Pi plays their best response on all links given the
         cards that are played on all the links that they have
      (3)Pj chooses their best response to the cards
         that are being played on their link
      (4)num ← num + 1
    until an equilibrium is reached for that link
        or num = n2
    if the joint actions = joint actions in the previous round
    then return equilibria
    else return -1
```

5 Experimental Studies

The Gini coefficient (also known as the Gini index or Gini ratio) [7, 30] is a measure of statistical dispersion, it is named after the Italian sociologist Corrado Gini.[2] It measures the ratio of areas above the Lorenz curve which plots the

---

[2] http://en.wikipedia.org/wiki/Gini_coefficient

proportion of the total income of the population that is cumulatively earned by the bottom x % of the population. The Gini coefficient measures the inequality among values of a frequency distribution. A Gini coefficient of zero expresses perfect equality, while the Gini coefficient of one expresses maximal inequality among values (for example where only one person has all the income in a society), i.e., larger Gini coefficients indicate greater income disparity. In this chapter, the Gini coefficient is used as a measure of inequality.

In our experiments, we considered a society of consisting 12 players. In each round, each player was given a number of cards chosen uniformly from one to five:

$$|C_i| \sim U(1, 5)$$

We had three card types: *green*, *red*, and *yellow*. Card colors were selected independently for each card using the following probabilities:

$$P([\text{green red yellow}]) = [0.20\,0.40\,0.40]$$

5.1 Reward Functions

There were two methods for selecting reward functions to generate the payoff matrix: (1) In the baseline method, each reward for links with two cards on them were chosen randomly:

$$R(c1, c2) \sim U(1, 1000) \quad \text{for } c1, c2 \in T \tag{5}$$

Links with one or zero cards gave zero reward to both players. (2) In the alternate method, the reward for an arbitarily chosen link (e.g. green–green) is replaced with 100 times the value of the maximum of all the rewards in the baseline method. The latter is to investigate a society where there is a significantly outlying reward available to a small number of people if they make the right connections. We can exaggerate the variance of rewards in such a way we can observe the game dynamics more clearly.

It is for this reason that the green cards occur at lower likelihood than the others. For a given game card and reward structure, we would run our various network formation algorithms and generate graphs of increasing size. Each network formation algorithm was run 10 times, thus generating 10 graphs with the same number of edges for each process. For each game structure (cards, rewards and graph) that resulted, we would find the set of equilibria. For each graph, the equilibrium-finding algorithm was run 40 times and each run was ended if the algorithm didn't terminate in 15 rounds.

For any single equilibrium, we calculated the social welfare as the sum of all the rewards to all players and the Gini coefficient. For each game, we calculated an associated social welfare with the weighted average of social welfares of equilibria of that game, where weights were the number of times the equilibrium was

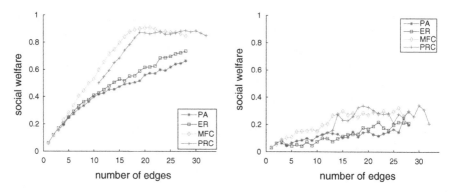

Fig. 5 Social welfare of given structure models with increasing number of edges. The *left-hand side* figure is with baseline rewards and the *right-hand side* one is with alternate rewards

discovered. We calculated associated Gini coefficients for each game structure similarly.

The Gini coefficient is normalized between zero (everyone has equal wealth) and one (one person has all the wealth), but social welfare for each game is a function of the payoff matrix. We use the following way to normalize the social welfare. First, we need to calculate the maximum value of all possibly generated social welfare by:

$$\max \sum_{(c_1, c_2) \in C_2} n_{c_1, c_2} (R(c_1, c_2) + R(c_2, c_1)) \qquad (6)$$

$$\text{such that} \sum_{\tilde{c} \in T} n_{c, \tilde{c}} \leq n_c \;\; \forall c \in T, \quad n_{c, \tilde{c}} \geq 0, \;\; \forall c, \tilde{c} \qquad (7)$$

This considers all possible combinations of cards on a link $(c_1, c_2) \in C_2$ and maximizes the reward obtained for having a particular number of card combinations on the graph (n_{c_1, c_2}) with the rewards obtained for that card combination $(R(c_1, c_2) + R(c_2, c_1))$, such that the number of card combinations of the graph does not violate the card constraints, i.e., the number of cards of a particular type (n_c) and non-negativity of the number of combinations. This yields an upper bound on the social welfare because it allows multiple links between players and links between cards of the same player. We use this to normalize social welfares across different card and reward structures.

5.2 Experimental Results

Figure 5 shows how social welfare changes as a function of network formation algorithm and graph size. We did not show the error bars for clarity in presentation

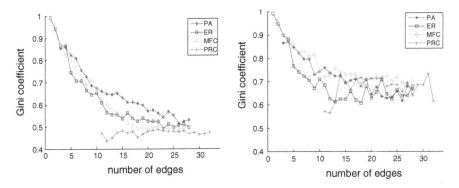

Fig. 6 Gini coefficient of given structure models with increasing number of edges. The *left-hand side* figure is with baseline rewards and the *right-hand side* one is with alternate rewards

but we discuss significance below. We see that social welfare improves as the society gets more connected for all algorithms. MFC and PRC are significantly better than ER and PA. ER is slightly better than PA but the result is not statistically significant. These results hold in both reward scenarios. For baseline rewards, MFC and PRC both reach about 0.9 efficiency in social welfare at about 18 links and do not improve much beyond that. We also see the impact of network structure as the 28-link ER and PA graphs are less efficient that MFC and PRC graphs that are half the size. For alternate rewards, the efficiency is significantly smaller than the baseline word, this could be the result of two factors: there are green–green links that are not being formed, and our normalization could be overcounting the number of potential green–green links.

Figure 6 shows how Gini coefficients change as a function of network formation algorithm and graph size. Inequality decreases as the network sizes increase. For the baseline reward structure, MFC, PRC and ER are significantly better than PA. The key change is that ER has jumped from the PA equivalence class to the MFC/PRC equivalence class. We note that the Gini coefficient is relatively flat after about 18 links. For the alternate reward structure, all the algorithms are in the same equivalence class. This is because once a few green–green links are formed, it is difficult to change the inequality of the world.

We then investigated the number of wasted cards in equilibrium, i.e., the number of cards that did not yield any reward to the player holding it. Figure 7 shows the number of wasted cards as a percentage of the total number of cards in a society. We see that wasted cards explains a lot of the phenomena in social welfare. The MFC and PRC algorithm, which has an MFC component, waste the fewest cards because that is part of their process. The others form links that are not as useful in allowing players to use their cards. ER performs slightly better that PA because it does not overload particular users with large numbers of links. Thus as fewer cards are wasted, social welfare improves. This similarly explains the Gini coefficient because as more cards are used, we have fewer users with low or no

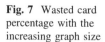

Fig. 7 Wasted card percentage with the increasing graph size

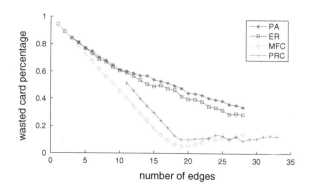

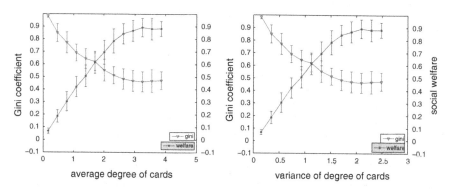

Fig. 8 Social welfare and Gini coefficient by average and variance of degree in graph with baseline rewards

rewards. Nevertheless, it is interesting to note that while ER wastes more cards than MFC and PRC, it does not perform worse in terms of inequality. This remains an open question. Interestingly, with half the possible links (33), we still have about 10% of cards being wasted.

We also looked at the impact of network properties on outcomes. Figure 8 shows social welfare and Gini coefficient as a function of the average and variance of the degrees of the nodes in the graph. Clearly, this will depend on the card and reward structure. In our case, both average and variance of degree showed similar curves in increasing social welfare and decreasing inequality. The inequality curves are similar in both reward structures and the social welfare curves are close to the best performing algorithms as a function of graph size. We believe the Networked Resource Game is a good starting point for modeling and investigating the complexities and design of economies of resource-bounded and socially networked agents (Fig. 9).

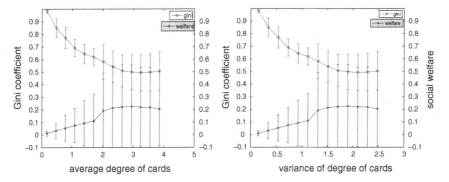

Fig. 9 Social welfare and Gini coefficient by average and variance of degree in graph with alternate rewards

6 Positive Feedback

In this section, we introduce a new model that we can use to simulate the both cooperations and competitions in a game. In each game, we like to reassign the cards to players based on his previous rewards. This is like that the rich people are likely to get more resources while such resources may help him become richer. This phenomenon is modeled be the following the following NRG model with positive feedback (p-NRG).

Given a game structure (cards, rewards and a graph), each game is conducted as a sequence of sub-games. In a sub-game, the agents only play their cards once. Like in the classical NRG, each player is first given a number of cards chosen uniformly from 1 to 5:

$$|C_i| \sim U(1,5)$$

We calculated each player's (p_i) welfare: W_i at the end of each sub-game. At the beginning of a sub-game, each player will be re-assigned with new cards provided by a *card-pool* wit infinite number of cards. At each round of the sub-game, N cards will be drawn from the pool based on a given probability distribution on card types. For a particular player p_i, the number of new cards he can get is depending on the welfare W_i he obtained at this round. In other others, the ones with large welfare values tends to get more new cards at the next round of sub-game. The probability for getting more new cards is calculated by

$$P_i = \frac{W_i + \theta}{\sum_{i=1}^{N} W_i + N\theta} \tag{8}$$

where $\theta > 0$ is smoothing factor used to avoid getting zero probabilities. In the following experiments, we first assign the card-pool distribution by:

$$P([\text{green red yellow}]) = [1/6 \ 2/6 \ 3/6]$$

Fig. 10 Gini coefficient comparison between NRG and p-NRG in the ER model

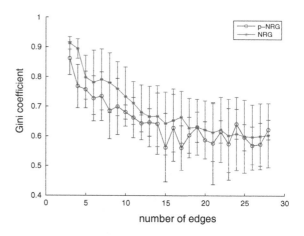

Fig. 11 Social welfare comparison between NRG and p-NRG in the ER model

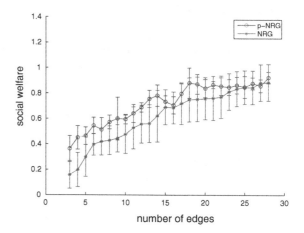

Each game is consisting 5 sequential sub-games. In this new game, the Gini coefficient of each game is the average value of the five sub-games, and social welfare is the sum of five sub-games.

The experimental results are presented in Figs. 10, 11, 12, 13, 14, 15, 16 and 17. In comparisons of NRG with and without positive feedback, we found that the results are similar for the ER and MFC models in both social welfare and Gini coefficient. However, we found there is a significant variance in the experiments with the PRC model. As we can see from Figs. 14 and 15, the p-NRG model is less than the classical NRG model in both the social welfare value and Gini coefficient. By introducing the positive feedback, the system becomes less productive in terms of social welfare. It hurts the economy somehow by introducing such circular causations. However, the social inequality is roughly the same comparing to the classical NRG. Based on our observation, positive feedback may hurt social welfare but won't influence the social inequality significantly.

Fig. 12 Gini coefficient comparison between NRG and p-NRG in the MFC model

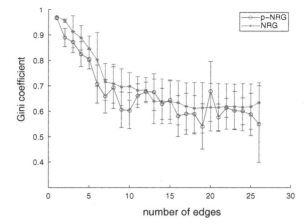

Fig. 13 Social welfare comparison between NRG and p-NRG in the MFC model

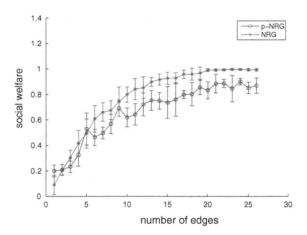

Fig. 14 Gini coefficient comparison between NRG and p-NRG in the PRC model

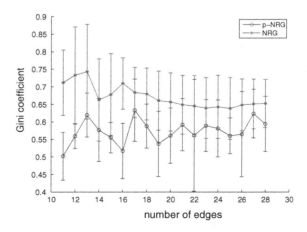

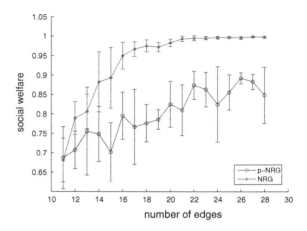

Fig. 15 Social welfare comparison between NRG and p-NRG in the PRC model

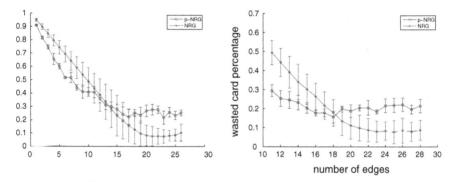

Fig. 16 Wasted card percentage, comparison of the baseline and positive-feedback NRG model given with the ER (*left*) and MFC (*right*) structure

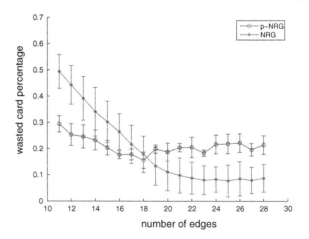

Fig. 17 Wasted card percentage, comparison of the baseline and positive-feedback NRG model given with the PRC structure

7 Conclusions and Future Work

In this chapter, we studied the Networked Resource Game and we investigated how network structure may influence the social welfare and inequality measured by the Gini coefficient. Based on empirical evaluations, we found that different network structure may lead to different game dynamics in terms of increasing social welfare and decreasing Gini Coefficient. Efficient interactions of players will increase both the social welfare as well as social equality. For the NRG with positive feedback, we found that introducing of positive feedback may lead to a less productive society but it won't cause significant social inequality comparing to the classical NRG model.

One potential future direction is using these properties as part of the network formation process because they may be more easily estimated than the requirements of the processes we presented. We also plan on investigating games where more than two players can collaborate. It is also a challenge to investigate appropriate outcomes for graphs as the scale of the society grows as equilibrium discovery will become more computationally demanding.

Acknowledgments This research is partly funded by the NCET Program, Ministry of Education, China, and the National Natural Science Foundation of China (NSFC) under Grant No. 61305047.

References

1. Albert, R., Barabasi, A.-L.: Statistical mechanics of complex networks. Rev. Mod. Phys. **74**, 47–97 (2002)
2. Axelrod, R., Hamilton, W.D.: The evolution of cooperation. Science **211**, 1390–1396 (1981)
3. Bowling, M.: Convergence and no-regret in multiagent learning. In: Advances in Neural Information Processing Systems 17 (NIPS), pp. 209–216, 2005. A longer version is available as a University of Alberta Technical Report, TR04-11
4. Duong, Q., Vorobeychik, Y., Singh, S., Wellman, M.P.: Learning graphical game models. In: IJCAI (2011)
5. Elkind, E., Goldberg, L., Goldberg, P.: Nash equilibria in graphical games on trees revisited. In: Proceedings of the 7th ACM Conference on Electronic Commerce, pp. 100–109. ACM, New York (2006)
6. Erdo, P., Renyi, A.: On random graphs. Mathematicae **6**, 290–297 (1959)
7. Gini, C.: On the measure of concentration with special reference to income and statistics. Colorado College Publication (1936)
8. Greenwald, A., Hall, K.: Correlated q-learning. In: 20th International Conference on Machine Learning, pp. 242–249 (2003)
9. Groh, G., Lehmann, A., Reimers, J., Friess, M., Schwarz, L.: Detecting social situations from interaction geometry. In: 2010 IEEE Second International Conference on Social Computing (SocialCom), pp. 1–8 (2010)
10. Heckerman, D., Geiger, D., Chickering, D.M.: Learning bayesian networks: the combination of knowledge and statistical data. In: Machine Learning, pp. 197–243 (1995)
11. Hsieh, H.-P., Li, C.-T.: Mining temporal subgraph patterns in heterogeneous information networks. In: 2010 IEEE Second International Conference on Social Computing (SocialCom), pp. 282–287 (2010)

12. Irving, R.W.: An efficient algorithm for the stable roommates problem. J. Algorithms. **6**(4), 577–595 (1985)
13. Jackson, M.O.: Allocation rules for network games. Games Econ. Behav. **51**(1), 128–154 (2005)
14. Jackson, M.O., Watts, A.: The evolution of social and economic networks. J. Econ. Theor. **106**(2), 265–295 (2002)
15. Kakhbod, A., Teneketzis, D.: Games on social networks: on a problem posed by goyal. CoRR. abs/1001.3896 (2010)
16. Kearns, M., Littman, M., Singh, S.: Graphical models for game theory. In: Conference on Uncertainty in Artificial Intelligence, pp. 253–260 (2001)
17. E. Kim, L. Chi, R. Maheswaran, and Y.-H. Chang. Dynamics of behavior in a network game. In IEEE International Conference on Social Computation (2011)
18. Li, Z., Chang, Y.-H., Maheswaran, R.T.: Graph formation effects on social welfare and inequality in a networked resource game. In: SBP, pp. 221–230 (2013)
19. Luca, M.D., Cliff, D.: Human-agent auction interactions: adaptive-aggressive agents dominate. In: IJCAI (2011)
20. Luo, L., Chakraborty, N., Sycara, K.: Prisoner's dilemma in graphs with heterogeneous agents. In: 2010 IEEE Second International Conference on Social Computing (SocialCom), pp. 145–152 (2010)
21. Nisan, N., Roughgarden, T., Tardos, E., Vazirani, V.V. (eds.): Algorithmic Game Theory. Cambridge University Press, New York (2007)
22. Ortiz, L., Kearns, M.: Nash propagation for loopy graphical games. In: Neural Information Processing Systems (2003)
23. Erdos, P., Renyi, A.: On random graphs. Publicationes Mathematicae **6**, 290–297 (1959)
24. Qiu, B., Ivanova, K., Yen, J., Liu, P.: Behavior evolution and event-driven growth dynamics in social networks. In: 2010 IEEE Second International Conference on Social Computing (SocialCom), pp. 217–224 (2010)
25. Shoham, Y.: Computer science and game theory. Commun. ACM **51**(8), 74–79 (2008)
26. Vickrey, D., Koller, D.: Multi-agent algorithms for solving graphical games. In: National Conference on Artificial Intelligence (AAAI) (2002)
27. von Neumann J., Morgenstern, O.: Theory of Games and Economic Behavior. Princeton University Press, Princeton (1944)
28. Wani, M.A., Li, T., Kurgan, L.A., Ye, J., Liu, Y. (eds.): In: The Fifth International Conference on Machine Learning and Applications, ICMLA 2006, Orlando, Florida, USA, 14–16 December 2006. IEEE Computer Society (2006)
29. Yang, X.-S. (ed.): Artificial Intelligence, Evolutionary Computing and Metaheuristics—In the Footsteps of Alan Turing, Studies in Computational Intelligence. vol. 427, Springer, Berlin (2013)
30. Yitzhaki, S.: More than a dozen alternative ways of spelling gini economic inequality. Econ. Inequality **8**, 13–30 (1998)

Genetic Algorithms for Multi-Objective Community Detection in Complex Networks

Ahmed Ibrahem Hafez, Eiman Tamah Al-Shammari,
Aboul ella Hassanien and Aly A. Fahmy

Abstract Community detection in complex networks has attracted a lot of attention in recent years. Communities play special roles in the structure–function relationship. Therefore, detecting communities (or modules) can be a way to identify substructures that could correspond to important functions. Community detection can be viewed as an optimization problem in which an objective function that captures the intuition of a community as a group of nodes with better internal connectivity than external connectivity is chosen to be optimized. Many single-objective optimization techniques have been used to solve the detection problem. However, those approaches have drawbacks because they attempt to optimize only one objective function, this results in a solution with a particular community structure property. More recently, researchers have viewed the community detection problem as a multi-objective optimization problem, and many approaches have been proposed. Genetic Algorithms (GA) have been used as an effective optimization technique to solve both single- and multi-objective community detection problems. However, the most appropriate objective functions to be used with each other are still under debate since many similar objective functions have been proposed over the years. We show how those objectives correlate, investigate their performance when they are used in both the single- and multi-objective GA, and determine the community structure properties they tend to produce.

A. I. Hafez (✉)
Department of Computer Science, Minia University, Minya, Egypt
e-mail: ah.hafez@gmail.com

E. T. Al-Shammari
Faculty of Computing Science and Engineering, Kuwait University, Kuwait, Kuwait

A. I. Hafez · A. e. Hassanien
Scientific Research Group in Egypt (SRGE), Cairo, Egypt
e-mail: aboitcairo@gmail.com

A. e. Hassanien · A. A. Fahmy
Faculty of Computers and Information, Cairo University, Giza, Egypt
e-mail: aly.fahmy@gmail.com

W. Pedrycz and S.-M. Chen (eds.), *Social Networks: A Framework*
of Computational Intelligence, Studies in Computational Intelligence 526,
DOI: 10.1007/978-3-319-02993-1_8, © Springer International Publishing Switzerland 2014

Keywords Social networks · Community detection · Genetic algorithms · Communities' quality measures

1 Introduction

Networks are used to model and represent many real world systems. This fact has made complex network analysis a popular research area. Collaboration networks, the Internet, biological networks, communication networks, and social networks are just some examples of such complex networks. A common feature of complex networks is community structure [1], i.e. groups of nodes in the network that are more densely connected internally than with the rest of the network. Communities play special roles in the structure–function relationship, and detecting communities (or modules) can be a way to identify substructures that may correspond to important functions in the network. Therefore, uncovering or detecting community structure is one of the most important problems in the field of complex network analysis.

Many methods have been developed for the community detection problem. These methods use tools and techniques from disciplines like physics, biology, applied mathematics, computer and social sciences. Results of a recent survey can be seen in [2].

One of the most popular algorithms proposed so far is the Girvan-Newman algorithm, which introduces a divisive method that iteratively removes the edge with the greatest betweenness value [3]. Some improved algorithms have also been proposed [4, 5]. These algorithms are based on a foundational measure criterion of community, Modularity which measures the number of within-community edges relative to a null model, which is a popular quality function that was proposed by Newman [3]. The larger the Modularity value, the more accurate the community partition. Consequently, community detection becomes a Modularity optimization problem. Because the search for the optimal (largest) Modularity value is an NP-complete problem, many heuristic search algorithms, such as extremal optimization, simulated annealing, and genetic algorithms (GA), have been applied to solve the optimization problem [2].

The remainder of this chapter is organized as follows. In Sect. 2 we define the community problem as a single-objective and multi-objective optimization problem and introduce the objective functions used in our GA. In Sect. 3 we discuss related work. We describe our GA, its genetic representation, and the operations used by the algorithm in Sect. 4. Section 5 reviews our experimental results for the single-objective and multi-objective cases on real life social networks and a synthetic network and apply the GA on a case study from Facebook social network. We then offer conclusions and suggestions for future work.

2 The Community Detection Problem

A social network SN can be modeled as a graph $G = (V, E)$ where V is a set of vertices, and E is a set of edges that connect two elements of V. A community structure S in a network is a set of groups of vertices having a high density of edges among the vertices and a lower density of edges between different groups. The problem of detecting k communities in a network, where the number k is unknown, given a quality measure of communities $F(S)$, can be formulated as finding a partitioning of the nodes in k subsets that best satisfy the quality measure $F(S)$. The problem can be viewed as an optimization problem in which one usually wants to optimize the given quality measure $F(S)$. We will describe the most popular quality measures that have been used to detect communities in the following subsection.

2.1 Single Objective Community Detection

We can view the community detection problem as a single objective optimization problem.

A single objective optimization problem $(\Omega; F)$ is defined as

$$\min F(S), \text{ s.t } S \in \Omega \tag{1}$$

where $F(S)$ is an objective function that needs to be optimized and $\Omega = \{S_1, S_2, ..., S_k\}$ is the set of feasible community structures in a network. We assume that all quality measures need to be minimized without loss of generality. So we want to find community structure S that maximizes or minimizes the quality measure $F(S)$ according to the definition of the quality measure. A formal definition of the optimization problem is given in [6].

2.2 Multi-Objective Community Detection

Recently, researchers have treated the community detection problem as a multi-objective optimization problem where, given a set of quality measures $F_1(S)$, $F_2(S),..., F_t(S)$, we want to find community structure S that simultaneously optimizes each quality measure.

A multi-objective optimization problem $(\Omega; F_1, F_2, ..., F_t)$ is defined as

$$\min F_i(S), i = 1, 2, ..., t; \text{ s.t } S \in \Omega \tag{2}$$

Since the goal is to optimize a set of competing objectives optimized simultaneously, there is not one unique solution to the problem. A set of solutions is found through the use of the Pareto optimality theory [7]. Given two solutions S_1

and $S_2 \in \Omega$, solution S_1 is said to dominate solution S_2, denoted as $S_1 \prec S_2$, if and only if

$$\forall i : \mathbf{F}_i(S_1) \leq \mathbf{F}_i(S_2) \text{ and } \exists i \text{ s.t. } \mathbf{F}_i(S_1) < \mathbf{F}_i(S_2) \tag{3}$$

Multi-objective optimization aims to generate and select non-dominated solutions, which are called Pareto-optimal solutions. It is worth noting that Pareto-optimal solutions, as outlined in [8], usually include the optimal solutions obtained by single-objective GA when applied to the clustering problems. This will be explained in more detail in Sect. 5.

Genetic algorithm first proposed in [9] is an optimization method applied to artificial intelligence problems that mimics the process of natural evolution. Genetic algorithms belong to a larger class of evolutionary algorithms, which generate solutions to optimization problems using techniques inspired by natural evolution, such as inheritance, mutation, selection, and crossover. It is a practical method, particularly when the solution space of a problem is very large and an exhaustive search for the exact solution is impractical. In GAs, potential candidate solutions in the solution set should be represented in a suitable data representation. Each candidate in the solution set, which is called a chromosome, represents a possible solution to the problem. The algorithm tries to find the candidate solution with the best fit. To improve the quality of the candidate solutions, the algorithm uses genetic operations, such as point mutation (random variations of some parts of the chromosome) and crossing over (generating new chromosomes by merging parts of existing chromosomes), on possible candidate solutions for a predefined number of iterations. At the outset, the algorithm randomly initializes the chromosomes. Then, for a number of iterations, it uses an objective function to assign a fitness value to evaluate each candidate solution's relative ability to solve the problem. The GA then reproduces candidate solutions for a new population that will be used in the next iteration by performing crossover between candidates selected according to their fitness values. It also applies some random mutation to candidates. GAs are fast algorithms for converging a problem to a smaller solution space, and if the algorithm has a good objective function, it produces near optimal solutions. The power of a GA is the cross-over mechanism, which produces better next generation of candidate solutions.

The above description of a GA only employs one objective function to describe how good a solution is. Hence, only one objective can be optimized. However, most real-world problems involve simultaneous optimization of several and often competing objectives. For example, in the community detection problem we want to find communities that contain a high density of internal connections inside each community and have a low number of external connections between nodes from different communities, i.e., low Cut Ratio. GAs have also been applied to multi-objective optimization problems, and many algorithms have been proposed to deal with the problem using the Pareto optimality theory [7]. When there are many, possibly conflicting, objectives to be optimized simultaneously, there is no longer a single optimal solution but rather a whole set of possible solutions of equivalent quality, and as we state before Multi-objective optimization aims to generate and select non-dominated Pareto-optimal solutions.

Due to the ability of multi-objective optimization to find multiple Pareto-optimal solutions in a single run, a number of multi-objective GAs have been suggested [10–14] over the past decade. The Non-dominated Sorting Genetic Algorithm (NSGA) proposed by Srinivas and Deb [15] was one of the first multi-objective GA algorithms. An improved and faster algorithm was proposed in [11]; the Fast Elitist Non-Dominated Sorting Genetic Algorithm for Multi-Objective Optimization: NSGA-II, which we will use in this work, overcomes some of the problems associated with the NSGA. Other algorithms in [12, 13] use a Strength Pareto Approach, such as the Strength Pareto Evolutionary Algorithm (SPEA) proposed in [12] and the Pareto Envelope-based Selection Algorithm (PESA) proposed in [14].

2.3 Objectives

In a GA, the objective function plays an important role in the evolution process. It is the "steering wheel" in the process that leads to good candidate solutions. For the community detection problem, many objective functions have been proposed to capture the intuition of communities, and there is no straightforward way to compare these objective functions based on their definitions.

Here we state objective functions that capture this intuition and/or are popular in the literature, and can potentially be used for community detection. A detailed description of objective functions can be found in [16] and a similarity comparison can be found in [16, 17].

In the following quality measures, the lower the value of $F(S)$ the better the community structure:

- Conductance [16] measures the fraction of total edge volume that points outside the cluster.
- Expansion [16] measures the number of edges per node that point outside the cluster.
- Internal Density [16] is the internal edge density of the cluster.
- Cut Ratio [16] is the fraction of all possible edges leaving the cluster.
- Normalized Cut [16, 18] is the normalized fraction of edges leaving the cluster.
- Maximum-Out Degree Fraction (ODF) [16, 19] is the maximum fraction of edges of a node pointing outside the cluster.
- Average-ODF [16, 19] is the average fraction of node edges pointing outside the cluster.
- Flake-ODF [16, 19] is the fraction of nodes in S that have fewer edges pointing inside than outside of the cluster.

In the following quality measures, the higher the value of $F(S)$, the better the community structure:

- Modularity [3] measures the number of within-community edges relative to a null model of a random graph with the same degree distribution.
- Community Score [20] measures the density of sub-matrices based on volume and row/column means.
- Community Fitness [21] is the ratio between the total internal degrees of the nodes belonging to that community and the sum of the total internal and external degrees of the nodes belonging to that community.
- Surprise [22] compares the number of links within and between communities in a partition with the expected number of links in a random network with the same distribution of nodes per community.

All but two of the quality measures are parameter free, i.e., calculation only depends on the network. Community Score and Community Fitness both have a positive real-valued parameter that controls the size of the communities.

3 Related Work

GAs have been used as an effective optimization technique for community detection. [23, 24] used GAs to optimize the network modularity proposed by Girvan and Newman [3]. Pizzuti proposed another GA to optimize the Community Score criterion [20, 25] and used the locus-based adjacency representation and uniform crossover for the genetic representation and the genetic operation. These algorithms have the advantage of automatically determining the number of communities during the evolutionary process. However, they also have a resolution limit, since a single objective is optimized.

A different approach is described in [26]; a random walk distance measure between graphs is integrated in a GA to cluster social networks. They use the k-medoids as the genetic representation in which each community center is represented by one of the nodes of the network. This means that k, the number of communities, must be known in advance.

More recently, researchers have applied multi-objective optimization techniques to the community detection problem using multi-objective evolutionary algorithms [6, 27–29]. Pizzuti [27] proposed a Multi-objective Genetic Algorithm (MOGA) for community detection in networks (MOGA-Net) based on NSGA-II [11] that simultaneously optimizes the Community Score and Community Fitness. Agrawal [28] proposed a bi-objective community detection method also based on NSGA-II [11]. However, Agrawal used the Community Score and Modularity as the two objectives. Shi et al. [6] proposed a new multi-objective evolutionary algorithm based on the Pareto Envelope-based Selection Algorithm version 2 (PESA-II) [30] and used the modularity objective to drive two new objectives and try to minimize the two objectives to find community structure.

4 Genetic Algorithms for Community Detection

In this section we describe the GAs, genetic representation, and genetic operations used in this work. For both single- and multi-objectives we have adopted the genetic representation and genetic operations proposed in [20]. We describe the various stages of the GA in the following subsections.

4.1 Genetic Representation

The algorithm uses the locus-based adjacency representation proposed in [31] In this representation, each individual chromosome consists of n genes $g_1, g_2,...,g_n$ and each gene can take values j in the range $\{1,...,n\}$. Where $n = |V|$ is the number of nodes in the network, and a value j assigned to the i-th gene is interpreted as a link between the nodes i and j. This means that, in the detected community structure, i and j will be in the same community. A further decoding step is necessary to identify all communities. The advantages of this representation are that the number k of communities is automatically determined by the number of components contained in an individual and determined by the decoding step, and the decoding step can be achieved in linear time, as mentioned in [32].

4.2 Initialization

A random generation of individuals could generate components that are discon-nected in the original graph. Pizzuti [20] proposed the term safe initialization to describe a process where each gene i is assign to a value j from the i-th node's neighbors. Here we use the same initialization steps used in [20]. Then, with a probability of $(1 - \text{mutation Rate})$ of the population size, we select a η percent of the genes, and for each selected gene i we assign it to itself, and for all its neighbors we assign the value i. Consequently, node i and its neighbors will be in the same community. After a series of trials, we found that a value of 0.5 for η achieved a good result.

4.3 Uniform Crossover

Given two parents, a random binary vector is created. Uniform crossover then selects the genes where the vector is a 1 from the first parent and where the vector is a 0 from the second parent. These genes are then combined to form the new child.

4.4 Mutation

The mutation operator that randomly changes the value of a randomly chosen gene causes a useless exploration of the search space. Therefore, as in the initialization step, we randomly select a m percent of the genes and for each gene i we randomly change its value to j such that node i and j are neighbors. After a series of trials, we found that a value of 0.1 for m achieved a good result.

4.5 Objectives

As mentioned in Sect. 2, many community quality measures have been proposed over the last several years, and some are similar in behavior. We apply the algorithm to each quality measure as the optimization objective and compare the results to gain more insight into the definition of each objective and its properties.

5 Experimental Results

In our experimental setup, we employed a standard single-objective GA algorithm with a roulette selection function and elitism. The algorithm was implemented in a .NET environment using C# in the single-objective GA. In the multi-objective case we used NSGA-II [11] implemented in the MATLAB Genetic Algorithm and Direct Search Toolbox as the multi-objective GA. For more description reader may see [15].

We tested the algorithm on a synthetic data set and two real social networks and finally we show the effect of the algorithm in detecting communities in online social network. To compare the accuracy of the resulting community structures, we used Normalized Mutual Information (NMI) to measure the similarity between the true community structures and the detected ones. NMI is a similarity measure proved to be reliable by Danon et al. [33]. The NMI similarity measure is inspired from information theory and is based on defining a confusion matrix $\mathbf{N}$, where the rows correspond to the real communities $\mathbf{A}$ and the columns correspond to the found communities $\mathbf{B}$. The members of $\mathbf{N}$, N_{ij}, are simply the number of nodes in the real community i that appear in the founded community j. The number of real communities is denoted as c_A and the number of founded communities is as denoted c_B. The sum over row i of matrix N_{ij} is denoted as $N_{.i}$, and the sum over column j is denoted as $N_{j.}$. Based on information theory, a measure of similarity between the partitions is then:

$$NMI(A,B) = \frac{-2\sum_{i=1}^{c_A}\sum_{j=1}^{c_B} N_{ij} \log\left(\frac{N_{ij}N}{N_{i.}N_j}\right)}{\sum_{i=1}^{c_A} N_{i.} \log\left(\frac{N_i}{N}\right) + \sum_{j=1}^{c_B} N_j \log\left(\frac{N_j}{N}\right)} \tag{4}$$

We calculated the average NMI for many runs of the algorithm for each objective and compared the results.

Real Social Network We tested the algorithm for each objective on two real life data sets, the Zachary's Karate Club and the Bottlenose Dolphins. Both data sets have been well studied in the literature (see [34]). The Zachary Karate Club data, which was first analyzed in [35], contains the community structure in a karate club. The network consists of 34 vertices and 78 edges. The network is divided into two approximately equal groups. The Bottlenose Dolphin data was compiled by Lusseau [36] and is based on observations over a period of seven years of the behavior of 62 bottlenose dolphins living in Doubtful Sound, New Zealand. A relationship between two dolphins was established by their statistically significant frequent association. The network split naturally into two large groups.

Synthetic network We used the benchmark proposed by Girvan and Newman in [1]. The network consists of 128 nodes divided into four equal-size communities. Edges are placed between vertex pairs at random such that $z_{in} + z_{out} = 16$, where z_{in} and z_{out} are the internal and external degrees of a node with respect to its community respectively. If $z_{in} > z_{out}$, the neighbors of a node inside its group are greater than the neighbors belonging to the other three groups then the network has a strong community structure. Thus, a good algorithm should discover the communities up to $z_{out} = 8$.

5.1 Single Objective Community Detection

We began by applying the single objective GA to the community detection problem. We ran each objective separately in the GA.

The results for the real life social networks, Zachary's Karate Club and the Bottlenose Dolphins, are presented in Fig. 1. The graph shows the average NMI over 30 runs of the GA for each objective. We employed standard parameters for the GA: a crossover rate of 0.9, a mutation rate of 0.4, the elite reproduction was 10 % of the population size, the population size was 60, and the number of iterations was 35. Three objectives, Conductance, Average-ODF, and Modularity, achieved a NMI value above 0.8 for both social networks. Two objectives, Community Score and Community Fitness, also achieved a good NMI value above 0.5 for both networks. For the synthetic network, we generated 15 networks for each value of z_{out} in a range from 0 to 6 and ran the algorithm 10 times. We calculated the average NMI value for the 10 runs and then calculate the average NMI value for each value of z_{out} for the 15 networks. Figure 2 shows the NMI values for the synthetic network for each objective separately. We used the same values for the standard parameters for the GA, except for the number of iterations (generation). Because the synthetic network is larger than the social networks, we set the number of iterations to 100.

We found that the Community Score objective achieved a good result until $z_{out} = 5$. The Community Fitness and Modularity objectives achieved a good

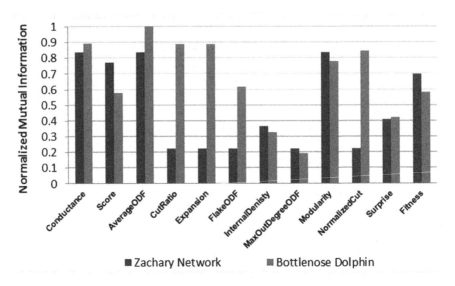

Fig. 1 NMI values obtained by each objective on Zachary Karate Club Network and Bottlenose Dolphin Network

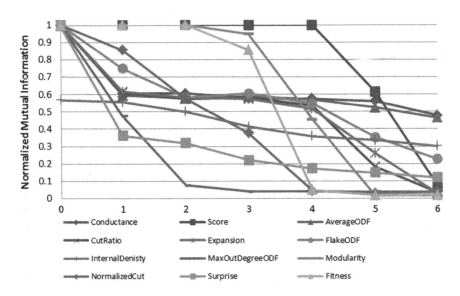

Fig. 2 NMI values for each objective on the GN Benchemark for z_{out} range from 0 to 6

result until $z_{out} = 3$. The Internal Density objective failed to detect the correct community structure for the network when $z_{out} = 0$; at this value the network consisted of four disconnected communities. We found that all objectives, except Conductance, Average-ODF, and Internal Density, decreased to a NMI value less than 0.1 when z_{out} reaches 6.

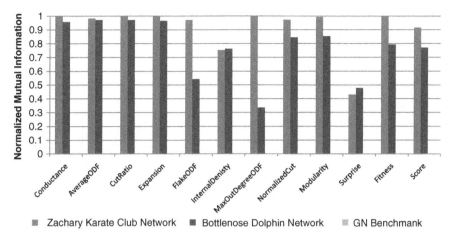

Fig. 3 Objectives Stabilities in term of NMI obtained by running GA over real life data sets; Zachary Karate Club Network and Bottlenose Dolphin Network, and GN benchmark for $z_{out} = 3$

5.1.1 Objectives Stabilities

GA is a non-deterministic algorithm, due to the random nature of the GA i.e., it may produce a different solution each run. The objective function is responsible for directing the algorithm to global/local maxima or global/local minima based on the definition of the objective over the solution space, usually we want the objective to direct the algorithm to a global maxima but it may fail sometimes to do that. A good objective function should assign the same fitness score to two solutions if they are the same or very similar. If two completely different solutions have the same fitness score obtained by the objective function, the algorithm will tend to produce a different solution in each run.

We define a stable objective as an objective that tends to produce similar community structures over many runs of the GA. Figure 3 shows the average similarity of solutions obtained by an objective in term of the NMI similarity measure. A low value indicates that, over many runs, the solutions vary considerably; a high value indicates that the solutions are much more similar to each other.

We ran the algorithm many times on each network, Zachary's Karate Club, the Bottlenose Dolphins, and the synthetic network, for a value of $z_{out} = 3$ and calculated the average NMI for each pair of solutions for each network. We observed that Community Score, Community Fitness, and Modularity achieved high values for all networks. This indicates that they were more stable than the rest of objectives because they produced similar results for different runs. Most of the objectives achieved high values for the two real life social networks, as shown in Fig. 3.

(a) (b)

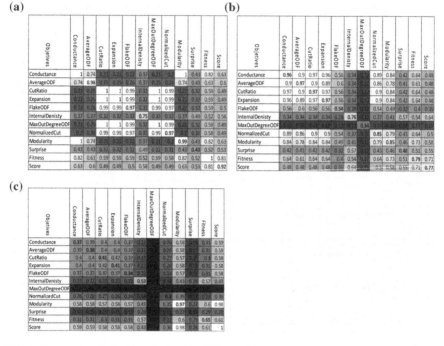

(c)

Fig. 4 Community similarities obtained by different objectives: **a** Zachary Karate Club Network. **b** Bottlenose Dolphin Network and **c** GN-Benchmark for zout = 3. Diagonal entries correspond to objective stabilities.

5.1.2 Objectives Similarities

In Sect. 2, we showed and defined many objectives, and stated that they are somehow similar in definition. Additionally, when deployed with the GA, some tend to produce equal or similar results.

To understand these similarities among objectives, we compared the solutions obtained by each objective with the solutions obtained by the remainder of the objectives. The average NMI values are shown in Fig. 4. A low value (dark cells) indicates low similarity, and a high value (white cells) indicates high similarity.

For the Zachary Karate Club Network, Community Fitness produced a result similar to Modularity, Conductance, and Community Score. Additionally, Expansion, Cut Ratio, Max-ODF, Flake-ODF, and Normalized Cut also show similar results. Conductance and Modularity produce the exact solution. For the Bottlenose Dolphin network, we found there is high similarity between Conductance, Average-ODF, Cut Ratio, Expansion, Normalized Cut, and Modularity. For the GN Network, there is a lot of variation in the output of all objectives except Community Score and Modularity, which produced similar results and good community structure.

Table 1 Modularity value for optimal community structure of the GN Network for different z_{out} values

Z_{out}	0	1	2	3	4	5	6	7
Modularity	0.4374	0.3975	0.3662	0.3391	0.3036	0.2696	0.238	0.2104

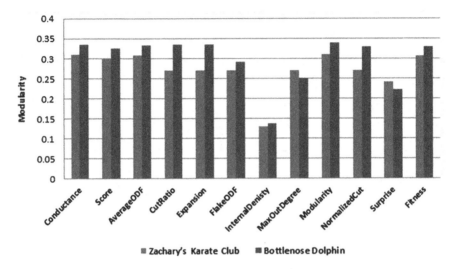

■ Zachary's Karate Club   ■ Bottlenose Dolphin

Fig. 5 Modularity values achieved by each objective for Zachary Karate Club network and Bottlenose Dolphin network

5.1.3 Network Modularity and Other Objectives

We also compared objectives based on network Modularity because it is a popular community quality measure used extensively in community detection.

First, it should be noted that the Modularity of the optimal community structure of Zackary is 0.3 and is 0.33 for Dolphin. For the GN benchmark, we show the average Modularity of the optimal community structure of the generated network for different z_{out} values in Table 1. We observed that when the z_{out} value increased, the Modularity of the original community structure decreased. When the z_{out} value increased, the number of external connections for each node increased, which led to increased randomness in the network. Therefore, the original community structure was no longer a strong structure, and there was a high probability that other community configurations with high Modularity value existed.

The same experiment was conducted as before, and for each run we calculated the Modularity of the community structure returned by each objective. We then calculated the average Modularity for all runs. In Fig. 5, we show the Modularity of the community structure returned by different objectives for the real social network. From the results, it can be seen that most objectives returned community structures with good Modularity values.

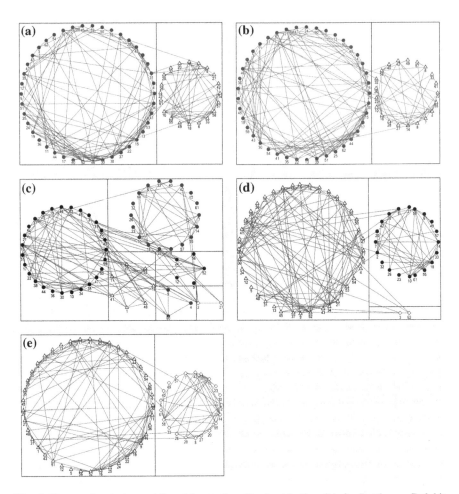

Fig. 6 Community strutures obtianed by appling Single-objective GA for Bottlenose Dolphin Network: **a** Average-ODF : NMI = 1 , Modularity = 0.333. **b** Modularity : NMI = 0.82, Modularity = 0.338. **c** Surpirse : NMI = 0.45 , Modularity = 0.22. **d** Normalized Cut : NMI = 0.9, Modularity = 0.329 and **e** CutRatio and Expansion : NMI = 0.88, Modularity = 0.335.

For the Dolphin network, only the Average-ODF objective detected the optimal community structure of the network with a Modularity value of 0.333. Additionally, the Modularity objective found a community structure with a Modularity value of 0.338, which is better than the optimal community structure of the network. Both Cut Ratio and Expansion found community structures with a Modularity value of 0.335, which is also better than the optimal solution in terms of Modularity. In Fig. 6, we visualize[1] the solution returned by Average-ODF, which

---

[1] Community Visualization has been done using NodeXL plugin for Microsoft Excel from http:// snap.stanford.edu/index.html.

Objectives	AverageODF	Conductance	CutRatio	Expansion	Fitness	FlakeODF	InternalDenisty	MaxOutDegree	Modularity	NormalizedCut	Score	AverageODF	Conductance	CutRatio	Expansion	Fitness	FlakeODF	InternalDenisty	MaxOutDegree	Modularity	NormalizedCut	Score
AverageODF	0.84	0.84	0.84	0.84	0.84	0.84	0.83	0.84	0.84	0.84	1	1	1	1	1	0.87	0.99	0.78	0.99	1	1	0.94
Conductance	0.83	0.84	0.84	0.84	0.84	0.84	0.87	0.84	0.84	0.84	1	0.89	0.89	0.89	0.89	0.82	1	0.84	0.89	0.89	0.89	0.95
CutRatio	0.25	0.23	0.23	0.23	0.84	0.23	0.52	0.25	0.84	0.23	1	0.89	0.89	0.89	0.89	0.85	1	0.85	0.9	0.89	0.89	0.9
Expansion	0.25	0.25	0.25	0.23	0.84	0.25	0.51	0.23	0.84	0.25	1	0.88	0.89	0.89	0.89	0.88	0.99	0.84	0.89	0.88	0.89	0.91
Fitness	0.45	0.45	0.23	0.24	0.7	0.72	0.43	0.84	0.84	0.82	0.58	0.35	0.35	0.35	0.31	0.58	0.8	0.33	0.82	0.75	0.8	0.38
FlakeODF	0.24	0.26	0.24	0.24	0.23	0.23	0.57	0.23	0.84	0.25	0.82	0.6	0.54	0.56	0.6	0.33	0.62	0.74	0.89	1	1	0.85
InternalDenisty	0.14	0.14	0.15	0.14	0.38	0.2	0.37	0.51	0.92	0.39	0.6	0.12	0.09	0.09	0.08	0.32	0.15	0.33	0.42	0.86	0.85	0.39
MaxOutDegree	0.24	0.23	0.24	0.23	0.23	0.23	0.14	0.23	0.84	0.23	1	0.37	0.14	0.26	0.19	0.23	0.15	0.06	0.19	0.89	0.89	0.94
Modularity	0.84	0.84	0.26	0.23	0.46	0.25	0.41	0.23	0.84	0.84	1	0.79	0.77	0.79	0.8	0.37	0.67	0.38	0.26	0.78	0.89	0.74
NormalizedCut	0.29	0.23	0.23	0.22	0.22	0.23	0.15	0.23	0.25	0.23	0.96	0.81	0.8	0.81	0.82	0.35	0.57	0.29	0.26	0.76	0.84	0.86
Score	0.58	0.6	0.23	0.23	0.44	0.23	0.43	0.23	0.59	0.23	0.84	0.42	0.41	0.4	0.4	0.35	0.36	0.34	0.28	0.42	0.4	0.58

(a) NMI of the best obtained solution for Zachary Karate Club (b) NMI of the best obtained solution for Bottlenose Dolphin

Objectives	AverageODF	Conductance	CutRatio	Expansion	Fitness	FlakeODF	InternalDenisty	MaxOutDegree	Modularity	NormalizedCut	Score	AverageODF	Conductance	CutRatio	Expansion	Fitness	FlakeODF	InternalDenisty	MaxOutDegree	Modularity	NormalizedCut	Score
AverageODF	0.31	0.31	0.31	0.31	0.31	0.31	0.31	0.31	0.31	0.31	0.31	0.33	0.33	0.33	0.33	0.33	0.33	0.32	0.33	0.33	0.33	0.33
Conductance	0.31	0.31	0.31	0.31	0.31	0.31	0.31	0.31	0.31	0.31	0.31	0.34	0.34	0.34	0.34	0.33	0.33	0.33	0.34	0.34	0.34	0.33
CutRatio	0.27	0.27	0.27	0.27	0.31	0.27	0.25	0.27	0.31	0.27	0.31	0.34	0.34	0.34	0.34	0.33	0.33	0.33	0.34	0.34	0.34	0.33
Expansion	0.27	0.27	0.27	0.27	0.31	0.27	0.23	0.27	0.31	0.27	0.31	0.36	0.34	0.34	0.34	0.33	0.33	0.32	0.34	0.34	0.34	0.33
Fitness	0.2	0.19	0.27	0.27	0.21	0.31	0.15	0.31	0.31	0.31	0.25	0.22	0.21	0.22	0.23	0.33	0.32	0.16	0.33	0.34	0.33	0.25
FlakeODF	0.27	0.27	0.27	0.27	0.27	0.27	0.25	0.27	0.31	0.27	0.31	0.44	0.3	0.3	0.3	0.22	0.29	0.31	0.33	0.33	0.33	0.33
InternalDenisty	0.25	0.25	0.25	0.25	0.13	0.25	0.13	0.22	0.31	0.19	0.26	0.25	0.25	0.25	0.25	0.14	0.24	0.14	0.23	0.33	0.33	0.25
MaxOutDegree	0.27	0.27	0.27	0.27	0.27	0.27	0.25	0.27	0.31	0.27	0.31	0.3	0.26	0.27	0.26	0.23	0.26	0.25	0.25	0.34	0.33	0.34
Modularity	0.31	0.31	0.27	0.27	0.19	0.27	0.19	0.27	0.31	0.31	0.34	0.34	0.34	0.34	0.34	0.23	0.32	0.21	0.27	0.34	0.33	0.34
NormalizedCut	0.27	0.27	0.27	0.27	0.26	0.27	0.25	0.27	0.27	0.27	0.31	0.33	0.33	0.33	0.33	0.21	0.3	0.21	0.27	0.33	0.33	0.33
Score	0.27	0.27	0.27	0.27	0.18	0.27	0.18	0.27	0.28	0.27	0.3	0.28	0.28	0.28	0.29	0.21	0.29	0.19	0.27	0.28	0.28	0.33

(c) the corresponding modularity value for Zachary Karate Club (d) the corrospending modularity value for Bottlenose Dolphin

Fig. 7 NMI values and Modularity values for the best and the worst community structure in the Pareto-optimal set for (**a**) and (**c**) the Zachary Karate Club Network, (**b**) and (**d**) the Bottlenose Dolphin Network. The *upper triangular area* of the matrix shows the values for the best solution for each pair. The *lower triangular area* shows the values for the worst solution for each pair. The diagonal entries correspond to the single objective case

is also the optimal solution of the network, and solutions returned by the Modularity, Cut Ratio, Expansion, Normalized Cut, and Surprise objectives.

5.2 Multi-Objective Community Detection

We applied the multi-objective GA to the community detection problem. We restricted the number of objectives used in the algorithm to two and studied the algorithm's performance for each pair of objectives. We excluded the Surprise objective because it is computationally expensive as it requires many binomial coefficient calculations and does not scale well for large networks. Additionally, the Surprise objective was excluded because it did not achieve good results in the single objective scenario.

Figure 7 shows the average NMI over 15 runs of the GA for each pair of objectives and the corresponding average Modularity value of each solution for the Zachary's Karate Club and the Bottlenose Dolphins social networks. We employed standard parameters for the GA; a crossover rate of 0.8, a mutation rate of 0.2, the elite reproduction rate was 10 % of the population size. We also employed a binary tournament selection function. The population size was 100, and the number of generations was 30.

For each run, the Pareto-optimal set returned by the algorithm was investigated. The best community structure was selected and compared to the optimal known solution of the network in terms of the NMI similarity measure. Additionally, the worst community structure in the set was also compared to the optimal solution of the network. Modularity values for both best and worst structures were also calculated.

Figure 7a, b show the NMI values for the best and worst community structures in the Pareto-optimal set when we applied the multi-objective GA using two objectives for the Zachary Karate Club Network and the Bottlenose Dolphin Network, respectively. The NMI values for the best solution are in the upper triangular area of the matrix. The NMI values for the worst solution compared to the optimal solution are in the lower triangular area of the matrix. The diagonal entries correspond to the single objective case. Figure 7c, d show the corresponding Modularity values of the best and worst solutions in terms of the NMI for the Zachary Karate Club Network and the Bottlenose Dolphin Network respectively. Modularity values for the best solutions are in the upper triangular area of the matrix, and Modularity values for the worst solutions compared to the optimal solution in term of NMI are in the lower triangular area. The diagonal entries correspond to the Modularity values for the single-objective case.

Figure 8 visualizes some of the results of the multi-objective case for the Bottlenose Dolphin Network. Figure 8a shows a result determined by the Modularity and Normalized Cut objectives with a Modularity value of 0.338, and Fig. 8b shows the result found by the Score and Fitness objectives with a Modularity value of 0.329. Figure 8c shows the result found by the Score and Modularity objectives with a Modularity value of 0.336.

When used with other objectives, Community Score achieved good NMI results for both networks, except when used with Community Fitness and Internal Density objectives. In addition, Community Fitness achieved good NMI results for both networks, except when used with Community Score and Internal Density objectives. Average-ODF and Conductance achieved very promising NMI results for both networks when used with any objective.

The NMI results for the best and worst solution for the synthetic network are shown in Figs. 9 and 10, we show the corresponding Modularity values for these solutions. Here, we employed the same standard parameter values used for the GA, except for the number of iterations, which was set to 100. We applied each pair of objectives to the network for z_{out} values ranging from 0 to 7. As was done previously, we generated different networks for each z_{out} value and took the average result from each run. We only show the results for z_{out} values from 1 to 7.

For $z_{out} = 0$, all pairs were able to detect the correct network community structure, except for Internal Density with Community Fitness, which achieved NMI values of 0.6. For z_{out} values from 1 to 3, when Community Score was applied with all objectives, the optimal community structure was detected, as it did for the single objective case. In addition, Community Fitness correctly detected the

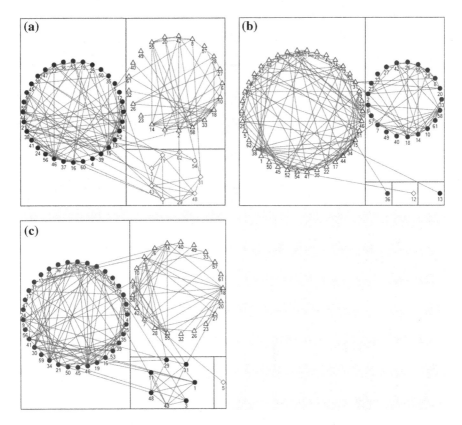

Fig. 8 Community strutures obtianed by appling Multi-objective GA for Bottlenose Dolphin Network found by: **a** Modularity and Normalized Cut : NMI = 0.6975, Modularity = 0.3384. **b** Score and Fitness : NMI = 0.8387, Modularity = 0.3332 and **c** Score and Modularity : NMI = 0.6861, Modularity = 0.3358.

optimal community structure when applied with all objectives except for Internal Density. Modularity achieved good NMI results with values above 0.9 when applied with other objectives, except for Normalized Cut. For values of $z_{out} > 4$, the performance of most objective pairs began to decrease; however, it was determined that Community Fitness achieved good results until $z_{out} = 6$ when applied with other objectives. Community Fitness with Flake-ODF and Average-ODF achieved good results for $z_{out} = 7$.

Figure 10e shows some of the solutions having low NMI values (lower triangular area of the matrix) have a better Modularity value. This indicates that such solutions have good community structure. This is due to high z_{out} values, which led to increased randomness in the network and the low Modularity value of the original community structure of the generated network (Table 1). This further indicates that the GN network benchmark is not sufficient for adequate testing.

(a) Zout = 1

Objectives	AverageODF	Conductance	CutRatio	Expansion	Fitness	FlakeODF	InternalDenisty	MaxOutDegree	Modularity	NormalizedCut	Score
AverageODF	-	0.58	0.67	0.67	1	0.58	1	0.58	1	0.86	1
Conductance	0.58	-	0.67	0.66	1	0.58	1	0.58	1	0.86	1
CutRatio	0.58	0.58	-	0.67	1	0.67	1	0.66	1	0.86	1
Expansion	0.58	0.58	0.67	-	1	0.67	1	0.66	1	0.86	1
Fitness	0.58	0.58	0.58	0.58	-	1	0.58	1	1	1	1
FlakeODF	0.58	0.58	0.67	0.67	0.73	-	1	0.58	1	0.86	1
InternalDenisty	0.05	0.29	0.52	0.09	0.54	0.67	-	1	1	0.92	1
MaxOutDegree	0.58	0.58	0.58	0.58	0.58	0.58	0.04	-	1	0.86	1
Modularity	0.58	0.58	0.67	0.67	0.73	1	0.65	0.58	-	1	1
NormalizedCut	0.58	0.58	0.58	0.58	0.58	0.7	0.85	0.51	0.58	-	1
Score	0.58	0.58	0.58	0.58	0.73	1	0.65	0.58	1	0.86	-

(b) Zout = 2

Objectives	AverageODF	Conductance	CutRatio	Expansion	Fitness	FlakeODF	InternalDenisty	MaxOutDegree	Modularity	NormalizedCut	Score
AverageODF	-	0.64	0.66	0.63	1	0.65	0.62	0.63	1	0.77	1
Conductance	0.64	-	0.65	0.64	1	0.65	0.67	0.63	1	0.7	1
CutRatio	0.66	0.65	-	0.66	1	0.66	0.89	0.62	1	0.79	1
Expansion	0.62	0.64	0.66	-	1	0.64	0.65	0.61	1	0.7	1
Fitness	0.65	0.66	0.62	0.63	-	1	0.56	1	1	1	0.99
FlakeODF	0.64	0.64	0.66	0.64	0.7	-	0.97	0.58	1	0.77	1
InternalDenisty	0.06	0.05	0.15	0.03	0.53	0.11	-	0.43	1	0.62	0.99
MaxOutDegree	0.6	0.58	0.59	0.58	0.58	0.05	0.03	-	1	0.63	1
Modularity	0.67	0.67	0.67	0.67	0.69	0.86	0.62	0.58	-	1	1
NormalizedCut	0.58	0.57	0.58	0.58	0.67	0.77	0.2	0.57	0.86	-	1
Score	0.67	0.67	0.65	0.67	0.71	0.86	0.66	0.58	0.99	0.86	-

(c) Zout = 3

Objectives	AverageODF	Conductance	CutRatio	Expansion	Fitness	FlakeODF	InternalDenisty	MaxOutDegree	Modularity	NormalizedCut	Score
AverageODF	-	0.59	0.59	0.59	1	0.58	0.5	0.56	0.94	0.57	1
Conductance	0.59	-	0.58	0.6	1	0.59	0.49	0.58	0.99	0.59	1
CutRatio	0.59	0.58	-	0.59	1	0.59	0.51	0.57	0.97	0.58	1
Expansion	0.59	0.6	0.59	-	1	0.6	0.52	0.57	0.98	0.57	1
Fitness	0.59	0.58	0.59	0.58	-	1	0.55	1	1	1	1
FlakeODF	0.58	0.59	0.59	0.6	0.71	-	0.52	0.57	1	0.58	1
InternalDenisty	0.06	0.06	0.03	0.04	0.52	0.05	-	0.31	1	0.41	1
MaxOutDegree	0.32	0.48	0.42	0.47	0.58	0.52	0.04	-	0.9	0.4	1
Modularity	0.58	0.64	0.65	0.64	0.72	0.99	0.62	0.58	-	0.91	1
NormalizedCut	0.57	0.57	0.58	0.57	0.57	0.57	0.05	0.1	0.58	-	1
Score	0.58	0.58	0.58	0.58	0.69	1	0.66	0.58	1	0.58	-

(d) Zout = 4

Objectives	AverageODF	Conductance	CutRatio	Expansion	Fitness	FlakeODF	InternalDenisty	MaxOutDegree	Modularity	NormalizedCut	Score
AverageODF	-	0.58	0.59	0.55	1	0.58	0.44	0.49	0.62	0.52	1
Conductance	0.58	-	0.59	0.58	1	0.6	0.44	0.53	0.6	0.55	0.99
CutRatio	0.59	0.59	-	0.52	1	0.57	0.43	0.44	0.57	0.49	0.99
Expansion	0.55	0.58	0.52	-	1	0.59	0.4	0.45	0.6	0.5	0.99
Fitness	0.58	0.58	0.57	0.58	-	1	0.55	1	1	1	0.98
FlakeODF	0.58	0.58	0.55	0.57	0.72	-	0.43	0.5	0.6	0.53	1
InternalDenisty	0.04	0.04	0.03	0.03	0.52	0.04	-	0.27	0.53	0.27	0.98
MaxOutDegree	0.11	0.11	0.03	0.03	0.58	0.25	0.03	-	0.45	0.06	0.96
Modularity	0.59	0.57	0.57	0.55	0.71	0.59	0.14	0.07	-	0.54	1
NormalizedCut	0.52	0.5	0.48	0.38	0.56	0.51	0.04	0.05	0.5	-	0.99
Score	0.58	0.58	0.58	0.58	0.7	0.86	0.67	0.56	0.99	0.57	-

(e) Zout = 5

Objectives	AverageODF	Conductance	CutRatio	Expansion	Fitness	FlakeODF	InternalDenisty	MaxOutDegree	Modularity	NormalizedCut	Score
AverageODF	-	0.58	0.44	0.38	0.98	0.56	0.32	0.39	0.35	0.35	0.62
Conductance	0.57	-	0.48	0.48	0.95	0.55	0.36	0.44	0.37	0.39	0.6
CutRatio	0.29	0.38	-	0.15	0.96	0.47	0.25	0.23	0.06	0.13	0.5
Expansion	0.22	0.38	0.15	-	0.98	0.51	0.23	0.11	0.02	0.08	0.44
Fitness	0.55	0.55	0.52	0.54	-	1	0.5	0.91	0.95	0.63	0.92
FlakeODF	0.49	0.48	0.33	0.43	0.54	-	0.4	0.39	0.4	0.39	0.65
InternalDenisty	0.05	0.05	0.02	0.02	0.47	0.05	-	0.23	0.22	0.24	0.4
MaxOutDegree	0.04	0.1	0.02	0.02	0.36	0.08	0.03	-	0.08	0.05	0.43
Modularity	0.05	0.18	0.02	0.02	0.64	0.15	0.02	0.02	-	0.04	0.4
NormalizedCut	0.04	0.04	0.02	0.02	0.02	0.05	0.04	0.03	0.02	-	0.46
Score	0.57	0.56	0.38	0.3	0.63	0.48	0.13	0.03	0.21	0.02	-

(f) Zout = 6

Objectives	AverageODF	Conductance	CutRatio	Expansion	Fitness	FlakeODF	InternalDenisty	MaxOutDegree	Modularity	NormalizedCut	Score
AverageODF	-	0.57	0.27	0.27	0.81	0.55	0.26	0.29	0.29	0.3	0.32
Conductance	0.55	-	0.28	0.3	0.82	0.6	0.31	0.3	0.3	0.34	0.43
CutRatio	0.04	0.04	-	0.03	0.75	0.34	0.21	0.06	0.03	0.06	0.11
Expansion	0.04	0.03	0.03	-	0.71	0.36	0.21	0.07	0.03	0.06	0.14
Fitness	0.46	0.48	0.17	0.05	-	0.96	0.46	0.73	0.63	0.46	0.8
FlakeODF	0.49	0.55	0.07	0.08	0.52	-	0.35	0.28	0.32	0.35	0.49
InternalDenisty	0.06	0.06	0.01	0.01	0.44	0.06	-	0.24	0.21	0.23	0.23
MaxOutDegree	0.07	0.06	0.02	0.02	0.09	0.03	0.05	-	0.04	0.06	0.07
Modularity	0.01	0.02	0.02	0.02	0.02	0.02	0.02	0.02	-	0.04	0.1
NormalizedCut	0.05	0.04	0.02	0.02	0.05	0.05	0.04	0.04	0.02	-	0.15
Score	0.27	0.4	0.02	0.03	0.61	0.41	0.07	0.03	0.02	0.04	-

(g) Zout = 7

Objectives	AverageODF	Conductance	CutRatio	Expansion	Fitness	FlakeODF	InternalDenisty	MaxOutDegree	Modularity	NormalizedCut	Score
AverageODF	-	0.43	0.2	0.22	0.8	0.37	0.24	0.21	0.21	0.22	0.22
Conductance	0.4	-	0.22	0.22	0.6	0.36	0.23	0.24	0.24	0.25	0.25
CutRatio	0.02	0.02	-	0.02	0.54	0.22	0.21	0.05	0.02	0.03	0.07
Expansion	0.02	0.02	0.02	-	0.54	0.26	0.21	0.04	0.02	0.03	0.08
Fitness	0.36	0.32	0.02	0.02	-	0.76	0.44	0.53	0.44	0.39	0.5
FlakeODF	0.28	0.29	0.02	0.03	0.34	-	0.25	0.14	0.25	0.3	0.25
InternalDenisty	0.04	0.04	0.02	0.02	0.42	0.04	-	0.21	0.2	0.21	0.22
MaxOutDegree	0.04	0.07	0.02	0.02	0.07	0.02	0.03	-	0.05	0.05	0.07
Modularity	0.02	0.02	0.02	0.02	0.02	0.02	0.02	0.02	-	0.03	0.06
NormalizedCut	0.03	0.03	0.02	0.02	0.03	0.03	0.03	0.03	0.02	-	0.08
Score	0.05	0.1	0.02	0.02	0.19	0.04	0.08	0.03	0.02	0.03	-

Fig. 9 NMI values for the best and worst community structure in the Pareto-optimal set for GN benchmark for z_{out} values from 1 to 7. The *upper triangular* of the matrix show the NMI values for the best solution for each pair. The *lower triangular* show the NMI values for the worst solution for each pair

(a) Zout = 1

Objectives	AverageODF	Conductance	CutRatio	Expansion	Fitness	FlakeODF	InternalDensity	MaxOutDegree	Modularity	NormalizedCut	Score
AverageODF	-	0.33	0.36	0.36	0.41	0.33	0.41	0.33	0.41	0.38	0.41
Conductance	0.33	-	0.36	0.35	0.41	0.33	0.41	0.33	0.41	0.38	0.41
CutRatio	0.33	0.33	-	0.36	0.41	0.36	0.41	0.35	0.41	0.38	0.41
Expansion	0.33	0.33	0.36	-	0.41	0.36	0.41	0.35	0.41	0.38	0.41
Fitness	0.33	0.33	0.33	0.33	-	0.41	0.09	0.41	0.41	0.41	0.41
FlakeODF	0.33	0.33	0.36	0.36	0.21	-	0.41	0.33	0.41	0.38	0.41
InternalDenisty	0.24	0.28	0.29	0.25	0.06	0.21	-	0.41	0.41	0.38	0.41
MaxOutDegree	0.33	0.33	0.33	0.33	0.33	0.33	0.24	-	0.41	0.38	0.41
Modularity	0.33	0.33	0.36	0.36	0.22	0.41	0.18	0.33	-	0.41	0.41
NormalizedCut	0.33	0.33	0.33	0.33	0.19	0.38	0.23	0.33	0.38	-	0.41
Score	0.33	0.33	0.33	0.33	0.22	0.41	0.17	0.33	0.41	0.38	-

(b) Zout = 2

Objectives	AverageODF	Conductance	CutRatio	Expansion	Fitness	FlakeODF	InternalDensity	MaxOutDegree	Modularity	NormalizedCut	Score
AverageODF	-	0.34	0.34	0.33	0.38	0.34	0.32	0.33	0.38	0.35	0.38
Conductance	0.34	-	0.34	0.34	0.38	0.34	0.31	0.33	0.38	0.34	0.38
CutRatio	0.34	0.34	-	0.34	0.38	0.34	0.36	0.33	0.38	0.36	0.38
Expansion	0.33	0.34	0.34	-	0.38	0.34	0.31	0.33	0.38	0.34	0.38
Fitness	0.3	0.26	0.33	0.29	-	0.38	0.08	0.38	0.38	0.38	0.38
FlakeODF	0.34	0.34	0.34	0.34	0.18	-	0.37	0.32	0.38	0.35	0.38
InternalDenisty	0.24	0.24	0.22	0.25	0.06	0.25	-	0.18	0.38	0.28	0.38
MaxOutDegree	0.32	0.32	0.32	0.32	0.32	0.32	0.24	-	0.38	0.33	0.38
Modularity	0.34	0.34	0.34	0.34	0.18	0.36	0.17	0.32	-	0.38	0.38
NormalizedCut	0.32	0.31	0.32	0.32	0.19	0.35	0.2	0.31	0.36	-	0.38
Score	0.34	0.34	0.34	0.34	0.2	0.36	0.18	0.32	0.38	0.36	-

(c) Zout = 3

Objectives	AverageODF	Conductance	CutRatio	Expansion	Fitness	FlakeODF	InternalDensity	MaxOutDegree	Modularity	NormalizedCut	Score
AverageODF	-	0.3	0.3	0.3	0.34	0.3	0.28	0.3	0.34	0.3	0.34
Conductance	0.3	-	0.3	0.3	0.34	0.3	0.29	0.3	0.34	0.3	0.34
CutRatio	0.3	0.3	-	0.3	0.34	0.3	0.29	0.3	0.34	0.3	0.34
Expansion	0.3	0.3	0.3	-	0.34	0.3	0.29	0.3	0.34	0.3	0.34
Fitness	0.3	0.3	0.3	0.3	-	0.34	0.07	0.34	0.34	0.34	0.34
FlakeODF	0.3	0.3	0.3	0.3	0.17	-	0.29	0.3	0.34	0.3	0.34
InternalDenisty	0.24	0.24	0.25	0.24	0.05	0.24	-	0.15	0.34	0.2	0.34
MaxOutDegree	0.27	0.29	0.28	0.29	0.3	0.29	0.24	-	0.33	0.28	0.34
Modularity	0.3	0.31	0.31	0.31	0.19	0.34	0.19	0.3	-	0.34	0.34
NormalizedCut	0.3	0.3	0.3	0.3	0.29	0.3	0.24	0.25	0.3	-	0.34
Score	0.3	0.3	0.3	0.3	0.16	0.34	0.16	0.3	0.34	0.3	-

(d) Zout = 4

Objectives	AverageODF	Conductance	CutRatio	Expansion	Fitness	FlakeODF	InternalDensity	MaxOutDegree	Modularity	NormalizedCut	Score
AverageODF	-	0.28	0.29	0.28	0.31	0.29	0.26	0.27	0.29	0.28	0.31
Conductance	0.28	-	0.29	0.29	0.31	0.29	0.26	0.28	0.29	0.28	0.31
CutRatio	0.29	0.29	-	0.28	0.31	0.29	0.27	0.27	0.29	0.28	0.31
Expansion	0.28	0.29	0.28	-	0.31	0.29	0.26	0.27	0.29	0.28	0.31
Fitness	0.29	0.29	0.29	0.29	-	0.31	0.07	0.31	0.31	0.31	0.31
FlakeODF	0.29	0.29	0.28	0.28	0.17	-	0.26	0.28	0.29	0.28	0.31
InternalDenisty	0.24	0.24	0.25	0.25	0.05	0.24	-	0.15	0.28	0.15	0.31
MaxOutDegree	0.25	0.25	0.25	0.25	0.28	0.26	0.24	-	0.27	0.24	0.31
Modularity	0.29	0.29	0.29	0.28	0.17	0.29	0.22	0.25	-	0.28	0.31
NormalizedCut	0.28	0.28	0.28	0.27	0.27	0.28	0.24	0.24	0.28	-	0.31
Score	0.29	0.29	0.29	0.29	0.17	0.31	0.17	0.29	0.31	0.29	-

(e) Zout = 5

Objectives	AverageODF	Conductance	CutRatio	Expansion	Fitness	FlakeODF	InternalDensity	MaxOutDegree	Modularity	NormalizedCut	Score
AverageODF	-	0.26	0.25	0.24	0.27	0.26	0.22	0.25	0.24	0.24	0.25
Conductance	0.26	-	0.25	0.26	0.27	0.26	0.24	0.25	0.25	0.25	0.26
CutRatio	0.25	0.26	-	0.25	0.27	0.25	0.2	0.25	0.25	0.24	0.25
Expansion	0.25	0.26	0.25	-	0.27	0.26	0.17	0.24	0.25	0.25	0.25
Fitness	0.26	0.25	0.26	0.26	-	0.27	0.06	0.26	0.26	0.16	0.26
FlakeODF	0.26	0.26	0.26	0.26	0.26	-	0.25	0.24	0.25	0.25	0.26
InternalDenisty	0.24	0.24	0.25	0.25	0.05	0.24	-	0.17	0.17	0.17	0.21
MaxOutDegree	0.24	0.24	0.24	0.24	0.24	0.24	0.24	-	0.24	0.24	0.24
Modularity	0.25	0.25	0.25	0.25	0.16	0.25	0.25	0.25	-	0.24	0.24
NormalizedCut	0.24	0.24	0.25	0.25	0.24	0.24	0.24	0.24	0.25	-	0.25
Score	0.26	0.26	0.25	0.25	0.14	0.26	0.22	0.24	0.25	0.25	-

(f) Zout = 6

Objectives	AverageODF	Conductance	CutRatio	Expansion	Fitness	FlakeODF	InternalDensity	MaxOutDegree	Modularity	NormalizedCut	Score
AverageODF	-	0.25	0.23	0.22	0.25	0.2	0.23	0.22	0.23	0.23	0.24
Conductance	0.25	-	0.22	0.23	0.23	0.25	0.21	0.23	0.23	0.23	0.24
CutRatio	0.24	0.24	-	0.24	0.21	0.22	0.17	0.24	0.25	0.24	0.24
Expansion	0.24	0.24	0.24	-	0.2	0.23	0.18	0.24	0.25	0.24	0.24
Fitness	0.21	0.17	0.22	0.24	-	0.24	0.05	0.22	0.18	0.11	0.21
FlakeODF	0.25	0.24	0.24	0.24	0.22	-	0.25	0.24	0.25	0.25	0.24
InternalDenisty	0.24	0.24	0.25	0.25	0.05	0.24	-	0.17	0.17	0.15	0.17
MaxOutDegree	0.24	0.24	0.24	0.24	0.24	0.22	0.24	-	0.24	0.24	0.24
Modularity	0.25	0.25	0.25	0.25	0.25	0.25	0.25	0.24	-	0.24	0.24
NormalizedCut	0.24	0.24	0.25	0.25	0.24	0.24	0.24	0.24	0.25	-	0.24
Score	0.24	0.24	0.25	0.25	0.13	0.24	0.24	0.24	0.25	0.24	-

(g) Zout = 7

Objectives	AverageODF	Conductance	CutRatio	Expansion	Fitness	FlakeODF	InternalDensity	MaxOutDegree	Modularity	NormalizedCut	Score
AverageODF	-	0.23	0.23	0.22	0.18	0.22	0.17	0.23	0.22	0.22	0.23
Conductance	0.23	-	0.21	0.21	0.18	0.22	0.2	0.21	0.21	0.22	0.22
CutRatio	0.25	0.25	-	0.25	0.17	0.21	0.18	0.24	0.25	0.25	0.24
Expansion	0.25	0.25	0.25	-	0.17	0.21	0.18	0.24	0.25	0.25	0.24
Fitness	0.22	0.22	0.25	0.25	-	0.19	0.05	0.16	0.13	0.1	0.13
FlakeODF	0.21	0.22	0.25	0.25	0.2	-	0.19	0.2	0.21	0.22	0.22
InternalDenisty	0.24	0.24	0.25	0.25	0.05	0.24	-	0.18	0.17	0.17	0.18
MaxOutDegree	0.24	0.23	0.25	0.25	0.24	0.2	0.24	-	0.24	0.24	0.24
Modularity	0.25	0.25	0.25	0.25	0.25	0.25	0.25	0.25	-	0.25	0.24
NormalizedCut	0.25	0.25	0.25	0.25	0.25	0.24	0.25	0.25	0.25	-	0.24
Score	0.24	0.24	0.25	0.25	0.21	0.24	0.24	0.24	0.25	0.25	-

Fig. 10 Corresponding Modularity values for the best and worst community structures in the Pareto-optimal set from the GN benchmark for z_{out} values from 1 to 7. The *upper triangular* areas of the matrices show the Modularity values for the best solution for each pair. The *lower triangular* areas show the Modularity values for the worst solution for each pair of objectives for the solution selected in Fig. 9

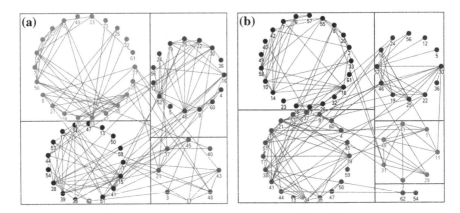

Fig. 11 Community struture founded by **a** Clauset, Newman and Moore and **b** Girvan-Newman algorithms for Bottlenose Dolphin Network

One advantage of a multi-objective GA is that it returns a set of Pareto-optimal solutions. There is no best or worst solution; the most appropriate solution can be selected according to specific needs or other criteria. For example, the algorithm could run using Community Fitness and Flake-ODF, and the solution with the maximum network Modularity could be selected from the Pareto-optimal solutions set. Modularity has been selected as our reference when comparing determined community structures because it is a popular measure and most recent work in community detection is based on this quality measure.

5.3 Comparing with Other Algorithms

As mentioned before the Girvan-Newman (GN) algorithm [3] and an improved GN algorithm proposed by Clauset, Newman, and Moore [5] are from the first community detection algorithms proposed to date. In this section we compare the results of those algorithms with the best result obtained by the GA in the single-objective case and the multi-objective case.

First we show the community structure found by the two algorithms for the Bottlenose Dolphin Network in Fig. 11. Results were obtained using NodeXL [37], a plugin for Microsoft Excel. However, the NMI calculations were done in a program that we wrote in C# to ease the comparison process.

Figure 12 shows the NMI values of the solutions found by the two algorithms compared with some of our results using the GA for both real social networks.

In the single-objective case, we observe that the GA produced better results than the GN and Clauset algorithms for both networks for most of the objectives. In the multi-objective case, the multi-objective GA with the proper objective function also produced a very promising result for both networks.

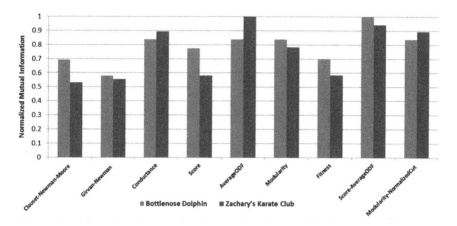

Fig. 12 NMI comparision between best of GA results, Clauset-Newman-Moore algorithm, and Girvan-Newman algorithm

5.4 Case Study: Communities in Online Social Network

Online social network such as Facebook or Google+ is a platform to build social networks or social relations among people who, for example, share interests, activities, backgrounds, or real-life connections. A social network service consists of a representation of each user (often a profile), his/her social links, and a variety of additional services. Social networking sites allow users to share ideas, pictures, posts, activities, events, and interests with people in their network due such online social behavior online communities are formed where online users tend to form communities that group users who share some comment interest. Users usually form groups or circles in online social network. Leskovec [38] collects some data for the Facebook website—10 ego networks—. The data was collected from survey participants using a Facebook application [39]. The dataset includes node features (profiles), circles, and ego networks. The ago network consist of a user's—the ego node—friends and their connections to each other. Each ego network has a ground truth community structure that has been identified by the ego used himself by just tagging each of his friends. The problem with the ego network ground truth is that it was identified by the ego user himself, making it valid form the user point of view and most participants users miss classify some of his/her friends.

We combine the 10 Facebook ego networks from [38] into one big network. We remove the ego nodes. The result network is undirected network which contain 3,959 nodes and 84,243 edges. There is no clear community structure for the network so we compare the result of the GA only in term of Modularity.

First we applied single-objective GA to the network for each objective. We employed standard parameters for the GA: a crossover rate of 0.9, a mutation rate of 0.4, the elite reproduction was 10 % of the population size, the population size was 100, and the number of iterations was 400. Figure 13 shows the result

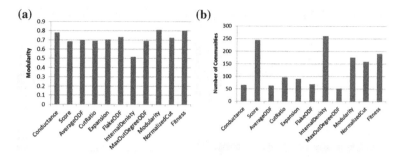

Fig. 13 a Modularity value for the detected community structure for each objective and the **b** Number of communities per each community structure

Modularity value for the detected community structure of each objective and the number of communities per each community structure. Almost all Objectives produce a good community structure in term of Modularity for Facebook dataset. First Modularity and Community Fitness objectives produce a good community structure with a high Modularity value above 0.8. Then there are Conductance and Normalized Cut objectives produce community structures with Modularity value about 0.75. Only Internal density achieves a low modularity value as observed in Fig. 13a. The number of communities' k in each community structure produced by each objective is different. Figure 13b shows the number of communities for the result of each objective. The community structures returned by Community Score and Internal density objective have more than 240 communities. The community structures returned by Modularity, Community Fitness and Normalized cut objectives have approximately the same size which in the range from 150 to 190. The rest of the objectives produce a community structure which has less than 90 communities.

Despite that the number of communities seems somehow large, the distribution of nodes in each community is not uniform, we observe that a small subset of community tend to be large—contain most of the nodes—and other is very small which have less than 10 nodes. In Fig. 14 we show the nodes distribution over communities for the result of two objectives Modularity and Conductance. We can observe that result community structure has about 5 large communities and 10 medium size communities and the rest of communities are very small which contain less than 15 nodes.

We visualize the best results of the single-objective GA for the Facebook dataset which are the result obtained by Modularity, Community Fitness and Conductance objectives. The community structure returned by Modularity is approximately the same as Community Fitness, the NMI comparison of both community structures obtained by Modularity and Community Fitness was 0.85. So we visualized the result for Modularity and Conductance only in Fig. 15. Visualization of this network has be done using Gephi [40] via utilizing Force-Atlas2 layout algorithm [41]. The network layout shown in Fig. 15 highlights the network community structure.

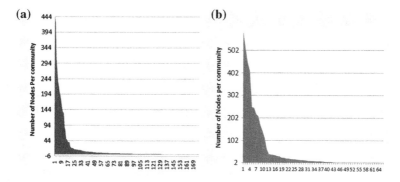

Fig. 14 Nodes distribution over communities for the result of two objectives **a** Modularity and **b** Conductance

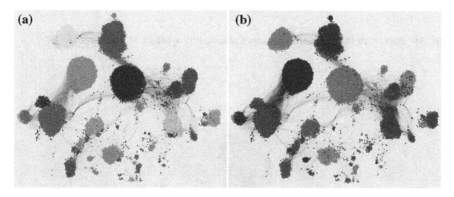

Fig. 15 Community structures of Facebook data sets as found by **a** Modularity and **b** Conductance objectives

From Fig. 15 we can observe the existence of strong communities in the network in which almost all nodes are connected to each other. The community structure returned by Modularity objective is shown in Fig. 15a; we can observe a clear community structure. The major difference between the two community structures shown in Fig. 15a, b is that Modularity objective was able to break about 4 large communities in Fig. 15b into smaller communities which is better.

Second we applied the Multi-Objective GA for each pair of objective on the Facebook dataset. We employed standard parameters for the GA; a crossover rate of 0.8, a mutation rate of 0.2, the elite reproduction rate was 10 % of the population size, population size was 100, and the number of generations was 300. We show only the best result for a few pairs which prove to produce a good result when applied with each other. Fig. 16 summarizes the best result of Multi-Objective GA when applied to the Facebook dataset.

The result of Multi-objective GA when applied using two objectives inherits some properties of the result of each objective when used in single objective GA.

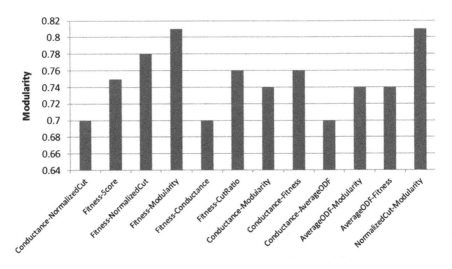

Fig. 16 Best result for the Facebook dataset obtained by applying Multi-Objective GA

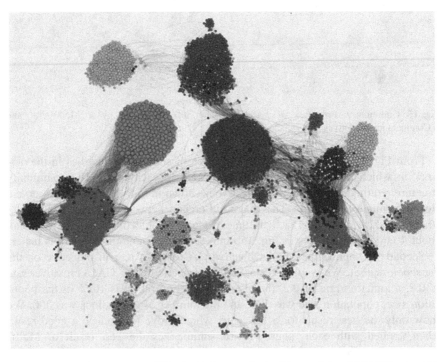

Fig. 17 Community structure of Facebook dataset obtained by applying Multi-Objective GA using Conductance and Modularity objectives

For example when applied Conductance and Modularity the result community structure has 122 communities, however the community structure of Conductance only has 66 communities and the community structure of Modularity only has 180 communities. The fact that GA try to optimize the solution for both objective, lead to a result that average the result of both objective. We show the result community structure obtained by Conductance-Modularity in Fig. 17.

6 Conclusion and Future Works

A GA as an optimization technique works effectively for the community detection problem in single-objective and multi-objective cases. However, performance is influenced directly by the objective function used in the optimization process. For the multi-objective case, we found that some objectives worked well together while others did not. However, it would be interesting to design new objectives, different from those used in this study, that are more suitable for the multi-objective genetic algorithm. Additionally, we could explore the performance of a different multi-objective optimization technique, such as multi-objective Bayesian optimization, for the community detection problem.

References

1. Girvan, M., Newman, M.: Community structure in social and biological networks. Proc. Natl. Acad. Sci. **99**(12), 7821–7826 (2002)
2. Fortunato, S.: Community detection in graphs. Phys. Rep. **486**(3–5), 75–174 (2010)
3. Newman, M.E.J., Girvan, M.: Finding and evaluating community structure in networks. Physics Rev. E **69**(2), 026113 (2004)
4. Radicchi, F., Castellano, C., Cecconi, F., Loreto, V., Parisi, D.: Defining and identifying communities in networks. Proc. Nat. Acad. Sci. U.S.A. **101**(9), 2658–2663 (2004)
5. Clauset, A., Newman, M., Moore, C.: Finding community structure in very large networks. Phys. Rev. E **70**(6), 066111 (2004)
6. Shi, C., Zhong, C., Yan, Z., Cai, Y., Wu, B.: A multi-objective optimization approach for community detection in complex network. In: IEEE Congress on Evolutionary Computation (CEC), Barcelona, pp.1–8 (2010)
7. Ehrgott, M.: Multicriteria Optimization, 2nd edn. Springer, Berlin (2005)
8. Handl, J., Knowles, J.: An evolutionary approach to multiobjective clustering. IEEE Trans. Evol. Comput. **11**(1), 56–76 (2007)
9. Holland, J.: Adaptation in Natural and Artificial Systems. University of Michigan Press, Michigan (1975)
10. Srinivas, N., Deb, K.: Multiobjective optimization using nondominated sorting in genetic algorithms. Evolitionary Comput. **2**(3), 221–248 (1994)
11. Deb, K., Agrawal, S., Pratap, A., Meyarivan, T.: A fast elitist non-dominated sorting genetic algorithm for multi-objective optimization: NSGA-II. In: Parallel Problem Solving from Nature PPSN VI Lecture Notes in Computer Science, vol. 1917, pp. 849–858 (2000)
12. Zitzler, E., Thiele, L.: Multiobjective evolutionary algorithms: a comparative case study and the strength pareto approach. IEEE Trans. Evol. Comput. **3**(4), 257–271 (1999)

13. Zitzler, E., Deb, K., Thiele, L.: Comparison of multiobjective evolutionary algorithms: empirical results. Evolutionary Comput. **8**(2), 173–195 (2000)
14. Corne, D., Knowles, J., Oates, M.: The pareto envelope-based selection algorithm for multiobjective optimization. In: Proceedings of the Parallel Problem Solving from Nature VI Conference, pp.839–848 (2000)
15. Deb, K.: Multi-objective optimization using evolutionary algorithms. Wiley, England (2001)
16. Leskovec, J., Lang, K., Mahoney, M.: Empirical comparison of algorithms for network community detection. In: Proceeding of 19th International Conference on World Wide Web (ACM WWW), pp. 631–640 (2010)
17. Shi, C., Zhong, C., Yan, Z., Cai, Y., Wu, B.: On selection of objective functions in multi-objective community detection. In: Proceedings of the 20th ACM International Conference on Information and Knowledge Management, Glasgow, pp. 2301–2304 (2011)
18. Shi, J., Malik, J.: Normalized cuts and image segmentation. IEEE Trans. Pattern Anal. Mach. Intell. **22**, 888–905 (1997)
19. Flake, G., Lawrence, S., Giles, C.: Efficient identification of web communities. In: Sixth ACM SIGKDD International Conference on Knowledge Discovery and Data Mining, pp. 150–160 (2000)
20. Pizzuti, C.: Ga-net: a genetic algorithm for community detection in social networks. In: Proceedings of the 10th International Conference on Parallel Problem Solving from Nature: PPSN X, Dortmund, pp. 1081–1090 (2008)
21. Lancichinetti, A., Fortunato, S., Kertesz, J.: Detecting the overlapping and hierarchical community structure of complex networks. arXiv:0805.4770v2 (2008)
22. Aldecoa, R., MarÕn, I.: Deciphering network community structure by surprise. PLoS ONE **6**, e24195 (2011)
23. Tasgin, M., Bingol, H.: Community detection in complex networks using genetic algorithm. arXiv:cond-mat/0604419 (2006)
24. Shi, C., Zhong, C., Yan, Z., Cai, Y., Wu, B.: A new genetic algorithm for community detection. Complex Sci. **5**(2), 1298–1309 (2009)
25. Pizzuti, C.: Community detection in social networks with genetic algorithms. In: Proceedings of the 10th Annual Conference On Genetic And Evolutionary Computation, Atlanta, pp.1137–1138 (2008)
26. Firat, A., Chatterjee, S., Yilmaz, M.: Genetic clustering of social networks using random walks. Comput. Stat. Data Anal. **51**(12), 6285–6294 (2007)
27. Pizzuti, C.: A multi-objective genetic algorithm for community detection in networks. In: 21st International Conference on Tools with Artificial Intelligence, pp.379–386 (2009)
28. Agrawal, R.: Bi-objective community detection (bocd) in networks using genetic algorithm. Contemp. Comput. Commun. Computer Inf. Sci. **168**, 5–15 (2011)
29. Hafez, A.I., Ghali, N.I., Hassanien, A.E., Fahmy, A.A.: Genetic algorithms for community detection in social networks. In: 12th International Conference on, Intelligent Systems Design and Applications (ISDA), 2012, pp. 460–465 (2012)
30. Corne, D., Jerram, N., Knowles, J., Oates, M.: PESA-II: region-based selection in evolutionary multiobjective optimization. In: Proceedings of the Genetic and Evolutionary Computation Conference, pp.283–290 (2001)
31. Park, Y., Song, M.: A genetic algorithm for clustering problem. In: Proceedings of the 3rd Annual Conference on Genetic Programming, pp.568–575 (1998)
32. Cormen, T., Leiserson, C., Rivest, R., Stein, C.: Introduction to Algorithms. MIT Press, Massachusetts (2001)
33. Danon, L., Diaz-Guilera, A., Duch, J., Arenas, A.: Comparing community structure identification. J. Stat. Mech: Theory Exp. **9**, 09008 (2005)
34. Network DataSets. http://www.personal.umich.edu/mejn/netdata. Accessed 2013
35. Zachary, W.W.: An information flow model for conflict and fission in small groups. J. Anthropol. Res. **33**(4), 452–473 (1977)
36. Lusseau, D.: The emergent properties of dolphin social network. Proc. R. Soc. Lond. B Biol. Sci. **270**(Suppl 2), S186–S188 (2003)

37. Stanford Large Network Dataset Collection. http://snap.stanford.edu/data/index.html. Accessed 2013
38. McAuley, J., Leskovec, J.: Learning to discover social circles in ego networks. In : NIPS, pp.548–556 (2012)
39. Leskovec, J.: Social Circles in Ego Networks. http://snap.stanford.edu/socialcircles/. Accessed 2013
40. Bastian, M., Heymann, S., Jacomy, M.: Gephi: an open source software for exploring and manipulating networks. In: International AAAI Conference on Weblogs and Social Media, http://www.aaai.org/ocs/index.php/ICWSM/09/paper/view/154 (2009)
41. Jacomy, M., Heymann, S., Venturini, T., Bastian, M.: ForceAtlas2, A continuous graph layout algorithm for handy network visualization. Medialab center of research (2011)

Computational Framework for Generating Visual Summaries of Topical Clusters in Twitter Streams

Miray Kas and Bongwon Suh

Abstract As a huge amount of tweets become available online, it has become an opportunity and a challenge to extract useful information from tweets for various purposes. This chapter proposes a novel way to extract topical structure from a large set of tweets and generate a usable summarization along with related topical keywords. Our system covers the full span of the topical analytics of tweets starting with collecting the tweets, processing and preparing them for text analysis, forming clusters of relevant words, and generating visual summaries of most relevant keywords along with their topical context. We evaluate our system by conducting a user study and the results suggest that users are able to detect relevant information and infer relationships between keywords better with our summarization method than they do with the commonly used word cloud visualizations.

Keywords Automated summarization · Clustering · Data mining · Twitter · Social networks · Keyword extraction · Topic modeling

1 Introduction

The area of computational intelligence traditionally refers to the nature-inspired computational techniques that attempt to address high-complexity real world problems, which cannot necessarily be addressed by classical statistical approaches or first-fit modeling. Well-known computational intelligence technologies

M. Kas (✉)
Electrical and Computer Engineering, Carnegie Mellon University,
Pittsburgh, PA 15213, USA
e-mail: miraykas@gmail.com

B. Suh
Advanced Technology Labs, Adobe Systems Inc, San Jose, CA 95110, USA
e-mail: bongwon@adobe.com

W. Pedrycz and S.-M. Chen (eds.), *Social Networks: A Framework* 173
of Computational Intelligence, Studies in Computational Intelligence 526,
DOI: 10.1007/978-3-319-02993-1_9, © Springer International Publishing Switzerland 2014

benefit from techniques like fuzzy sets, neural and evolutionary computing, which are techniques that offer adaptive mechanism that enable intelligent behavior in complex and dynamically changing environments [1]. Computational intelligence techniques have found application in several areas of research and engineering, including performance evaluation of weapon systems [2], fault diagnosis [3], temperature forecasting [4], design decisions in software processes and products [5, 6].

More recently the intelligence aspect of computational intelligence has also started being attributed to discovery science and knowledge mining [7], which can be referred to as a trendy and hot topic at the time of writing. Analyzing dynamically changing, complex social networks that can be obtained from online social networking websites is one of the areas that attract significant attention from the knowledge discovery and data mining community. Another area of computational intelligence that has connections with the social network analysis is swarm intelligence, which refers to the collective behavior of self-organized, decentralized natural behavior such as fish schooling or bird flocking. The term was originally coined by Beni and Wang in the context of cellular robotics systems in 1989 [8]; and nowadays its use has been extended to other areas beyond robotics such as understanding collective behavior, group mind or social adaptation to knowledge [9].

In the last decade, activity and user engagement in online social media have increased substantially, and a huge volume of user generated data becomes available online every day; which opens doors for the analysis of collective intelligence behavior in a very large user population. Among many popular social platforms, Twitter, with roughly 150 million active users and 340 million tweets per day, makes an enormous volume of users' activities (i.e. tweets) publicly available. This opens up great opportunities to investigate interesting communication patterns and content shared within the social network.

The 140-character limit and immediate visibility on followers' timelines propel information shared on Twitter to be short and timely. This, in turn, allows users to disseminate timely information about the breaking news all around the world. For example, Kwak et al. [10] showed how the news of a recent airplane crash propagated over the network of Twitter users. However, as Naaman et al. [11] point out, not all tweets have such informational value. They report that the majority of tweets are rather personal status updates. Many other researchers also found that users make use of Twitter for many different purposes including promoting political views, marketing, tracking real time events, and so on [10, 12, 13].

Due to such diverse user behavior and mixed usages, it is not an easy task to digest tweets and get a general sense of what is going on in Twitter. However, as Twitter becomes more widely used, more researchers become interested in taking advantage of the massive user-generated data available in Twitter. One natural but promising direction is to examine tweets to measure general audience interests. Many researchers investigate novel ways of processing the massive number of tweets to accurately measure what the current trending topics are and how popular they are at each moment. For instance, Bollen et al. [14] proposed a method that can make accurate predictions on stock markets. The authors used sentiment analysis on tweets to extract overall socio-economic mood such as fatigue, confusion, depression or anger, and relate the collective emotive trends with social and economic indicators to predict the stock market behavior. Another interesting Twitter-based prediction example is Asur and Huberman [15] where the authors attempt to forecast future box-office revenues of movies by analyzing the tweets about them even before they are released. The proposed model achieved higher accuracy in prediction than any other known systems. They also discuss that sentiments extracted from Twitter can be further utilized to improve the prediction of general trends.

In addition to tracking general trends, tweets can also be used when one wants to track a specific topic of interest. Since the discussion in Twitter covers a very wide spectrum of topics that people care about, it is often possible to find tweets relevant to any reasonable topic. In this context, researchers have focused on issues including selecting relevant tweets, identifying topical structures, and visualizing the results. Tweet Motif [16] is an example of how users can be served with only relevant tweets of a specific topic of interest. The system is a navigation tool for the searched tweets. Once relevant tweets are returned as a result of a search on Twitter, unlike traditional search results presented in a sequential list, Tweet Motif identifies subtopics of tweets and lays out tweets by the subtopics, so that the users can navigate and drilldown through a faceted search interface.

Nokia Internet Pulse [17] is another system focusing on visualizing discussions around a particular topic on Twitter. It visualizes terms in relevant tweets as a stacked set of tag clouds. Words are sized in proportion to their frequencies and colored according to the emotional content of the tweet allowing users to investigate the social activity around the topic.

Although the above-mentioned systems show that tools for topic based summarization of tweets have great potential, there are also a number of challenges involved in processing a massive number of tweets and generating a topical summary on relevant information.

First, a tweet is a short piece of text and it is hard to detect topics from it primarily because it does not provide enough information on the relationships between the words and their contexts. In addition, people use Twitter for various purposes, resulting in tweets having very diverse topics. Ramage et al. [18] propose an LDA based model to map tweet contents into several dimensions such as substance, style, status, and social characteristics. While LDA based models have been used for several years for longer texts, how well it works with short texts such as tweets is still an open question. Second, it is still a challenge to process a very large number of tweets in real-time. Extracting relevant topical structure and digesting a continuous stream of tweets often requires scalable computer frameworks such as Apache Hadoop. Furthermore, the problem is exacerbated when it is difficult to present a complex topical structure in a comprehendible and easy-to-analyze fashion, which calls for novel visualization techniques.

Among the main data mining tasks that are of particular interest to computational intelligence community are data dimensionality reduction, classification, and rule extraction [7]. This chapter focuses more on reduction of data to a succinct visual summarization form, which can be understood by an analyst and enables intelligent decision making at business level. In particular, this chapter proposes a method and presents a system to summarize tweets around a specific topic by identifying relevant keywords and providing visual summaries using those keywords.

Our contributions are threefold:

1. We design and prototype a system that automates the entire tweet summarization process, starting with fetching tweets all the way up to providing visual summarizations as results at the end.
2. We present a method of forming topical clusters of words in tweets and grouping relevant information together.
3. We conduct a user study to evaluate the method for summarizing daily tweet contents and show that people perform better with the proposed summarization than with word cloud visualizations, which is one of the most popular ways of summarizing Twitter content.

We structure the rest of the chapter as follows. After briefly reviewing related work, we first discuss the details of our system and explain our design choices and the algorithms used. Next, we describe our dataset and move on to the user study and the results. We discuss our user study results as well as insights obtained in detail. In our discussion section, we specifically focus on several potential future research problems that can be solved by computational intelligence techniques. Finally, we conclude the chapter highlighting our key findings and contributions.

2 Related Work

Recently, Twitter has become a popular platform where people share opinions, news, and breaking stories on any topic that is of interest to them, resulting in a wide variety of topics and users mingling together in a unified, social platform. Twitter is primarily used by people to talk about their daily activities and to seek or share information and news about almost any topic [13]. This also allows its users to participate in conversations with individuals, groups, or the public at large [12].

Due to the enormous diversity of tweet contents and its user profile, Twitter is now home to a very large-scale dataset. This has resulted in numerous research studies utilizing Twitter data including disaster management [19], crime reporting [20], analysis and prediction on political campaigns [21], financial forecasts [12, 15], and other everyday activities such as information seeking and sharing related to local news consumption, shopping, and recommendation making [22].

In this chapter, our major contribution is the summarization of tweet contents and presenting our results with a user-friendly visualization. There have been many attempts to summarize tweets in a usable fashion. The rest of this section briefly reviews notable examples of research related to our work that we have not already mentioned so far.

Eddi, a system developed by Bernstein et al. [23], groups tweets in a user's feed into topics by clustering them via linguistic syntactic transformation and callouts to a search engine. The tool also provides interactive visualizations of the clusters. Theme Crowds [24] is another system that summarizes what Twitter users are saying about certain topics over time. The system generates hierarchical, multi-resolution cluster visualizations based on the similarity of users' tweet profiles each day. Liu et al. describe another interactive, visual text analysis tool called TIARA, whose text summarization results are based on the Latent Dirichlet Allocation (LDA) model [25]. The visualization results are represented as ordered layers of topics over time [26].

Our work is different from these aforementioned systems for a number of reasons. First, all of the tweet summarization systems mentioned above use some form of word clouds. In this work, we use treemaps to visually summarize our output, which we later found to be more successful than the word cloud visualizations. Another differentiator is the fact that we use a graph-based approach to identify topic structure from tweets. We use clustering on co-occurrence graphs for the words that appear in tweets for a given day and topic. Other studies employ different techniques such as LDA models [26], callouts to external search engines [23], or clustering users and their texts based on their similarities measured by metrics like cosine similarity. In this work, we attempt to design and prototype an entire tweet summarization system that generates summaries based on clusters of words on co-occurrence graphs and produces output in the form of treemap visualizations.

There are also other, recent studies that emerge from the computational intelligence community that specifically examines community detection and clustering on social networks. For instance, [27] uses Genetic Algorithms (GA) as an effective optimization technique to solve the community detection problem. The authors examine solutions with single-objective and multi-objective functions and show how those objective correlate with each other. Another chapter, [28], examines community detection in complex networks using genetic algorithms to optimize the modularity as defined by Newman. Another attempt along these lines is [29], which focuses on detecting closely connected communities that are sparsely connected with one another via optimizing a fitness function. There exist other computational intelligence studies that specifically target social networks and converting the acquired intelligence into predictions for government policies [30] and potentially profitable solutions via optimizing sales on directed social networks where ties between social agents are not necessarily reciprocal [31].

Fig. 1 Main steps of the
Twitter summarization
prototype system. Search
keywords are the inputs to the
system and visual cluster
summaries are the outputs of
the system

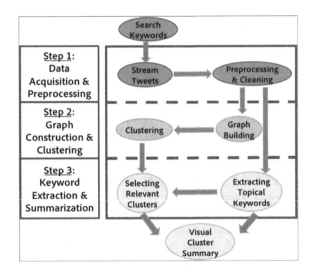

3 Methodology

In this chapter, we describe a novel method of extracting topical keywords and
providing topic-relevant and easy-to-analyze visual summaries of tweets on a
certain topic. In this section, we introduce the overall architecture of our prototype
system.

The key idea of the proposed system is to periodically form a co-occurrence
graph of words that appear in the tweets and form clusters of words that have
strong co-existence relationships in the graph. Then, central keywords are systematically selected and clusters around those keywords are elicited. The final
output is a visual summary that describes the major events and news of the day.
Figure 1 presents a high-level overview of the main steps of the system. Next, we
explain each step in more detail.

3.1 Twitter Data Stream

We use the Twitter Streaming API to track tweets that contain specific keywords.
Our system requires the users to specify an initial set of keywords (one or more
words) for the topics to be tracked. We pass this parameter to Twitter Streaming
API as a comma-separated list, which in turn determines the set of tweets to
download. Twitter API returns all tweets that contain one or more of the specified
words, regardless of order and ignoring case. We periodically form tweet files (a
new file every 10 min), which are then aggregated to a single daily file for analysis. We extract tweets using Java JSON parser library.

3.2 Preprocessing

We first preprocess our data to transform tweets into a clean, easy-to-analyze format. Preprocessing involves cleaning tweets through several steps. We convert all words to lower case, and remove urls, user mentions (e.g. @username), and a twitter specific stop word "RT", which are less useful for topical analysis. We also clean punctuation, including the removal of "#", the hashtag marker. However, we do not remove terms used in hashtags; we only remove the # sign. Hence, after our cleaning process SYRIA, Syria, syria, and #syria are all converted to syria for further processing. We also remove the commonly used stop words such as 'a, an, the' to reduce the noise introduced by frequently occurring, non-topical words. We also remove number-only words such as 123. However, we keep words that are alphanumeric combinations such as html5 because a lot of them correspond to technical terms. Finally, we apply the initial steps of the Porter Stemming algorithm from the Ling Pipe library [32] to remove inflectional affixes such as the plural affix and the tense affixes to reduce redundancy in the number of entities to be analyzed. We keep derivational affixes as they may result in changes in meanings or classes of words. For instance, if we were to remove derivational affixes the words 'useful' and 'useless' would be converted to 'use', which would result in very misleading and inaccurate analysis. Hence, we only remove inflectional affixes.

3.3 Building the Co-Occurrence Graph

After preprocessing the data and preparing it for analysis, we construct a word co-occurrence graph. In a word co-occurrence graph, a node denotes a word in tweets and a link between two nodes denotes that the corresponding two words co-occur together within a single tweet. Links in the co-occurrence graph are undirected, and link weights are determined by the co-occurrence frequency of two words. In a co-occurrence graph, if two words appear together more frequently, the link weight (strength) between those two words increases and the distance (cost) decreases. As a result, similar words or words that belong to similar concepts tend to have lower costs between them.

When constructing the co-occurrence graph, we look for co-occurrences of words within each tweet. In traditional text mining, many systems often look for co-occurrences in a sliding window of usually 5–7 words or within a single sentence or paragraph. Since a single tweet is short enough and self-contained, we use a single tweet as a unit for finding co-occurrences.

In this research, we use only unigrams as our nodes in the graph. Once we have the co-occurrence graph, we perform pruning and remove nodes and links that are below certain thresholds. We empirically set the node threshold as 3 and the link threshold as 10. In other words, nodes that appear less than 3 times and links that appear less than 10 times are removed from the graph. However, both threshold values are design parameters that can be adjusted as required.

Fig. 2 Sample co-
occurrence graph constructed
from two tweets

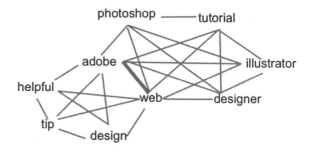

3.3.1 Example for Co-Occurrence Graph Construction

In this section, we walk through a simple illustration of data cleaning and construction of a co-occurrence graph. Consider the following two tweets on digital art/technology:

"Helpful tips from Adobe on web design!"

"30 Adobe Photoshop and Illustrator tutorials for Web Designers"

After the preprocessing step, we transform the tweets into the following forms:

"helpful tip adobe web design"

"adobe photoshop illustrator tutorial web designer"

Using the cleaned versions, we construct a co-occurrence graph as shown in Fig. 2. Since 'adobe' and 'web' co-occur twice, the link between them is stronger with a cost of 0.5, while all other links have a cost of 1.

The co-occurrence graphs generated in this way share a couple of properties. First, since we use the entire tweet to look for co-occurrences, there are several very dense groups in the form of overlapping cliques. This leads to a locally-dense, globally-sparse structure overall. Second, since we form undirected links between the words (i.e. if t1 co-occurs with t2, then t2 co-occurs with t1 as well), a lot of words (nodes) are reachable from one another. Hence, trying to get groups of similar words will not be as successful if we try to identify groups based on connectivity or through the identification of major, disconnected subcomponents within the co-occurrence graph.

3.4 Clustering of Words on Co-Occurrence Graphs

To discover groups of words that co-occur very frequently, we use clustering algorithms. Clustering is defined as the process of organizing objects into groups whose members are similar in some way [33, 34]. In our context, the words that frequently appear together on a certain topic or concept are the objects that belong to the same group/cluster.

Clustering is a very widely studied research area. To date, many clustering algorithms have been proposed [35]. One plausible way is classifying clustering algorithms based on their outputs. The output of a clustering algorithm can be hard or fuzzy clusters. Clustering is also of significant interest to computational intelligence community, especially when the increasing interest in fuzzy modeling and fuzzy clustering are considered [36, 37]. In hard clustering, an object/item belongs to a single cluster. In fuzzy clustering, objects have varying degrees of membership in different groups and can be members of several of different clusters to different extents.

Another way of classifying clustering algorithms is to consider how different groups are formed: hierarchical or partitional. Hierarchical algorithms produce dendrograms by splitting or merging groups based on a similarity criteria. To simply put, a dendrogram is a hierarchical tree formed of nested series of partitions. Hierarchical clustering algorithms that form clusters by merging subgroups are called agglomerative algorithms (e.g. single link clustering [38], group-averaging clustering, complete link clustering [39]) while the algorithms that work by splitting larger groups into smaller groups are called divisive algorithms (e.g. Girvan-Newman [40]). Partitional clustering (e.g. k-means) algorithms produce unnested grouping, which simply divides data into several groups based on a clustering criterion.

In our study, we use the complete link (max) clustering algorithm. This algorithm is an agglomerative, hierarchical clustering algorithm that forms clusters by merging subgroups. In the complete link clustering, initially each object is a unique cluster on its own and the closest clusters are successively merged until only a single cluster remains. The distance between two nodes x and y is equal to the cost of the shortest path between x and y. Initially, all nodes start as individual clusters and the ones that are closest to one another form a cluster. In our example, the words 'adobe' and 'web' are merged at the first level. To keep merging in the next level, the distances between the clusters are measured again. To combine another word with the 'adobe-web' cluster, the distance of the new word is checked against both words and the distance between the two clusters is set as the maximum of these two distances. This way, 'design', 'helpful', and 'tip' are combined as one cluster while the remaining words 'designer', 'tutorial', 'Photoshop', and 'illustrator' are combined as another cluster. Because, all the words that end up in the same cluster are one hop away from one another. The final step of iteration combines the two bigger clusters and finalizes the algorithm. The final output of the complete link clustering is a dendrogram. A dendrogram is a hierarchical tree representing a nested series of partitions that are formed by splitting or merging groups based on some similarity criteria. Figure 3 presents a dendrogram created for the co-occurrence graph depicted in Fig. 2, using the complete link clustering algorithm.

In our case, the data that we need to find clusters in is presented in the form of a co-occurrence graph. The distance between two clusters on a graph can be defined

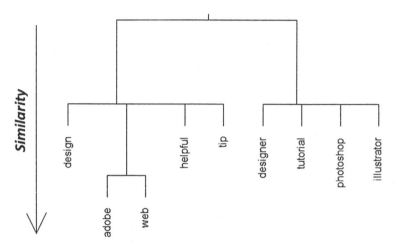

Fig. 3 Sample dendrogram for the co-occurrence graph depicted in Fig. 2

in a number of ways, and clustering based on minimum or maximum distance are the two most intuitive ones. For instance, in single link (min) clustering [38], the distance between two clusters is defined as the distance between the two closest points that are in different clusters. However, in the complete link (max) clustering, the distance between two clusters is defined as the maximum distance between a pair of objects, with the condition that the objects compared in the pair are in distinct clusters.

There are a number of reasons as to why we chose to use the complete link clustering in this work. As mentioned earlier, there are a number of common features of co-occurrence graphs and we want to exploit those properties to find of groups of similar concepts. One important feature of co-occurrence graphs that can be exploited by clustering algorithms is the existence of several, partially overlapping cliques. Using the farthest points' distance between two clusters, the complete link-clustering algorithm is good at finding cliques in a graph. This makes complete link clustering suitable for co-occurrence graphs. Since it is geared towards finding cliques, the complete link clustering is able to avoid the downsides of single link clustering which results in a chaining effect as it may keep linking an unrelated object to a cluster because that unrelated object happens to be very close to a single object in that cluster, while it may be far apart from all other objects in the cluster.

In the dendrogram representation, the vertical location of a branch is proportional to the value of the intergroup similarity between the subgroups of the branch. In other words, the closer a branch is to the root of the tree, the lower the similarity between the sub-groups.

Therefore, when a similarity threshold is given, it is rather straightforward to cut a dendrogram into many clusters by grouping nodes whose similarity is higher than the given threshold. However, finding a good threshold is a non-trivial problem. One other way to determine a good threshold is to use gap statistics and calculate

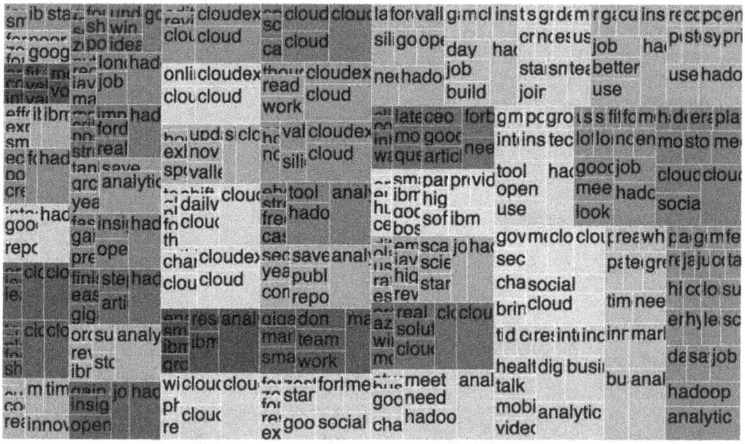

Fig. 4 Treemap visualization that illustrates the behavior of fuzzy clustering algorithms and LDA based schemes

the number of clusters at varying threshold level to see where the knee of the curve for the dendrogram size forms. Then, the threshold value around the steeper change in the number of clusters is used. In our experiments, we tried both methods and used the level whose number of clusters is closest to the half of the number of leaf nodes in the dendrogram. Note that this is a value that is empirically set after several experiments; a different threshold value might work better in other datasets.

Given that clusters of words are identified, the next step is to select the directly relevant clusters among all the clusters identified by the algorithm. To elicit relevant clusters, we first identify central topical keywords and then extract the clusters of keywords surrounding them. In the following section, we discuss how to extract the central topical keywords.

3.4.1 Discussion on the Design Choices

Before finalizing our decision on the clustering algorithm to be used in our system, we have implemented and experimented with several other algorithms for generating clusters including single link clustering, k-means, g-means [41], and fuzzy clustering algorithms. We obtained the best results with the complete link clustering due to the reasons explained above. For this particular problem, the problem with fuzzy clustering algorithms was that very popular terms usually appeared in several different clusters, and got replicated several times in the end-result summary visualization in our preliminary results. As a result, this reduced the recognizability and the readability of the words because each popular word had to replicate in different clusters, appearing in smaller separate boxes as shown in Fig. 4. This also

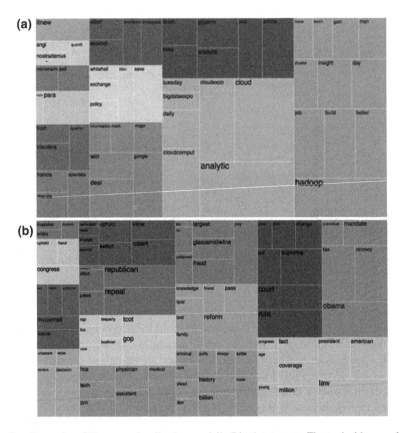

Fig. 5 a Examples of Treemap visualization on daily Big data tweets. The topical keywords are presented in a group based on their co-location relationships. The size of each rectangle represents the relative frequency of the corresponding keyword. **b** Examples of Treemap visualization on daily Healthcare tweets. The topical keywords are presented in a group based on their co-location relationships. The size of each rectangle represents the relative frequency of the corresponding keyword

reduces the possibility of correctly assessing a word's frequency (or importance) as the overall frequency has been partitioned across several clusters.

Similar problems occur for Latent Dirichlet Allocation (LDA) based techniques (i.e. the ability to associate the same word to different topics). LDA is one of the most commonly used text mining techniques and a generative probabilistic model for collections of discrete data such as text corpora that tries to allocate observations (e.g. words or phrases) to a set of unobserved groups (e.g. topics hidden in a document) [25]. Each word or phrase can be assigned to multiple topics with a certain degree of membership. Similar to our method, the LDA-based methods also build on the notion of co-occurrence. However, it would result in fuzzy clusters and a visualization that is unusable for human users as shown below. For instance, in Fig. 4, the words 'cloud' and 'hadoop' are replicated several times, but their importance can only be noticed after the figure is examined carefully for a long time. In our pilot user study, the users found that using Fig. 4, it is not easy to judge which word is more important while it is very to appreciate their importance for the topic considering its strict partitioning equivalent presented later in Fig. 5a. Moreover, statistical approaches such as LDA do not work well for short text because it is difficult to infer topic distributions from text when each document contains approximately ten words; resulting in low clustering accuracy as previously shown in [42]. Considering the reasons stated above, we preferred using strict partitioning clustering algorithms in our design over LDA-based schemes or fuzzy clustering algorithms, and did not include them in the rest of our user study. Thus, we do not provide further comparisons.

3.5 Keyword Extraction and Eliciting Relevant Clusters

Central keywords refer to the words that express a topic the best and are frequently used within that topic. In other words, central keywords are used for summarizing tweet contents in the most relevant way possible. In most cases, these central keywords are directly shown to the users, so it is important to focus on their quality.

In our prototype, we choose central keywords from words used in tweets. For a word to be a central keyword for a topic, it has to be a word that frequently appears in the stream of tweets about that topic while it is not all that popular in general. If a word also appears frequently in the generic set of tweets, it is more likely to be a word that is commonly used, regardless of the topic, such as stop words. Therefore, we are interested in identifying words that appear frequently in the set of topical tweets while they appear with relatively lower frequency in the set of generic tweets, which share the similar reasoning as Tf-Idf weighting used in document indexing to understand how important a term is for a document, considering all the documents in corpus. Tf-Idf (Term frequency—Inverse document frequency) is one of the most commonly used weighting schemes in automated text retrieval [43, 44]. The Tf-Idf weighting scheme returns the highest score when a term occurs many times within a small number of documents; thus lending high differentiating power. The result returned by the Tf-Idf weighting is lower for a term when the term occurs fewer times in a document, or occurs in many documents. The lowest Tf-Idf scores are observed when a term occurs in virtually all documents (e.g. stop words).

Similar to Tf-Idf weighting, we have designed a ranking scheme for selecting central keywords, which calculates the ratio of topical relevance over general popularity. For calculating topical relevance, we count how many times a word occurs in tweets about the topic. For the general popularity of a word, we examine how many times the word appears in the 1 % sample of all tweets posted on Twitter during the same day. Table 1 shows four attributes that are used to calculate the topical strength of each word.

Table 1 Attributes used for calculating the topical strength of each word

		Topical	
		Low Frequency	High Frequency
General	**Low Frequency**		★
	High Frequency		

Measure	Description
TWF (*word*)	Topical Word Frequency. Number of occurrences of a *word* in the topical tweet set. The topical tweet set is defined by tweets that contains a specific topical keyword e.g. Adobe
TTT	Total Topical Tweets. Number of tweets collected as the topical tweets set
GWF (*word*)	General Word Frequency. Number of occurrences of a *word* in the 1 % sample tweets. This measure is to assess the general popularity of a word
TGT	Total General Tweets. Number of tweets collected in the 1 % sample

$$TopicalStrength(word) = \frac{\frac{TWF(word)^2}{TTT}}{\frac{GWF(word)}{TGT}}$$

The weighting we use in this study is designed to penalize the words appearing in many non-relevant documents. In particular, the denominator of the formula above is designed to penalize globally popular words. However, we also boost the numerator by multiplying the 'Topical Word Frequency' of the word (*TWF* (*word*)). We experimented with several different ways for calculating the topical strength and found that the proposed measure produces the best results for identifying central topical keywords. After we calculate the topical strength for all the words that appeared in the topical tweet set, we rank them in ascending order. Then, we select the top-n words and choose the clusters that contain those words for visual summarization. In our design, we use $n = 20$. This threshold is set by examining the visualizations created at the end of the summarization process to ensure that there is enough information in the visualization while it is reasonably comprehensible when it is displayed in full screen mode in a laptop computer.

3.6 Summarization and Visualization

The final step in our system is to visually summarize the information we have extracted in earlier steps. In our system, we use treemap visualization [45, 46] to summarize our findings as we present in Fig. 5.

Fig. 6 **a** Example word cloud visualization on daily tweets on the same tweet set as Fig. 5a.
b Example word cloud visualization on daily tweets on the same tweet set as Fig. 5 b

Treemap visualization is a hierarchical (tree-structured) way of displaying data, where each node is represented as a rectangle and clusters are represented as bigger, bounding rectangles containing individual objects [46]. In our prototype, the area of a word is proportional to the square root of its frequency in the tweets. Words that belong to the same cluster have the same color, which makes it easier to identify groups of similar words. However, the colors for clusters are randomly chosen, and the proximity of two clusters does not imply any further relationship between those two clusters. Treemap visualization is well known to be good for detecting patterns that would otherwise be hard to notice due to correlations between the colors and dimensions. In addition to the treemap visualizations, we also use word clouds to visually summarizing the tweets, where the font sizes of the words are scaled with respect to their frequencies.

In our study, we also build word cloud summaries as presented in Fig. 6. In word clouds, the colors and positions of the words are chosen randomly to make

Fig. 7 Number of tweets per day on each topic (Healthcare, Big data, and Syria)

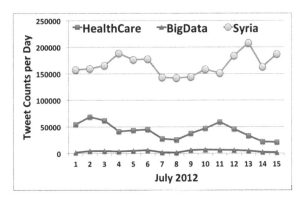

them visually more appealing. For the colors, we use the default color scheme in d3.js library [45]. One advantage of the word clouds is that the top 3–5 words usually stand out in the figure, making it easier to spot the most frequent keywords. Rivadeneira et al. discuss that font size has a robust effect while the layout of words in a word cloud had no effect on recognition [47]. In their experiments, users recognized words with larger fonts easier and quicker. In our evaluations, we compare user responses for treemap and word cloud visualizations.

3.7 Information on Coding Environment

The code for this prototype system is written using open-source Java libraries (Ling Pipe [32]) and open-source visualization library D3 JS [45]. The Streaming Twitter API that we used for streaming (downloading) tweets that contain a certain keyword is also available for developers for free. Given the level of detail provided in this chapter about the system design, this prototype system is easily replicable.

4 Dataset

Using the Twitter Streaming API, from July 1st to July 15th 2012, we collected tweets on three arbitrarily chosen domains: (i) Syria, (ii) Healthcare, and (iii) Big data. These three topics are selected because they are general and popular enough to generate many tweets suitable for analysis. As described earlier, tweets containing specific keywords are collected for the study. For Syria, we only use 'syria' as our search keyword and collect all the tweets containing that specific term. For healthcare, we have used both of its potential spellings: 'health care' and 'healthcare'. Similarly, for big data, we used 'big data' and 'big data'. While collecting these datasets, we did not observe any anomalous activities such as sudden increased or decreased volume of tweets or spam activity. The average numbers of tweets per day for each domain are as follows: Big data: 4,086.46, Healthcare: 42,093.53, and Syria: 166,930.2. Figure 7 shows the number of tweets per day on each topic during July 1st 2012–July 15th 2012.

5 User Study

In order to evaluate the effectiveness of our prototype system, we designed a 3-by-2 two-way within-subject factorial experiment (3 topics by 2 summarizations). We defined a set of tasks to measure the effectiveness of our method in terms of users' visual comprehension and understanding of the underlying information on three general topics. We then compared the users' performance with our method against the one with word cloud summarization, which is a popular way of summarizing information from Twitter [24, 46, 48, 49].

5.1 Participant

For the user study, we recruited fifteen participants from a research organization of a software company. They had a variety of professional backgrounds including computer science, marketing, business, physics, and neuropsychology. The average age of the participants is 32 (min: 24, max: 54) while four of the participants (26.7 %) are female.

To obtain a better insight on how participants use the summarizations, we observed four of participants while they were working through the given tasks. Some participants were uncomfortable about being monitored and timed, so we did not time how long each task takes, and we do not use time as one of independent measures. Typically, it took between 20 and 40 min for participants to complete the given tasks. Thirteen of them completed the study within the first 30 min.

5.2 Method

In our experiment, each participant is asked to evaluate six summarizations since we employed 3 × 2 within subject factorial experiment design. Each user is shown a total of six summarizations, each of which represents a combination of three topics (Syria, big data, and health care) and two summarization techniques (topical treemap and word cloud), Since each topic is visualized with two different types of summarization techniques (treemap and word cloud), we prepared two different sets of tweets for each topic to make sure that a user is not presented with two summarizations on the same tweet set. This approach is taken to avoid bias which might affect the evaluation results. Assume that tweets for Day-1 on Syria are summarized using both word clouds and treemaps. A user who has seen treemap visualization already will have some bias and pre-existing knowledge when he/she is presented with the word cloud visualization for the same set of tweets. Hence, the treemap and word cloud visualizations of Syria presented to a single user are drawn from two different sets of tweets. Another user is presented its counterpart. The order of summarizations and the topics have also been shuffled randomly at each experiment.

For each visual summary, we ask the participants to complete two tasks. In Task 1, we present six statements per summarization to measure how accurately users can identify information in the presented summarization. For each statement, participants were asked to choose one of the following options 'Probably Relevant', 'Probably Irrelevant', or 'Not Sure' upon their judgment on whether the given statement could be directly inferred from the presented visual summary. These statements were carefully prepared as a representative summary of the actual tweets on a given topic on a designated date.

The statements were independently written after one author read through all the tweets used for the user study. To prepare the statements that are "Irrelevant" to the presented visual summary, we have read tweets from random dates distant in time. In other words, instead of using fabricated fake statements, we use statements that summarize the tweets on a very distant day in time. As examples, the statements we used included the following:

- Cloudexpo is a big event on data analytics and cloudcomputing. (Topic: Big data on July 03, 2012, Relevant for Figs. 5a and 6a)
- Glaxosmithkline agrees to plead guilty and pay settlement fees. (Topic: Health care on July 02, 2012, Relevant for Figs. 5b and 6b)
- Troops set fire in Latakia mountains, destroying massive green area. (Topic: Syria on July 04, 2012)

After a user completes the study, we collect the number of correct answers and use it as a dependent variable for our model.

For Task 2, we ask the users to find the two major news/events that happened on the day summarized in the visualization. The goal of Task 2 is to measure how well users can spot the scale/importance of the events and news presented in the summarizations. When scoring users' answers, we consider two criteria: (i) whether they are commenting about an event relevant to the visualized day, and (ii) whether the event they comment is a major or minor event of the day. In this task, a user can get at most two points from each visual summary where each event he or she claims to be one of the two major events of the day is worth 1 point at most. If an event the user suggests as a major event of the day is related to the visualized day's events, but not a major one, then the user gets 0.5 points for that answer instead of a full 1 point. If the event the user has selected is not relevant to the visualized day's events at all, the user does not get any points for that answer. Finally, if the user is able to identify a major event of the day correctly, he or she gets 1 full point for each correct answer.

Table 2 Repeated measure ANOVA for Task 1 and Task 2

	Factors	Deg. F	F value	Sig. p
Task 1	Summarization	1	7.176	0.018[**]
	Topic	2	6.923	0.004[***]
	Interaction	2	2.347	0.114
Task 2	Summarization	1	4.035	0.066[*]
	Topic	2	4.675	0.018[**]
	Interaction	2	0.009	0.957

(Significance: *** $p < 0.01$, ** $p < 0.05$, * $p < 0.1$)

Fig. 8 The number of correct answers for Task 1. The score represents the accuracy of the detection of relevant statements

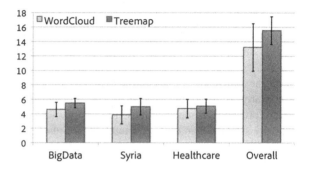

Fig. 9 The number of correct answers for Task 2. The score represents the accuracy of the detection of major events/news of a certain day

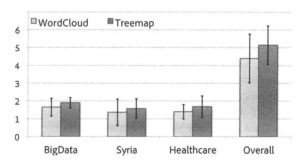

After processing the six visual summaries, we also asked participants to complete the post-experiment questionnaire. The questions were asked to evaluate overall subjective satisfaction, relevance of information, and ease-of-use for both summarization methods. We collected users' input in a 5-Point Likert scale (e.g. (*a*) Very relevant, (*b*) Somewhat relevant, (*c*) Neutral, (*d*) Not so relevant, and (*e*) Not relevant at all).

6 Results

This section reports our user study results and the statistical analysis performed on these results. We use two-way repeated-measure ANOVA to measure the effect of

Fig. 10 Users' ratings on
post-experiment
questionnaires. Each bar
represents the evaluation of
overall satisfaction, relevance
of information presented, and
ease-of-use for two
summarization techniques
(i.e. treemap and word cloud)

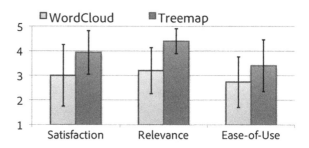

the two factors (topic and summarization) and the interaction between these
factors.

The statistical analysis results presented in Table 2 clearly shows that the two
factors have a significant effect on participants' performance both on Task 1 and
Task 2.

In the following set of figures (Figs. 8, 9 and 10), we present results from our
user study. In the figures, the error bars present the standard deviations while the
bars represent the mean values. Important mean values are also mentioned specifically when we analyze and interpret the user study results.

In Fig. 8, we present the result for Task 1, which is designed to measure how
accurately users can infer the relevancy of presented statements. Since each
summarization is presented with a set of six statements, in each task for each
method, users can get at most 6 points, resulting in at most 18 points per method.
Overall, the results we obtain with treemap are substantially better than word cloud
results (e.g. mean score 15.53 for treemap versus 13.2 for word cloud,
p value $= 0.018$ in Table 2).

Another interesting point to note is the marginal interaction between summarization and topic factors (p-value $= 0.114$ in Task 1: Interaction in Table 2). The
interaction indicates that the summarization can be particularly useful when the
users are examining tweets from an unfamiliar domain. When examined in detail,
we observed interesting patterns in user responses based on the topic. For instance,
with the healthcare topic, there was little difference in the accuracy of users'
performance obtained using treemap and word cloud visualizations. Our observation suggests that this result might be due to the impact of users' prior
knowledge on the topic since most users in our study have some prior knowledge
about the healthcare system to some extent. However, when the results for Syria
are considered, the impact of summarization becomes more obvious as most of our
users are less knowledgeable about Arabic keywords and Syria. For the case of
Syria, treemap provides substantially better results than word cloud visualizations.
The results hint that there may be an interesting interaction between topic and
summarization. Further research is required to explore the relationships between
prior knowledge on topics and the efficiency of summarization techniques.

Next, in Fig. 9, we present the results for Task 2, which aims for detection of
the two major event/news of the day. The results presented in Fig. 9 are similar to
those in Fig. 8. In this task, the average user score is 5.14 out of 6 with treemap

visualizations across all topics, while 4.39 point with word cloud visualization. The analysis in Table 2 shows that both factors, summarization and topic, have a significant effect on users' performance in detecting major events and news.

Finally, Fig. 10 reports results of our post survey questions regarding overall satisfaction, relevancy, and ease of use for the two summarization methods: treemaps and word clouds.

With the participants' ratings obtained in the post-experiment questionnaires, we performed a series of paired t-tests to measure the significance in differences in the users' responses. As is easily noticed in Fig. 10, the users' responses on overall satisfaction indicate that the participants preferred the treemap summarization over the word clouds (paired t-test, $t = 2.114$, $df = 14$, p-value $= 0.053$). A more noteworthy result is about the second question in the questionnaire, concerning information relevance. The relevance of information is important because it provides a subjective assessment on the effectiveness of our methods in keyword extraction and clustering techniques. In terms of information relevance, treemap summarization achieves markedly higher results: 4.4 out of 5 for treemaps versus 3.2 out of 5 for word cloud visualization. The paired t-test shows a very significant effect of our treemap method with $t = 4.58$, $df = 14$, and p-value $= 0.0004$.

However, we could not detect significant differences in the ease-of-use question ($t = 1.50$, $df = 14$, p-value $= 0.16$). The result shows that many users gave lower ratings overall on this question. Most users thought that they should have been able to find information more easily. Even though users were able to complete tasks with relatively high scores, this questionnaire result implies that the tasks were not trivial to complete.

7 Discussion

In this section, we discuss the lessons learned and points out a number of interesting research questions and potential use cases that are inspired by the research presented in this chapter and may benefit from various computational intelligence techniques.

7.1 Lessons Learned

One of the interesting comments from the participants was about the usefulness of treemap visualizations in identifying bi-grams. With our clustering techniques, the words that constitute a commonly used phrase usually co-occur together frequently and they end up in the same cluster in treemap. Due to the spatial proximity of words that are commonly used together, users are able to infer and suggest bi-grams based on the information presented in our treemap visualizations.

Another strength of treemaps is that many participants stated that the scaling of words with respect to their frequencies and clustering of related concepts helped them organize their thoughts and infer the relations among words by following the presented structure of the data. Most participants perceived this as an advantage over word clouds in which words are positioned to make the overall visualization more appealing, not for forming a structure.

Three out of fifteen participants completing our user study suggested that treemap summarizations and word cloud visualizations are complementary, and they would find relevant information quicker if they were presented both summarizations at the same time. This seems to be because word clouds help users to identify the top three outstanding terms quicker while treemap summarizations would be useful for identifying patterns in the presented information in an organized manner. This observation suggests that it might be a worthwhile effort to prepare both summarizations and present them to the users for the analysis of the data.

After conducting our pilot study, we were expecting that most users would perform better with treemap summarization in Task 1 because relevant words are clustered together. We were not sure if the users would perform better with our treemap method in Task 2 because the most frequent words stand out in word cloud visualizations. Surprisingly, users were able to perform better in both tasks using treemap summarization. We think that this is an effect of the way treemap summarization organizes information, which in turn helps users organize the information they inferred from the visualization.

Some participants pointed out minor usability issues and suggested improvements to the visual design of treemaps such as restricting the spectrum of colors to lighter, bolder colors and providing interactivity for the highlighted keyword and cluster. Another suggestion along these lines was to add numerical information to treemaps which indicate the area of each rectangle to enable direct quantitative comparison of rectangles that have similar areas.

7.2 Future Research

Next, we discuss a number of potential research directions on a broader spectrum, inspired by the users' comments.

7.2.1 Spamming Control

In our dataset, we did not observe any particular Twitter users dominating the entire discussion. However, it could have been the case if the discussion was under the influence of spamming parties. In that case, using the set of words that appear most frequently becomes totally susceptible to the influence of spammers and results in misleading conclusions. Therefore, one interesting research direction is to analyze the distribution of contributions from different users on a certain topic and design a model that filters out and/or weighs the information acquired with respect to the credibility of the users before generating summaries for tweets. There have been a number of research studies that touch upon computational intelligence techniques that especially focus on noise reduction for more intelligence clustering of unlabeled or semi-labeled data (e.g. new emails arriving) [50], spam filtering based on collective classification [51], and feature extraction for identifying and classifying the spammers on Twitter [52].

7.2.2 Evolution of Concepts over Time

In this chapter, we focus on summarizing tweets to provide daily digests on a certain topic. However, we do not examine how these summaries evolve over time. Or we do not analyze the popularity trends of topical keywords. 'Which words lose popularity?', 'What are the emerging concepts?', 'Are there any repeating trends or seasonal effects?' are among the questions that can be answered by performing over-time analysis either in the form of time series analysis [53], regression models [54], machine-learning based analysis techniques [55–57] or by using computational intelligence techniques. Such a problem can be very-well solved using techniques that inherently use evolutionary computation mechanisms. Evolutionary computation techniques are primarily inspired by the principals of natural evolution where the three mechanisms that drive the evolution are reproduction, mutation, and natural selection; geared towards the survival of fittest [58, 59]. Evolution of concepts or topics can be modeled as co-occurrence between different unigram concepts to generate new bigrams (reproduction), the survival of very popular and trendy concepts only (natural selection), and the emergence of new concepts that drive from earlier concepts with some randomness involved due to outside effects (mutation).

One direct business application of this kind of research is brand monitoring: 'What do people say about our product?'. By analyzing over-time evolution of concepts, it is possible to infer how a brand or a product is perceived in the society and how the discussions involving that brand/product change over time.

7.2.3 Adaptive Online Marketing

Another direct business application of analyzing real-time Twitter summaries is online marketing. For instance, it would be important for a marketer to monitor users' Twitter activities around a specific theme (e.g. a newly launched product). These days, many online marketing systems are based on a competitive marketplace for keywords, in which the highest bidder for a given keyword gets to display its advertisement when there is an activity for that keyword. Hence, choosing the right set of keywords and placing the right amount of budget on the profitable ones in auction based advertisement systems require timely analysis and the ability to make snap decisions. However, this is a dynamic game played by several different bidders simultaneously which would benefit from the support of the computational intelligence techniques especially with regards to forecasting how other players' strategies will evolve as time goes by. This is a business questions that is already under investigation by several researchers [60, 61].

7.2.4 Adaptive Keyword Expansion

The keyword extraction technique proposed in this chapter can also be extended to perform adaptive learning/expansion of the advertisement keyword set that can be used for bidding in online advertisement systems. With our system, we provide visual summarizations that only use the data collected during the period of analysis. The system proposed and described in this chapter does not incorporate feedback from the previous analysis results to understand which keywords proved to be useful for winning the online advertisement auction or turning the advertisement browsing into actual sales. It is possible to make the evolution of the set of keywords smarter by incorporating a feedback mechanism that is based on computational intelligence and soft computing techniques into the system such as fitness approximation [62, 63] to evaluate how well a term fits into the set of keywords to be used in online advertisement auction.

7.2.5 Co-Evolution of Text and Social Networks

This chapter focuses only on co-occurrence networks (i.e. content networks) that were derived from the actual tweets sent by the users on Twitter. However, there are a number of recent studies that investigate the co-evolution of complex content and user networks. For instance, [54] is one such paper where the authors study the temporal co-evolution of content and social networks on Twitter and bi-directional influences between them by using multilevel time series regression models. Understanding the co-evolution of social and content networks can provide valuable information on the factors that impact the activity and popularity of social media applications. This problem space might benefit substantially from computational intelligence techniques such as fuzzy sets and fuzzy time series analysis to forecast the direction the community will move and also from swarm intelligence to model collective behavior and its evolution better.

8 Conclusion

In this work, we have made several contributions. We built and evaluated a system that summarizes tweets on a given topic and provides topical keywords and visual summaries of clusters around those keywords as the final output. At the core of our tweet summary generation, we employ a graph-based clustering approach that models relationships of words and keyword extraction. In addition, by conducting a user study, we have shown that many users find information provided by treemap summarizations more relevant, and easier to work with than the information provided by corresponding word clouds. Considering the interest in analyzing large-scale online social data, we hope that the research presented in this study will inspire further research that aims to improve the state of the art for automated Twitter summarization models.

References

1. Engelbrecht, A.: Computational Intelligence: an Introduction. Wiley, Chichester (2007)
2. Chen, S.M.: Evaluating weapon systems using fuzzy arithmetic operations. Fuzzy Sets Syst. **77**(3), 265–276 (1996)
3. Palade, V., Bocaniala, C.D.: Computational Intelligence in Fault Diagnosis, 1st ed. Springer Publishing Company, New York (2003)
4. Hwang, S.M., Chen,J.R.: Temperature prediction using fuzzy time series. Trans. Syst. Man Cybern. Part B:Cybern. **30**(2), 263–275 (2000)
5. Pedrycs, W., Peters, J.F.: Computational intelligence in software engineering. In: Canadian Conference on Engineering Innovation: Voyage of Discovery, pp. 253–256. St. Johns (1997)
6. Pedrycs, W.: Computational intelligence as an emerging paradigm of software engineering. In: Proceedings of the 14th International Conference on Software Engineering and Knowledge Engineering (SEKE '02), pp. 7–14 (2002)
7. Wang, L.: Data Mining with Computational Intelligence. Springer, Heidelberg (2009)
8. Beni, G., Wang, J.: Swarm intelligence in cellular robotic systems. In: NATO Advanced Workshop on Robots and Biological Systems. Tuscany (1989)
9. Kennedy, J.: The particle swarm: social adaptation of knowledge. In: International Conference on Evolutionary Computation, (1997)
10. Kwak, H., Lee, C., Park, H., Moon, S.: What is Twitter, a social network or news media? In: WWW, pp. 591–600 (2010)
11. Naaman, M., Boase, J., Lai, C.H.: Is it really about me?: message content in social awareness streams. In: CSCW, pp. 189–192 (2010)
12. Boyd, D., Golder, S., Lotan, G.: Tweet, tweet, retweet: conversational aspects of retweeting on Twitter. In: HICSS, pp. 1–10 (2010)
13. Java, A., Song, X., Finin, T., Tseng, B.: Why we Twitter: understanding microblogging usage and communities. In: WebKDD & SNA-KDD, pp. 56–65 (2007)
14. Bollen, J., Mao, H., Zeng, X.J.: Twitter mood predicts the stock market. J. Comput. Sci. **2**(1), 1–8 (2011)
15. Asur, S., Huberman, B.A.: Predicting the future with social media. In: arXiv Preprint (2010)
16. O'Connor, B., Krieger, M., Ahn, D.: Tweet motif: exploratory search and topic summarization for Twitter. In: ICWSM, pp. 384–385 (2010)
17. Kaye, J.J., et al.: Nokia internet pulse: a long term deployment and iteration of a Twitter visualization. In: CHI EA, pp. 829–844 (2012)
18. Ramage, D., Dumais, S., Liebling, D.: Characterizing microblogs with topic models. In: ICWSM, pp. 384–385 (2010)
19. Acar, A., Muraki, Y.: Twitter for crisis communication: lessons learned from Japan's tsunami disaster. Int. J. Web Based Communities **7**(3), 392–402 (2011)
20. Li, R., Lei, K.H., Khadiwala, R., Chang, K.C.C.: TEDAS: a Twitter-based event detection system and analysis system. In: ICDE, pp. 1273–1276 (2012)

21. Shamma, D.A., Kennedy, L., Churchill, E.F.: Tweet the debates: understanding community annotation of uncollected sources. In: WSM, pp. 3–10 (2009)
22. Brooks, A.L., Churchill, E.F.: Tune in, tweet on, twit out: information snacking on Twitter. In: CHI, pp. 1–4 (2010)
23. Bernstein, M.S., et al.: Eddi: interactive topic-based browsing of social status streams Eddi: interactive topic-based browsing of social status streams. In: UIST, pp. 303–312 (2010)
24. Archambault, D., Greene, D., Cunningham, P., Hurley, N.: Theme crowds: multi resolution summaries of Twitter usage. In: SMUC, pp. 77–84 (2011)
25. Blei, D.M., Ng, A.Y., Jordan, M.I.: Latent dirichlet allocation. J. Mach. Learn. Res. **3**, 993–1022 (2003)
26. Liu, S., et al.: Interactive, topic-based visual text summarization and analysis. In: CIKM, pp. 543–552 (2009)
27. Hafez, A.I., Ghali, N.I., Hassanien, A.E., Fahmy, A.A.: Genetic algorithms for community detection in social networks. In: International Conference on Intelligent Systems Design and Applications (ISDA), pp. 460–465, Kochi (2012)
28. Pizzuti, C.: Boosting the detection of modular community structure with genetic algorithms and local search. In: Proceedings of the 27th Annual ACM Symposium on Applied Computing (SAC), pp. 226–231 (2012)
29. Pizzuti, C.: Mesoscopic analysis of networks with genetic algorithms. World Wide Web, pp. 1–21 (2012)
30. Brown, M.A., Alkadry, M.: Predictors of social networking and individual performance. In: Citizen 2.0: Public and Governmental Interaction through Web 2.0 Technologies. IGI Global, New York, p. 17 (2012) (Ch 8)
31. Wang, C.G., Szeto, K.Y.: Sales potential optimization on directed social networks: a quasi-parallel genetic algorithm approach. Appl. Evol. Comput. (LNCS) **7248**, 114–123 (2012)
32. Baldwin, B., Carpenter, B.: Ling Pipe. http://alias-i.com/lingpipe/ (2003)
33. Anderberg, M.R.: Cluster Analysis for Applications. Academic Press Inc., New York (1973)
34. Jain, A.K., Dubes, R.C.: Algorithms for Clustering Data. Prentice-Hall advanced reference series, Upper Saddle River (1988)
35. Jain, A.K., Murty, M.N., Flynn, P.J.: Data clustering: a review. ACM Comput. Surv. **31**(3), 264–323 (1999)
36. Pedrycz, W.: Knowledge based clustering in computational intelligence. In: Challenges in Computational Intelligence, pp. 317–341. Springer, Berlin (2007)
37. Xu, R., Wunsch, D.: Computational intelligence in clustering algorithms, with applications. In: Algorithms for Approximation, pp. 31–50. Springer, Berlin (2007)
38. Sibson, R.: SLINK: an optimally efficient algorithm for the single-link cluster method. Comput. J. **16**(1), 30–34 (1973)
39. Sorensen, T.: A method of establishing groups of equal amplitude in plant sociology. Vidensk. Selsk. Biol. Skr. **5**(4), 1 (1948)
40. Girvan, M., Newman, M.E.J.: Community structure in social and biological networks. Proc. Natl. Acad. Sci. **99**(12), 7821–7826 (2002)
41. Hamerly, G., Elkan, C.: Learning the k in k-means. In: NIPS, pp. 281–288 (2003)
42. Song, Y., Wang, H., Wang, Z., Li, H., Chen, W.: Short text conceptualization using a probabilistic knowledgebase. In: IJCAI, pp. 2330–2336 (2011)
43. Salton, G., McGill, M.J.: Introduction to Modern Information Retrieval. McGraw-Hill, New York (1986)
44. Salton, G., Buckley, C.: Term-weighting approaches in automatic text retrieval. Inf. Process. Manage. **24**(5), 513–523 (1988)
45. Ogievetsky, M., Heer, V., Bostock, J.; D3 data-driven documents. IEEE Trans. Vis. Comput. Graph. **17**(12), 2301–2309 (2011)
46. Shneiderman, B., Wattenberg, M.: Ordered treemap layouts. In: INFOVIS, pp. 73–78 (2001)
47. Rivadeneira, A.W., Gruen, D.M., Muller, M.J., Millen, D.R.: Getting our head in the clouds. In: CHI, pp. 995–998 (2007)

48. Carmel, D., Uziel, E., Guy, I., Mass, Y., Roitman, H.: Folksonomy-based term extraction for word cloud generation. In: CIKM, pp. 2437–2440 (2011)
49. Herring, S.R., Poon, C.M., Balasi, G.A., Bailey, B.P.: Tweet spiration: leveraging social media for design inspiration. In: CHI EA, pp. 2311–2316 (2011)
50. Lowongtrakool, C., Hiransakolwong, N.: Noise filtering in unsupervised clustering using computation intelligence. Int. J. Math. Anal. 6(59), 2911–2920 (2012)
51. Laorden, C., Sanz, B., Santos, I., Galan-Garcia, P., Bringas, P.: Collective classification for spam filtering. In: Computational Intelligence in Security for Information Systems, vol. 6694, pp. 1–8. Malaga (2011)
52. Benevenuto, F., Magno, G., Rodrigues, T., Almeida, V.: Detecting spammers on Twitter. In: Seventh annual Collaboration, Electronic messaging, Anti-Abuse and Spam Conference, pp. 1–9. Redmond (2010)
53. Mathioudakis, M., Koudas, N.: Twitter monitor: trend detection over the Twitter stream. In: SIGMOD, pp. 1155–1158 (2010)
54. Singer, P., Wagner, C., Strohmaier, M.: Understanding co-evolution of social and content networks on Twitter. In: WWW, pp. 57–60 (2010)
55. Cataldi, M., Di Caro, L., Schifanella, C.: Emerging topic detection on Twitter based on temporal and social terms evaluation. In: MDMKDD, vol. 4, pp. 4–10 (2010)
56. Jo, Y., Hopcroft, J., Lagoze, J.: The web of topics: discovering the topology of topic evolution in a corpus. In: WWW, pp. 257–266 (2011)
57. Lin, C.X., Mei, Q., Han, J., Jiang, Y., Danilevsky, M.: The joint inference of topic diffusion and evolution in social communities. In: ICDM, pp. 378–387 (2011)
58. Back, T., Fogel, D.B., Michalewicz, Z.: Handbook of Evolutionary Computation, 1st edn. IOP Publishing Ltd, Bristol (1997)
59. Raidl, G.: Evolutionary computation: an overview and recent trends. ÖGAI J. 24, 2–7 (2005)
60. Borgs, C., et al.: Dynamics of bid optimization in online advertisement auctions. In: WWW, pp. 531–540 (2007)
61. Yih, W., Goodman, J., Carvalho, V.R.: Finding advertising keywords on web pages. In: WWW, pp. 213–222 (2006)
62. Jin, Y.: A comprehensive survey of fitness approximation in evolutionary computation. Soft. Comput. 9(1), 3–12 (2005)
63. Jin, Y., Olhofer, M., Sendhoff, B.: A framework for evolutionary optimization with approximate fitness functions. IEEE Trans. Evol. Comput. 6(5), 481–494 (2002)

Granularity of Personal Intelligence in Social Networks

Shahrinaz Ismail, Mohd Sharifuddin Ahmad and Zainuddin Hassan

Abstract This chapter introduces and exposes the latest development of agent-modeling framework that emerges from understanding the human behaviour of personal knowledge management (PKM) in social networks. The growing interest in personal knowledge management in exposing personal intelligence has brought out the postulation of GUSC model, which becomes the basis of agent-mediated PKM. In micro view, this GUSC model is the significant tool in understanding the granularity of personal intelligence in social networks. It eventually leads to a theoretical treatment on the bottom-up manifestation of organisational knowledge management from individual PKM processes.

Keywords Software agent · Personal knowledge management · GUSC model · Personal intelligence · Social networks · Bottom-up approach

1 Introduction

There are many aspects and approaches to understand the concept of personal intelligence (PI). As covered by this chapter, most researchers agree that PI is electronic based or technologically enhanced facility that represents human

S. Ismail (✉)
Malaysian Institute of Information Technology, Universiti Kuala Lumpur,
1016 Jalan Sultan Ismail, 50250 Kuala Lumpur, Malaysia
e-mail: shahrinaz@miit.unikl.edu.my
URL: http://www.unikl.edu.my

M. S. Ahmad · Z. Hassan
College of Information Technology, Universiti Tenaga Nasional,
Jalan IKRAM-UNITEN, 43000 Kajang, Selangor, Malaysia
e-mail: sharif@uniten.edu.my

Z. Hassan
e-mail: zainuddin@uniten.edu.my
URL: http://www.uniten.edu.my

W. Pedrycz and S.-M. Chen (eds.), *Social Networks: A Framework*
of Computational Intelligence, Studies in Computational Intelligence 526,
DOI: 10.1007/978-3-319-02993-1_10, © Springer International Publishing Switzerland 2014

counterparts and extends human capabilities. This chapter does not look into the PI concept as how the rest of the researchers define them, but it is viewing the concept from the realm of artificial intelligence where the tasks of the human counterparts are being mediated by the intelligent software agents.

In facilitating our understanding on this aspect, the latest development of agent-modelling framework that emerges from the understanding of the human behaviour on personal knowledge management (PKM) in social networks is introduced and exposed. Three keywords to be pointed out in this topic are: granularity (how 'low' can we go); personal intelligence (how much of individual 'intelligence' can be presented); and social networks (where all these happen in real life).

1.1 Personal Knowledge Management

The concept of personal knowledge management (PKM) started off with the burst of Web 2.0 tools over the Internet and World Wide Web, where personalisation and customisation are possible at individual level. This concept challenges the long-lasting controversial concept of knowledge management, in particular organisational knowledge management (OKM), where the hype of knowledge management is being questioned by the pessimist researchers who doubt that knowledge can be managed. In the realm of information and communication technology (ICT), knowledge management is seen to be embedded within an integrated system that facilitates every level of employees and managers in an organisation to 'manage' what they know and need to know. At the back of the mind, there is always an issue of humanity and personal aspects on what knowledge to manage, who knows what and when knowledge is needed, and these cannot be denied or hundred percent solved by the integrated information system, which gives an ample room for user behavioural research domain.

The important aspect agreed by researchers in this PKM domain is the 'people factor'. PKM processes are defined in terms of networking (e.g. finding people who share the same social interest, sharing knowledge, collaborating, extending and extrapolating), with the core focus or the thing that 'moves' PKM to be 'personal inquiry' (i.e. a quest to find, connect, learn and explore) [1]. Undoubtedly, individual knowledge workers perform different processes of PKM and often with different approaches at different times, depending on the situations, the need, the experience, and many other factors. Overall, there are still similarities in the patterns of the processes, due to the goals of performing them to be towards the common goals. In other words, PKM can collectively contribute to OKM because knowledge is a source of competitive advantage at organisatonal level as well as at individual level [2].

Figure 1 shows the comparison of PKM processes across literature, with similarities found existing among them. Each column represents the PKM processes postulated by the respective researchers, while each row represents the similarities in terms of the meaning or process defined, except for Grundspenkis's [3]

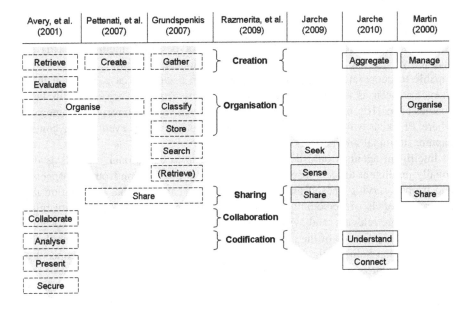

Avery, et al. (2001)	Pettenati, et al. (2007)	Grundspenkis (2007)	Razmerita, et al. (2009)	Jarche (2009)	Jarche (2010)	Martin (2000)
Retrieve	Create	Gather	Creation		Aggregate	Manage
Evaluate						
Organise		Classify	Organisation			Organise
		Store				
		Search		Seek		
		(Retrieve)		Sense		
	Share		Sharing	Share		Share
Collaborate			Collaboration			
Analyse			Codification		Understand	
Present					Connect	
Secure						

Fig. 1 Comparison of PKM processes flow among authors[1]

'retrieve' process, which is found to be in the different position compared to the processes of the similar terminology.

The three processes on the left side in Fig. 1 represent the theoretical side of PKM processes, whereas the three processes on the right side represent the technical side of PKM processes, both in proper order as suggested by the respective authors. The middle column shows the PKM processes found to exist over the Internet and Web 2.0 tools by Razmerita et al. [2], not in any particular order, but arranged in accordance to the sequence defined by other authors.

In general, PKM processes might be initiated from the social science or theoretical point of view, but the practice is seen to be very real over the virtual world—the Internet, Web 2.0 tools and the social network. In terms of granularity of PKM processes and how far refined this concept can go; the software agent technology is a way to look at.

1.1.1 Software Agent in Mediating PKM Processes

Software agent technology came into picture when the automation of PKM processes is deemed useful in enhancing the productivity and efficiency of knowledge workers at individual level. Proclaimed to be under artificial intelligence (AI)

---

[1] Avery et al. (2001) [4]; Pettenati et al. (2007) [5]; Grundspenkis (2007) [3]; Razmerita et al. (2009) [2]; Jarche (2009) [6]; Jarche (2010) [7]; Martin (2000) [8].

domain, software agents are expected to be 'intelligent' with capabilities to react and proact on given situations. Among the capabilities and features of software agents that are used to justify the emergence of personal intelligence (PI) are autonomy, reactive, proactive, able to communicate, adaptive, goal-oriented, capable to cooperate, reason, and flexible [9]. In a separate case, personal intelligence is seen as a layer that constitutes collective intelligence, with this layer enables users [10], but it is still similar to the software agent conceptual framework where PI exists within a restricted 'environment', such as event, user, content capture, terminal and network.

Intelligent agents would depend on the agents to be rational, as well as personally intelligent to recognise personally-relevant information from introspection and from observing oneself and others [11]. They can form that information into accurate models of personality, guide one's choices by using personality information where relevant, and systemise one's goals, plans, and life stories for good outcomes [11]. On top of that, the personal intelligence contributes to 'cumulative decisions' that helps in the individual's well-being, which reflects the concept of bottom-up approach in achieving OKM.

On a technical side, current research looks into the mediation processes in PKM, since the agents are claimed to be able to carry out all the actions and exhibit the behaviours within a knowledge flow [12]. In simplifying the concept, four processes of PKM are postulated to be mediated by software agents: get/retrieve, understand/analyse, share and connect [13]. These processes are aligned with the processes suggested by other researchers [2–5, 8, 14], but with more focus on PKM over computer and internet technologies.

PKM processes are enabled cognitively, in terms of method of performing tasks (method), how knowledge sources are identified (identify), how decision is made on the approach to take in seeking knowledge experts or sources (decide), and the drive for knowledge workers to seek knowledge experts (drive) [13, 15, 16]. These enablers co-exist with the processes to ensure the effectiveness of managing personal knowledge at human aspect, as shown in Fig. 2. Since it is proven to be part of the equation for effective PKM, these enablers fit in the overall picture of agent-mediated PKM processes, where they are treated as elements exist in the agent environment, with the help of the Belief-Desire-Intent (BDI) concept defined in software agents.

The effective PKM framework, as shown in Fig. 2, symbolises the interrelation among the PKM processes and cognitive enablers, postulated by Ismail and Ahmad [17].[2]

---

[2] (Ismail and Ahmad, 2012) It does not constitute that all cognitive enablers should exist in full form in order for each PKM process to work efficiently. In fact, there is a possibility of having the cognitive enablers carrying different percentages of weight in terms of importance and necessity, to support each PKM process. This may also varies across different background of individual knowledge workers, especially those from different industries. Yet, the whole concept of having these enablers and processes is valid across all individuals and industries. Hence the way the model is drawn is in a dynamic shape, to show fluidity of the existing elements within the concept.

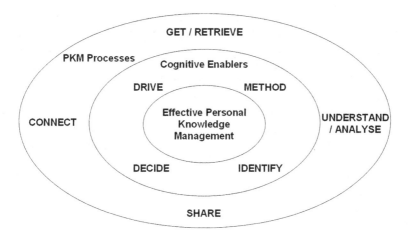

Fig. 2 Effective personal knowledge management: the human aspect

1.1.2 Social Networks: Where the PKM Intelligence Exists

The growth of Web 2.0 tools is seen as the major factor in fostering PKM processes and allowing people to be more effective in supporting knowledge management in diverse ways and scales [2]. Web 2.0 tools are used as PKM tools since they provide conditions where knowledge can be shared and new knowledge can be created or exchanged in social networks. This emphasises the strength in social network to be a good platform to prove the existence of personal intelligence in managing PKM processes.

Social network is modelled by a graph with nodes representing individuals and the links between the nodes indicating a discovered direct relationship between the individuals [18]. Part of PKM is to locate and connect to knowledge experts, and social network is a platform where this is done without the individuals realising it, and this whole process is done through the nodes (individuals) and the links (relationships). The referral or recommendation hops from one individual to another through the relationships, during the process of locating and finding the right knowledge experts within the social network. The difference between a social-network-based referral system and other recommender systems (where both systems could be agent-based) are the latter is designed to provide anonymous recommendations instead of providing referrals via chains of named individuals.

Supporting the creation of new knowledge, the concept of 'learning' has evolved with the boost of Web 2.0 technology and social networks, where 'learning' extends the common scene in classrooms, and this has expanded the environment into informal learning over the network outside the classrooms where it was found hard to 'reach' in traditional learning system. Learning, in this context, does not rely on formally registered students only, but also in daily adult life of knowledge workers, where informal learning picks up in the corporate world driven by the desire to capitalise on the intellectual assets of the workforce to manage organisational knowledge [19]. Learning environment exists in agent-mediated systems, where the concept is depicted by researchers to include informal learning and managing knowledge in the processes of learning, using tools made available by computer and internet technologies.

Social networks are Web-based services that allow individuals to construct a public or semi-public profile within a bounded system, articulate a list of other users with whom they share a connection, and view and traverse their list of connections and those made by others within the system [20]. On top of that, social networks reduce the barriers of trust among the individuals, since the concept started off by having known individuals to be connected based on relationships. The key point is 'connection', which social networks provide as a vital strength in making them significant to be the platform for agent-mediated PKM and learning processes.

In other words, social network is a governing term that covers many forms and functions, with each node having distinct relative worth, where nodes are sometimes used to represent events, ideas, or objects, with them behaving as filters, amplifiers, investors, providers, community builders or facilitators [21], on top of being referred to as gatekeepers and points of reference. These are the concepts within social networks, in which the PKM processes need to fulfill the goals in managing knowledge, with the help of software agents.

1.2 GUSC

The Get-Understand-Share-Connect (GUSC) model is suggested in recent re-search on agent-mediated PKM processes, where the processes are described as get or retrieve knowledge, understand or analyse knowledge, share knowledge, and connect to other 'knowledge' (i.e. knowledge expert, knowledge source, or other forms of knowledge) [17]. For the aspect of human PKM, the sequence of these processes is as shown in Fig. 3.

Comparing this GUSC model to renowned SECI model by Nonaka and Takeuchi [22], where SECI stands for socialisation-externalisation-combination-internalisation among individuals in an organisation, the PKM processes are found to be similar in nature where the GUSC represents the processes among individual and himself but in a different order of sequence. The link that ties the two models is the interchange between explicit and tacit knowledge during the processes, which is defined by both SECI and GUSC in a different manner based on the perspective (i.e. between individuals and within individual). The details of these processes are shown in Fig. 3.

At this point, the GUSC model has proven the granularity of knowledge management from organisational to personal. The next level of granularity is in terms of how the knowledge management can further be refined as practiced online and can be implemented at software agent level.

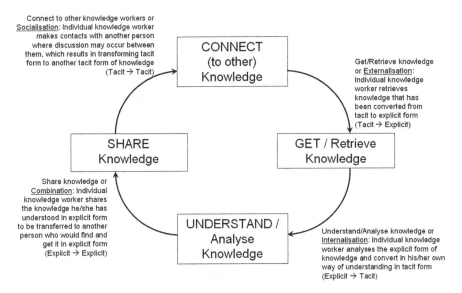

Connect to other knowledge workers or
Socialisation: Individual knowledge worker
makes contacts with another person
where discussion may occur between
them, which results in transforming tacit
form to another tacit form of knowledge
(Tacit → Tacit)

CONNECT
(to other)
Knowledge

Get/Retrieve knowledge
or Externalisation:
Individual knowledge
worker retrieves
knowledge that has
been converted from
tacit to explicit form
(Tacit → Explicit)

SHARE
Knowledge

GET / Retrieve
Knowledge

Share knowledge or
Combination: Individual
knowledge worker shares
the knowledge he/she has
understood in explicit form
to be transferred to another
person who would find and
get it in explicit form
(Explicit → Explicit)

UNDERSTAND /
Analyse
Knowledge

Understand/Analyse knowledge or
Internalisation: Individual knowledge
worker analyses the explicit form of
knowledge and convert in his/her own
way of understanding in tacit form
(Explicit → Tacit)

Fig. 3 GUSC PKM processes model within individuals

2 GUSC at Granular Level

In most literature, the 'granularity' can be seen when organisational concept of knowledge management (KM) is scoped down to individual level in PKM. Even in technology, a KM system is seen as a computer-based network-facilitated system that caters for the OKM, which is further narrowed down to PLE-based tools that caters for PKM. Little did we realise that the intelligence behind the scene that enables both these systems can be further detailed out as fine as a granule, if we further study the bots or agents that work in this realm. Let us phase in this scope to understand how GUSC is used to understand the granularity of the whole concept of personal intelligence.

As we know, OKM is where individuals connect to each other, and PKM happens when an individual is processing knowledge within himself/herself. With the help from artificial intelligence, a software agent can be assigned and programmed to mediate a human to manage knowledge through the processes of GUSC, and this is the level one of granularity, or we coin it as 'coarse granular' level. It is proven that agent-mediation system can be injected with capabilities to react and proact on its own, and can best work in an environment of multiple agents, known as multi-agent system (MAS). This multi-agent system can consist of agents that carry the roles of get, understand, share and connect (i.e. GUSC). In other words, this multi-agent system manifests the PKM processes among agents, which is finer than the human-agent mediation processes. In fact, the agent–agent processes can prove the intelligence of a person (human) when the agents duplicate the act of GUSC among them. This personal intelligence level is coined as 'fine granular' level in our postulation.

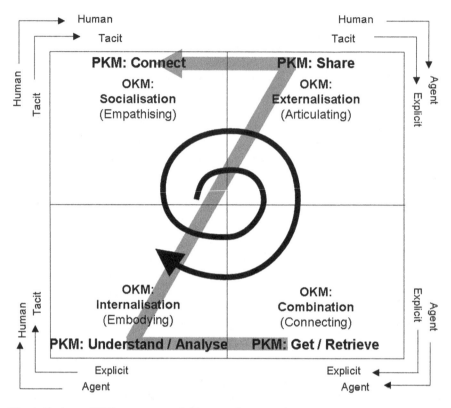

Fig. 4 Coshgean PKM processes model between human and agent

In proving this 'evolution', Fig. 4 shows the thin line that differentiates among OKM concept of SECI (shown in black spiral arrow), PKM concept of GUSC (shown in grey Z-arrow), and human-agent interaction (shown in thin arrows outside the main conceptual box). The agent–agent interaction is more flexible, where it can follow the GUSC processes or it can be structured depending on the need and situation of the multi-agent roles in the agent-based system development.

2.1 Agent-Mediated PKM Framework

In the early 2011, the agent-mediated PKM framework was introduced to understand and extract the general processes of effective PKM among individual knowledge workers, to be translated and applied to a software agent environment. The effective PKM processes are known as GUSC, as discussed earlier. Understanding that this concept is simplified to make it generalisable and implementable at agent level, it does not limit its capability to be expanded into a more complex

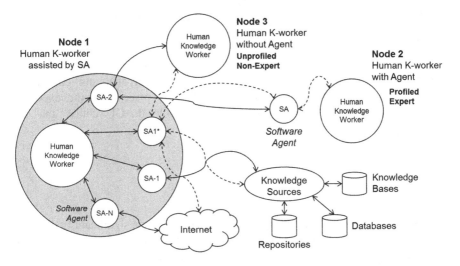

Fig. 5 Overview of an agent-mediated personal knowledge management

manner later on. The fundamental theory still relies on the concept of get, understand, share and connect (i.e. GUSC).

Focusing on the agent capabilities in performing the GUSC tasks on behalf of the human counterpart, Fig. 5 provides a graphical overview of the elements expected to exist in the agent-mediated PKM framework. This consists of three aspects related to the keywords mentioned in the beginning of this chapter: who (the actors or humans involved); how (the procedure of getting the task done); and where (the knowledge sources to be found or located).

In terms of 'who', there are two major nodes or human counterparts that are defined in the agent environment: Knowledge Seeker (or in Fig. 5 shown as Node 1 Human K-worker); and Knowledge Expert (shown as Node 2 and Node 3 in Fig. 5). The Knowledge Expert can be a Human K-worker (or Human Knowledge Worker) with or without a profile that 'advertises' himself/herself as an expert. In this framework, a 'profiled expert' is understood to have a software agent assisting it to profile or advertise his/her expertise. On the other hand, the 'unprofiled non-expert' does not necessarily mean that he/she is not an expert, but he/she could be a person who can refer the agent to another person who is more expert than him/her, since his/her expertise is not profiled.

The 'how' aspect in this framework is depending on the aspect of 'where' the knowledge is expected to be found. Three locations are defined for knowledge source (i.e. 'where'): the World Wide Web or the Internet (in general search); the knowledge experts; and other knowledge sources that are commonly found in the databases and repositories (as shown in Fig. 5). Depending on the situation and intelligence, the software agent will approach the knowledge sources according to the feasible methods, such as: proactively searching over the semantic Web; directly refer to the databases; and directly communicate with an identified knowledge expert.

This overview of agent-mediated PKM framework is based on the GUSC model processes, where the agents are mediating the tasks of get knowledge (from the knowledge sources), understand knowledge (in order to verify if the knowledge is what the agent supposed to be looking for), share knowledge (in terms of profiled expert advertising the expertise of the human counterpart), and connect to knowledge source (where the communication takes place once the expert is identified). The roles of the agent(s) are determined by the GUSC tasks they are assigned with.

2.2 Why GUSC

Recent studies, especially in educational scene, have shown significant findings on the existence of GUSC processes in a virtual learning environment, where social networks are used to facilitate students and adult learners in understanding subject-related topics outside of the common classroom environment. The significance does not end here; it brings out the curiosity on how much intelligence is generated from this online GUSC pattern.

In the said research conducted by Ismail, Abdul Latif and Ahmad [23],[3] content analysis was conducted by researchers on groups of students managing their learning environments in a social network. Out of a few groups identified, two were chosen for analysis due to the similarity of the topics and learning contents, which was on a final year project that basically requires similar patterns of learning across all universities. In order to ensure consistency in analysing and ease of data triangulation, a score sheet is developed based on the GUSC Model as the basis for further qualitative analysis. This study ensures that the researchers are clear on what they are looking for, before the analysis is conducted separately. The scores are shown in Table 1, which include group scores (i.e. by the case groups), and total scores (i.e. by the whole research).

On the other hand, a recent study on students' approach on using social network as learning tool has successfully model the GUSC in social network analysis diagram. Fig. 6 shows the preliminary findings on Connect relationship among the final year project students, based on the questionnaire survey answered by three respondents.[4] Each relationship or link has its own strength and value (i.e. from 0 to 1.0), indicating the significance of connection between two nodes (where a node indicates a person or member in the social network environment). The same diagram can be produced for Get, Understand and Share processes.

---

[3] The score sheet is based on the GUSC Model (that covers the PKM processes), but at the same time tries to analyse the cognitive enablers.

[4] This is an unpublished work as it is still in progress. The full result will include all the four GUSC relationships in social network analysis (SNA) diagram.

Table 1 Score sheet results on 2 case groups (in percentage)

Scores[a]	GE	UN	SH	CO	ME	ID	DE	DR
Group A	90.63	70.00	68.75	50.00	63.75	55.00	75.00	89.29
Group B	75.00	17.78	0	30.16	50.62	40.00	24.07	85.71
Total	82.35	42.35	32.35	39.50	53.53	47.06	75.49	87.39

[a] GE = Get/Retrieve knowledge; UN = Understand/Analyse knowledge; SH = Share knowledge; CO = Connect to knowledge source; ME = Method; ID = Identify; DE = Decide; DR = Drive

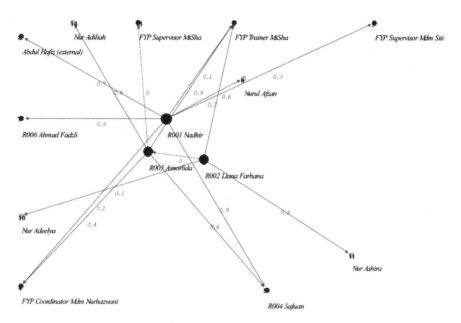

Fig. 6 Social network diagram based on preliminary findings from three respondents, showing the connect process only

Even though the GUSC score sheet and GUSC social network analysis are two different approaches in understanding the existence of GUSC processes in a network environment, the results prove that GUSC Model can be used to understand and analyse a network environment where connection exists among members, and standardisation in terms of measurement and analysis can be achieved. With these results, an agent environment can be further understood, and PKM at granular level can be proven and made standardised to be implemented in agent-based system.

3 Applying the Concept of GUSC

There is a need for agents to apply the GUSC processes in the environment that exists in social networks, especially in imitating and mediating human knowledge workers' way of managing personal knowledge, for example in deciding on the knowledge source. The strong notions of software agents can support this learning scheme. A case study is presented here to demonstrate how the GUSC concept can be applied in a real case scenario.

3.1 Case Study of PKM

An overview of a researcher's personal knowledge management (PKM) is illustrated in Fig. 7, according to research work processes that a researcher goes through over the computer and internet tools.[5] The flow of the research work processes includes four main processes: get/retrieve, understand/analyse, share/ publish, and connect. These processes are aligned with the PKM processes used in GUSC Model, and it is proven to be true among human knowledge workers across three industries (i.e. manufacturing, service and education) in Malaysia [25].

As shown in Fig. 7, the four work processes of a human knowledge worker with the role of a researcher are further broken down into lists of methods or ways the researcher performs. For example, the researcher would use general search (e.g. using online search engine), e-mail, RSS feed, aggregation tool, online library database search, search results that are auto-fed to email in preferred frequency, and 'follow' or subscribe to shared updates posted by other people. Of course, not every researcher uses all these methods to get or retrieve knowledge, but these are the possible ways a researcher may use. There are times when re-searchers tend to change the choice of method in retrieving knowledge at different time, depending on some factors like convenience, facilitating conditions, perceived usefulness, environment, and such.

In understanding and analysing the knowledge retrieved via computer and internet tools, the researcher may summarise, write reviews, comment, collect more data through research, write research paper, or decide who the knowledge is for and pass the knowledge to the person. These methods are shown in Fig. 7. A recent study found that some knowledge workers tend to share knowledge retrieved and 'tag' the shared knowledge to specific person or a group of people whom they know would need, appreciate and benefit from such knowledge. This is observed to be very common over Web 2.0 tools, such as blogs, micro-blogs and social networks. On top of just tagging, the knowledge worker may 'shove in' an idea or a review on what he/she understood from the knowledge tagged, as an act of starting a conversation on the topic and/or verifying the understanding. This is very tightly related to the sharing process.

From this case study, it is found that sharing and publishing personal knowledge are possible in different methods, such as blogging, RSS to blog, sending email, sharing RSS or hyperlinks (or URLs) with reviews, tagging people when sharing hyperlinks, updating wiki, social bookmarking, chatting, and publishing

---

[5] This case study is based on the preliminary findings from a data triangulation result, which is the base of the agent-mediated PKM framework formulated by Ismail and Ahmad (2011) [13].

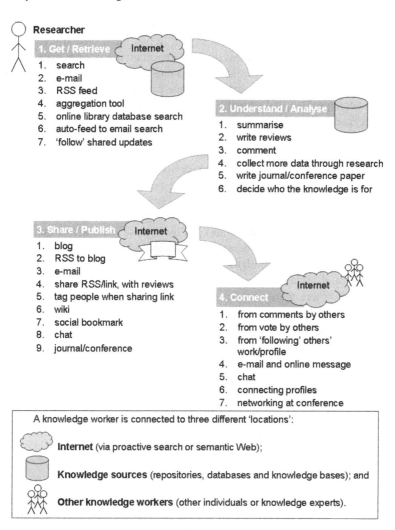

Fig. 7 Overview of a researcher's personal knowledge management (PKM)

through journals and/or conferences. It is a trend and habit to tag people we know in this current online environment that publishing research papers through journals and conferences are not the only means for everyone to do anymore. In fact, the act of 'publishing' and sharing thoughts or knowledge are no longer belong to re-searchers and journalists only, but to everyone who wish to do so over the Internet. Since the more we share, the more we can manage what we know better, then the ease of sharing eases the way we manage what we know.

People connection and networking is found to be vital in today's world, but it is found to be vital long before in a researcher's or academician's world. It is not just about expanding the network of people we know, but most importantly in locating knowledge experts, where even business people find critical at times since the world becomes smaller through globalisation. In this PKM case study, it is found that connecting to others is a process deemed important in research work processes, where a researcher may connect from comments posted by others, from vote by others, from 'following' others' work or online profile, email, and online message, chat, connected profiles, and networking at conferences (as shown in Fig. 7). Since the advancement of the Internet, even the researchers do not have to be physically present at the conference held many miles away, yet can still be connected to the people in the conference via web conference and other facilitating tools.

In general, a human knowledge worker (or in this case, a researcher), is connected to three different 'locations' when managing personal knowledge: Internet (via proactive search or Semantic Web), knowledge source (repositories, data-bases and knowledge bases), and another human knowledge worker (either a non-expert without profile, or an expert with profile). In Fig. 7, these locations are indicated with symbols of internet cloud, database symbols, and stickman figures, on the process headers. Even so, it does not limit the locations to be only for the indicated processes.

This case study is an example of the application of the PKM framework and GUSC Model, where there is a need to know and locate the knowledge experts (and knowledge source) within social networks (where most of the activities exist among the Internet tools and 'web'). The findings do not stop here; in fact, it proves the existence of socialisation embedded within the GUSC processes among the people (or the nodes, in networking terms) in the social networks. Seeing this existence, socialisation as part of computational intelligence will be discussed to support this case scenario.

3.2 The Nodal Approach

The basic concept of formalising software agents to mediate the work processes usually handled by the human counterpart is by having the human counterpart to delegate the tasks to a software agent or multiple software agents.[6] The conceptual model for agent-mediated research work processes commonly handled by human knowledge worker is shown in Fig. 5, with the locations of knowledge source illustrated, in a nodal approach format.

In the nodal approach, which corresponds to the model of social network, a knowledge worker manages his/her personal knowledge by working cooperatively with a software agent in a virtual workspace called a node. A node consists of a knowledge worker and one or more (role) agents, to perform some supporting roles of the knowledge worker. The knowledge worker has a set of functions for the PKM processes, some of which could be delegated to the agents.

---

[6] From this point to Fig. 8 and equation (1): This section on nodal approach is adapted from Ismail, Ahmad and Hassan (2012) [24], to fit the purpose of understanding the concept before applying the GUSC processes to role agents.

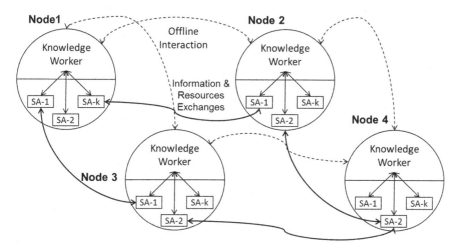

Fig. 8 Multiple nodes within a social network, with k agents

In understanding the application of a node in agent-mediation, a set of mundane housekeeping functions to which a software agent could be assigned, is identified. This is based on our analysis of human-agent and agent-agent cooperation, where agent functions to assist and/or complement humans (e.g. forward/receive information, communicate with other agents, calculate reputation/referral points, log actions, track process, information search, etc.).

Formally, a node can thus be specified as:

- a knowledge worker, KW;
- a set of the knowledge worker's functions, F;
- a set of one or more agents, A; and
- a set of agent's functions, G;
    i.e. a node N can be formulated as a four-tuple structure:

$$N = <KW, F, A, G> \tag{1}$$

With the formulated structure, a node can be replicated to instantiate another virtual workspace that consists of a knowledge worker, agents and their functions. Depending on the selected functions for a knowledge worker, each node represents a unique 'character' of its knowledge worker within the social environment.

In relating Fig. 5 to Fig. 8, the latter shows how the human knowledge worker connects to the three locations with the help of software agents for these different 'routes'. As an example, SA-2 (in Fig. 5) refers to 'Software Agent 1' that connects the human knowledge worker to another human knowledge worker who is an expert in the required field, who (in Fig. 8) is assisted by a software agent. Another case is SA-1, which refers to another software agent which focuses on mediating the human counterpart to search, retrieve and connect to the knowledge source (as shown in Fig. 5).

As an agent-mediated architecture, the model in Fig. 5 includes agent environments where a schedule is pertained. An agent's movement is determined by a schedule that guides it to perform its tasks, hence the need of the environment to be programmed into the agent-mediated system. Once triggered by the schedule, the agent performs its tasks step-by-step and follows the 'route' assigned to it. For a case of agent SA-N, the route is a proactive search or by using Semantic Web to search over the Internet. Any search results retrieved by the agent would be reported back to the human counterpart.

Multiple software agents (such as SA-1, SA-2 and SA-k shown in Fig. 8) can be assigned for different areas or locations (refer to SA-1, SA-2 and SA-N in Fig. 5). Another possible method is by having SA-1* where a software agent is assigned by 'role' to manage personal knowledge by connecting to all the three sources intelligently, for specific task. This method is more complex and requires the agent to have strong notions to support its full capabilities to perform.

In general, the agent-mediated PKM framework points out two different methods for a software agent to be used: location-based and role-based. In our granularity theory, we focus on the role-based mediation where the behaviour, diligence, intent and others (BDI) that include the strong notions of a software agent are highly required.

Let us zoom into the GUSC processes and visualise them as roles in agent-mediated system. In connecting the nodes or (role) agents, links of information and resources exchange are manipulated between the nodes, since agents are performing most of the necessary communication to complement the normal interactions between human knowledge workers. From the formulation of agent mediation across nodes (Fig. 8) and the conceptual model of locating knowledge experts in other nodes (Fig. 5) within social network (Fig. 6), agent-mediated PKM can be replicated to represent a function of individual knowledge worker's intelligence (or in other words, personal intelligence) in social network.

In the domain of social network, each node in the network (that represents a knowledge worker with profiles) has his/her own software agent to delegate the tasks to. On top of the tasks discussed earlier, these agents are also capable in accepting, receiving, and acknowledging the seekers' requests to be connected and communicated regarding the topics of expertise. Fig. 8 shows how the multiple (role) agents can be deployed within an ecosystem of multiple nodes, or in this case a social network ecosystem. Fig. 8 shows the intelligence manifested by the connections between the nodes, with information and resources intelligently exchanged via the agents.

Applying the GUSC processes as roles to the agents shown in the nodal concept presented in Fig. 8, each software agent is expected to carry a task of get, understand, share or connect. By focusing on the links between the nodes alone (i.e. putting aside the Internet cloud and databases), the multiple agents system with GUSC roles can be illustrated as in Fig. 9, where Get, Understand and Connect agents are used to seek knowledge (referred to as Node 1, Knowledge Seeker) while Share agent is used by a knowledge worker who wishes to make known of his/her expertise (referred to as Node 2, Knowledge Expert with profile).

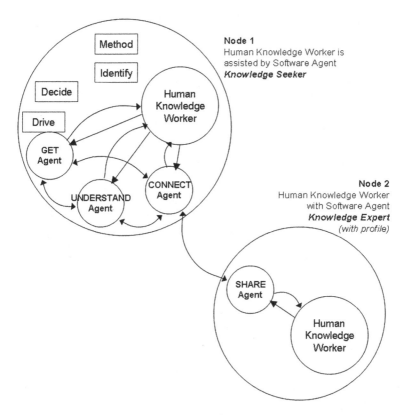

Fig. 9 Multiple agents with GUSC roles in nodal form

What about the cognitive enablers? A recent study proves that these enablers are the mediating variables for GUSC, thus they are needed within the GUSC role agents' environment. The four cognitive enablers are shown existed in a knowledge seeker's node (shown in Node 1, Knowledge Seeker, in Fig. 9) as part of the rules embedded in the agent environment.

Get, Understand and Connect are seen as significant tasks performed by a human knowledge worker to seek for a knowledge expert (i.e. being proactive), due to factors that exist in the person's environment or situation (referred to as cognitive enablers in this context). On the other hand, a knowledge expert who wishes to tell the world of his/her expertise, partly to be recognised by those who seek for his/her knowledge (like what researchers and academicians often do), would simply perform a Share task or publish his/her work over the Internet. If the knowledge expert already knows beforehand that a certain knowledge seeker would appreciate his/her share of updated knowledge, then the expert would be willingly let the seeker knows in advance (i.e. being reactive).

In reality and as shown in the replicated nodes in Fig. 8, the knowledge expert is also a knowledge seeker and vice versa, especially in a social network. Having said this, in ensuring that an agent can survive in a virtual world, it has to perform

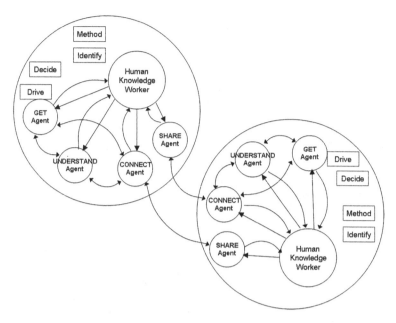

Fig. 10 Viable multi-agent system for PKM based on GUSC Model

both ways—as a proactive seeker and a reactive expert. When two nodes act as both seeker and expert, the viability of the multi-agent system can be illustrated as shown in Fig. 10.

Since one node (i.e. consist of a human knowledge worker, four GUSC agents and cognitive enablers in agent environment) reflects the personal intelligence of a human knowledge worker when interacting with another human knowledge worker, then this multi-agent system is a proof of granularity of personal intelligence, mainly in social networks.

4 Summary

The concept of personal intelligence is derived from the understanding that software agents can be used to demonstrate it. Since software agents are under the domain of artificial intelligence, they are expected to be 'intelligent' with the capabilities to react and proact on given situations within an environment. Up to this point, the intelligence of a personal human being has been zoomed in and replicated at granular level to be applied to an agent environment, thanks to the concept of personal knowledge management and GUSC Model.

Yet, at the end of the day it is about the contribution of this theory of granularity to the society and the real world. Let us zoom out from PKM to OKM to find out.

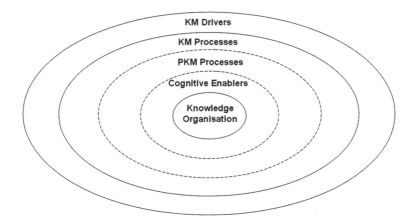

Fig. 11 Knowledge organisation theoretical framework (Bottom-up towards PKM-OKM approach)

4.1 Bottom-Up Approach

Knowledge management (KM) has been an interesting domain in information technology since the birth of the Internet. Renowned authors like Nonaka and Takeuchi [22] and Awad and Ghaziri [26] have postulated frameworks for KM to be implemented in organisations, and these have been referred to as fundamental knowledge by almost all KMers. We have discussed on how SECI Model by Nonaka and Takeuchi [22] being granulated to GUSC Model, so what about knowledge organisation theory by Awad and Ghaziri [26]?

In knowledge organisation theory, the main elements needed are KM drivers and KM processes. Compared to the PKM framework we have postulated, the elements required are PKM processes and cognitive enablers. Since it is agreed that individual knowledge workers are important to an organisation, hence the need to investigate the bottom-up approach of OKM [27, 28], with PKM supporting the processes of the OKM, then Fig. 11 can be illustrated as such embedment.

As shown in Fig. 11, the effective PKM framework (shown in Fig. 2) is embedded in the knowledge organisation framework proposed by Awad and Ghaziri [26], because the knowledge organisation depends on the ability of individual knowledge workers to manage their personal knowledge. The effective PKM framework (Fig. 2) is foreseen as the granular KM construed on an organisation, since the core player that makes the PKM effective is the 'people' or knowledge workers who exist and work within the organisation. With this fundamental concept in mind, KM processes suggested in Knowledge Organisation Model [26] can be further broken into PKM processes and their mediating cognitive enablers, as shown in Fig. 11.

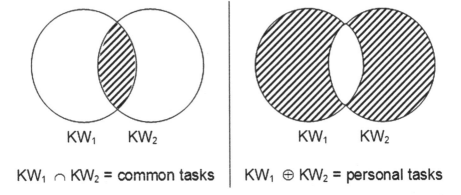

KW$_1$ ∩ KW$_2$ = common tasks │ KW$_1$ ⊕ KW$_2$ = personal tasks

Fig. 12 Common and personal tasks

It is proven that the final outcome of individuals' knowledge gain and use during the PKM processes is the achievement of their organisational goals and key performance indicators (KPI). It is believed that if knowledge workers could not manage their personal knowledge well, then they may not be able to achieve their personal goal or KPI, and the organisational goal will be in jeopardy. Mathematically, this can be formed in equations, supported by questionnaire and interview surveys conducted prior to this.

On a micro level, diverse tasks performed by individual knowledge workers overlap in an organisational setting and can be represented by a Venn diagram in Fig. 12, showing the common tasks and the individual or personal tasks. The overlapping tasks can be considered as the common tasks which all knowledge workers need to perform to achieve a common organisational goal. These tasks are to be performed concurrently or sequentially depending on the dependencies between tasks.

Fig. 13 supports that the replicated agent-mediated PKMs overlap each other to reveal tasks for a common goal. It also proves that the symmetric difference areas of the nodes as the personal tasks of both knowledge workers.

The common tasks of individual knowledge workers support their corresponding key performance indicators (KPI), which they have to achieve to fulfill a common goal of the organisation. The common tasks may be independent of each other (i.e. each knowledge worker needs to perform task A) or dependent on each other (i.e. task B must be completed before task C can be performed). The integration of all tasks leading to the achievement of each KPI can be manifested as an OKM process, as demonstrated in Fig. 13.

Based on the concept shown in Fig. 13, if the intersection of common organisational tasks of knowledge workers (as defined in the following equation):

$$KW_1 \cap KW_2 \cap KW_3 \cap \ldots \cap KW_{N-1} \cap KW_N \tag{2}$$

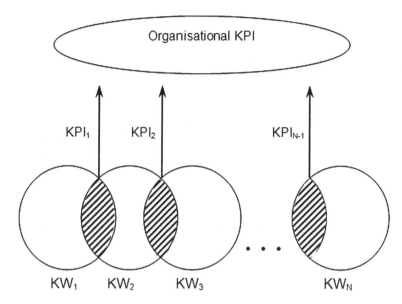

Fig. 13 KPIs manifests OKM

can be measured by their individual KPIs, as follows:

$$KPI_1 + KPI_2 + KPI_3 +\ldots + KPI_{N-1} \tag{3}$$

then the collective processes of achieving the organisational KPI, KPI_O, can be construed as the OKM process, as follows:

$$KPI_O = KPI_1 + KPI_2 +\ldots KPI_{N-1} \tag{4}$$

$$KPI_O = KW_1 \cap KW_2 \cap \ldots \cap KW_{N-1} \cap KW_N \tag{5}$$

The formulation of the equations support the overlapping replicated agent-mediated PKMs that can be integrated to manifest an agent-mediated OKM with KPI as the measurement metric. In other words, there is a tight-coupling between the PKM processes and knowledge management processes in achieving organisational goal and OKM, through the achievement of the KPIs at both personal and organisational level. The dependencies between the elements in the Knowledge Organisation Theoretical Framework (Fig. 11) can be seen in Fig. 14, where KM processes and PKM processes are dependent on each other.

Since the whole idea is determined on how the granularity of PKM processes can finally produce an outcome of a knowledge organisation, the Knowledge Organisation Theoretical Framework (Fig. 11) is also supporting the bottom-up approach to PKM-OKM, as shown in Fig. 14.

Knowledge Organisation (in achieving collective goals)

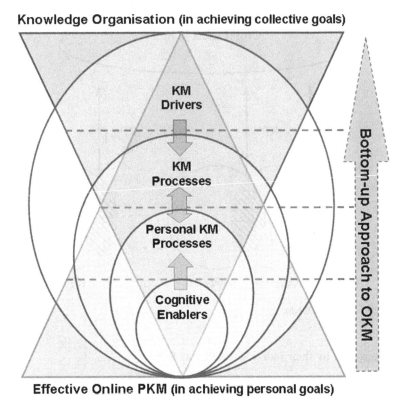

Effective Online PKM (in achieving personal goals)

Fig. 14 Bottom-up approach to OKM: the overlap between two models

4.2 Significance of the Granularity

We have covered the contribution of this granularity theory by looking from the management's view of the knowledge organisation. The significance of the granularity is actually within the scope of human's personal intelligence and software agent's personal intelligence. We coin this as coarse-grained and fine-grained, based on the interactions among the members of the system environment where the explicit knowledge and tacit knowledge are transformed in each interaction:

- Coarse-grained: the human-agent-agent-human interaction, as shown in Fig. 4 (agent-mediation for humans)
- Fine-grained: the agent-agent-agent-agent interaction (in terms of agent roles, and the intelligence will be further challenged especially for tacit knowledge)

The significance of the granularity theory lies in the postulated GUSC roles and personal intelligence in agent-mediation system, towards a viable multi-agent system for PKM.

References

1. Verma, S.: Personal knowledge management: a tool to expand knowledge about human cognitive capabilities. IACSIT Int. J. Eng. Tech. **1**(5), 435–438 (2009)
2. Razmerita, L., Kirchner, K., Sudzina, F.: Personal knowledge management: the role of Web 2.0 tools for managing knowledge at individual and organisational levels. Online Inf. Rev. **33**(6), 1021–1039 (2009)
3. Grundspenkis, J.: Agent based approach for organization and personal knowledge modelling: knowledge management perspective. J. Intell. Manuf. **18**(4), 451–457 (2007)
4. Avery, S., Brooks, R., Brown, J., Dorsey, P., O'Connor, M.: Personal knowledge management: Framework for integration and partnerships. In: Association of Small Computer Users in Education Conference. North Myrtle Beach, South Carolina, United States, pp. 29–39 (2001)
5. Pettenati, M.C., Cigognini, E., Mangione, J., Guerin, E.: Using social software for per-sonal knowledge management in formal online learning. Turk. Online J. Distance Educ. **8**, 52–65 (2007)
6. Jarche, H.: Sense-making with PKM. Harold Jarche—Life in Perpetual Beta (2009), http://www.jarche.com/2009/03/sense-making-with-pkm/18/Nov/2009
7. Jarche, H.: PKM in 2010. Harold Jarche—Life in Perpetual Beta (2010), http://www.jarche.com/2010/01/pkm-in-2010
8. Martin, J.: Personal knowledge management. In: Martin, J., Wright, K. (eds.): Managing knowledge: Case studies in innovation. Spotted Cow Press, Edmonton (2000), http://www.spottedcowpress.ca/KnowledgeManagement/pdfs/06MartinJ.pdf
9. Paprzycki, M., Abraham, A.: Agent systems today: Methodological considerations. In: International Conference on Management of e-Commerce and e-Government Conference, pp. 1–7 (2003)
10. Solachidis, V., Mylonas, P., Geyer-Schulz, A., Hoser, H., Chapman, S., Ciravegna, F., et al.: Collective intelligence generation from user contributed content. In: Fink, A., et al. (eds.) Advances in data analysis, data handling and business intelligence, studies in classification, data analysis, and knowledge organization, pp. 765–774. Springer, Berlin (2010)
11. Mayer, J.D.: Personal intelligence. Imagination, Cogn. Pers. **27**, 209–232 (2008)
12. Newman, B.D., Conrad, K.W.: A framework for characterizing knowledge management methods, practices, and technologies. In: 3rd International Conference on Practical Aspects of Knowledge Management (PAKM2000) Conference, Basel, Switzerland, pp. 30–31 Oct (2000)
13. Ismail, S., Ahmad, M.S.: Personal knowledge management among researchers: knowing the knowledge expert. In: QIK2011. Bandar Sunway, Malaysia, pp. 389–397 (2011)
14. Apshvalka, D., Wendorff, P.: A framework of personal knowledge management in the context of organizational knowledge management. In: D. Remenyi (ed.): 6th European Conference on Knowledge Management (ECKM). University of Limerick, Limerick, Academic Conferences Limited: Reading, pp. 34–41 (2005)
15. Agnihotri, R., Troutt, M.D.: The effective use of technology in personal knowledge management: a framework of skills, tools and user context. Online Inf. Rev. **33**(2), 329–342 (2009)
16. Schwarz, S.: A context model for personal knowledge management. Modeling and Retrieval of Context, Lecture Notes in Computer Science, Vol. 3946, pp. 18–33. Springer, Heidelberg (2006)
17. Ismail, S., Ahmad, M.S.: Effective personal knowledge management: a proposed online framework. World Academy of Science, Engineering and Technology (WASET) Vol. 72, pp. 542–550 (2012), https://www.waset.org/journals/waset/v72/v72-98.pdf
18. Kautz, H., Selman, B., Shah, M.: Referralweb: combining social networks and collaborative filtering. Commun. ACM **40**(3), (1997)

19. Attwell, G.: Personal learning environments—The future of e-learning. eLearning Pap. **2**(1), (2007)
20. Boyd, D.M., Ellison, N.B.: Social network sites: definition, history, and scholarship. J. Computer-Mediated Commun. **13**(1), (2007)
21. Serrat, O.: Social network analysis. Regional and Sustainable Development Department, Asian Development Bank, Mandaluyong (2009)
22. Nonaka, I., Takeuchi, H.: The knowledge creating company: how Japanese companies create the dynamics of innovation. Oxford University Press, New York (1995)
23. Ismail, S., Abdul Latif, R., Ahmad, M.S.: Learning environment over the social network in personal knowledge management. In: International Workshop on Collaboration and Intelligence in Blended Learning (CIBL-2012). Kuching, Malaysia, pp. 2–12 (2012)
24. Ismail, S., Ahmad, M.S., Hassan, Z.: Social intelligence in agent-mediated personal knowledge management. J. Inf. Sys. Res. Innov. (JISRI) **1**, pp. 1–10 (2012), http://seminar.spaceutm.edu.my/jisri
25. Ismail, S., Ahmad, M.S.: Emergence of social intelligence in social network: a quantitative analysis for agent-mediated PKM processes. In: ICIMμ 2011 Conference. UNITEN, Malaysia (2011)
26. Awad, E.M., Ghaziri, H.M.: Knowledge Management. Pearson Education Ltd., New Jersey (2004)
27. Myint, D.: PKM: the starting blocks for KM. Inside Knowl. Mag. **7**, (2004)
28. Zhang, Z.: Personalising organisational knowledge and organisationalising personal knowledge. Online Inf. Rev. **33**, 237–256 (2008)

Social Network Dynamics: An Attention Economics Perspective

Sheng Yu and Subhash Kak

Abstract Within social networking services, users construct their personal social networks by creating asymmetric or symmetric social links. They usually follow friends and selected famous entities, such as celebrities and news agencies. On such platforms, attention is used as currency to consume the information. In this chapter, we investigate how users follow famous entities. We analyze the static and dynamical data within a large social networking service with a manually classified set of famous entities. The results show that the in-degree of famous entities does not fit to a power-law distribution. Conversely, the maximum number of famous followees in one category for each user shows a power-law property. Finally, in an attention economics perspective, we discuss the reasons underlying these phenomena. These findings might be helpful in microblogging marketing and user classification.

Keywords Social network · Social network analysis · Network evolution · Attention economics

1 Introduction

A Social Networking Service (SNS) consists of online sites and applications that have three components: users, social links, and interactive communications. During recent years this kind of service has advanced greatly and changed our lives. Three worldwide popular SNS providers, Twitter, Facebook, and Tencent (qq.com), demonstrate the explosive growth and profound impact of this service. According

S. Yu (✉) · S. Kak
Department of Computer Science, Oklahoma State University,
Stillwater, OK 74078, U.S.A
e-mail: yshe@cs.okstate.edu

S. Kak
e-mail: subhashk@cs.okstate.edu

W. Pedrycz and S.-M. Chen (eds.), *Social Networks: A Framework*
of Computational Intelligence, Studies in Computational Intelligence 526,
DOI: 10.1007/978-3-319-02993-1_11, © Springer International Publishing Switzerland 2014

to Alexa ranking, these three providers are in the top 10 most-visited websites in the world. The least known of these three services, Tencent Inc., is one of the major social networking platforms in China. And Tencent Weibo, one major production of Tencent Inc, has 425 million registered users and 67 million daily users.

Within the social networking services, users mirror social relations in real life, build new social connections based upon interests and activities, or both. When building new social links, users typically adopt different kinds of famous entities. For example, on Twitter, a user might follow BBC Breaking News (@BBC-Breaking) for news and Johnny Depp (@J0HNNYDepp) for personal preference.

In the usage of social networking services, the common user, as information seeker, uses *attention* as *currency* where information from the followees represents the utility. Famous entities, as information providers, receive the attention to increase their influence. In such a system, *information* always tends to be excessive and *attention* is in scarcity. The economic theory that deals with this situation of excessive information and scarce attention is called *attention economics*.

In this chapter, based upon a large-scale dataset with 1.94 million users and 50.66 million directed links in a real social networking service, we analyze how users follow famous entities that have been put into human-chosen categories. The result confirms that there is power law phenomenon in users' adoption of categorized famous entities. Additionally, we discuss the reasons of these phenomena in an attention economic perspective. We expect the results to be useful for improving microblogging marketing and user classification.

The rest of this chapter is organized as follows: in Sect. 2, some technical background and related research works are introduced. Section 3 describes the dataset and discusses why the distribution of famous entities is not similar among different categories. In Sect. 4, static and dynamic analyses of the dataset are provided and further discussion is provided in Sect. 5. Conclusions are presented in Sect. 6.

2 Technical Background and Related Works

We begin with technical background and related research. Firstly, we introduce the concept of social network and its significant properties. Then we describe main features of a social networking service. Finally we discuss attention economics, which has the potential of explaining several phenomena related to social networking service.

2.1 Social Network

A social network represents the social structure associated with persons and/or organizations, which usually are represented as nodes, and where social relations correspond to the connections among nodes [1]. The social relation could be either explicit, such as kinship and classmates [2], or implicit, as in friendship and

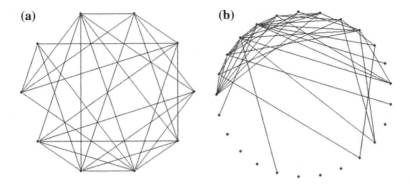

Fig. 1 Two major social network models. **a** Small world network. **b** Scale free network

common interest [3, 4]. As shown in Fig. 1, the small world and the scale free are two significant properties of social networks.

When a social network is viewed as a small world network, most nodes can reach every other node through a small number of links [5, 6]. In the real world, the famous theory of "six degree of separation" suggests that, on average, every two persons could be linked within six hops. The situation in online SNS is somewhat different. The average distance on Facebook in 2008 was 5.28 hops, while in November 2011 it was 4.74 [7]. In MSN messenger network, which contains 180 million users, the median and the 90th percent degree of separation are 6 and 7.8, respectively [8]. On Twitter, the median, average, and 90th percent distance between any two users are 4, 4.12 and 4.8 respectively [9]. In other words, the degree of separation varies on different SNS platforms and it changes with time.

At the same time, other researchers argue that the social network is not connected as strongly as indicated by a straightforward analysis [10]. In the measurement of users' distance, it is customary to include all connections. But if we only consider connections with some communications, the number of friends is approximately half of the total number of followees on Twitter. Thus in the friends' network, the degree of separation should be greater than that in the original network.

Many properties of social networks show the scale free property [11, 12], that is, the degree distribution asymptotically follows a power law. For example, on Twitter, the number of followees/followers fits to the power-law distribution with the exponent of about 2.276 [9]. In addition, the number of tweets being retweeted and mentioned by other users on Twitter also follows a power law [13]. If we take the number of followers as the sole indicator of influence, as power law implies, the information propagation and trend promotion in a social network are strongly influenced by a set of information "oligarchs".

2.2 Social Networking Service

Any social networking service embraces collections of online websites, applications, and platforms, which allow users to build social network and provide additional services [14, 15]. The social network could be symmetric or asymmetric. In symmetric SNS, such as Facebook, the undirected social relations must be confirmed by both peers. Conversely, in an asymmetric SNS, such as Twitter, the directed social link could be made without the explicit permission from the destination user.

Different users publish their opinions and experiences via SNS, which aggregates personal wisdom and different standpoints. If extracted and analyzed properly, the data on SNS can lead to successful predictions of certain human related events in a near-time horizon [16–18].

In this chapter, we focus on the structure of asymmetric SNS as microblogging service. First of all, we present some definitions.

Follow: user a follows user b means that there is a directed social link from a to b.

Follower/Followee: if user a follows user b, b is a followee of a, and a is a follower to b.

Famous entity: famous entities are specific users on social networking service that typically are celebrities, famous organizations, or some well-known groups. In this chapter, we focus on the followees being famous entities, which are named as *famous followees*.

Tweet: the form of messages. In our discussion, it is not specific to the messages on Twitter, but to all kinds of information pieces on microblogging serves.

Focusing on the directed relationships, we could convert the original social network $G = (V, E)$ into a bipartite graph $G' = (U', V', E')$. For a user u, who is a follower to others and followee of someone else, we "*divide*" user u into two nodes as u_1 and u_2 in G'. The u_1 represents the follower role of u, while u_2 denotes the followee role. In addition, the U' in G' includes all followers nodes, and V' contains all followees nodes. For each edge $\{a, b\}$ in G, we could construct an equivalent edge $\{a_1, b_2\}$ in G'. Figure 2 demonstrates the equivalent bipartite graph for a social network example.

Within a real social network service, an overwhelming proportion of users are followers and followees at the same time. Thus the bipartite graph is nearly a balanced bipartite graph.

In this chapter, we discuss how users adopt famous followees. Therefore, we exclude the nodes in V', which represent the non-famous entities. For example, in Fig. 2b, if user b was not in our famous entities dataset, the node b_2 and all directed edges that end with b_2 would be removed.

Furthermore, we group the remaining nodes in V' into categories according to a human-labeling dataset. For instance, users a and b are in the same category C_1, while a and c are in different groups.

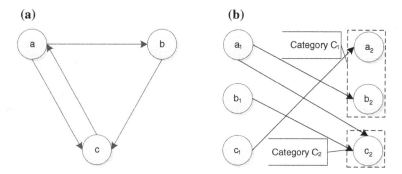

Fig. 2 a Social network example. **b** Corresponding bipartite graph

2.3 Attention Economics

Attention economics is a new branch of economics, which treats the individual's attention as a resource [19, 20]. The development of attention economics stems from the rise of information industry [21, 22].

In the pre-information age, most items of exchange in the economic system were physical. In the information era, we also exchange items of information. These items of information can lead to wealth that is beyond our expectation [23]. But production and consumption of information has some significant points of difference from the material world.

In the economy of material things, most of what is wanted is scarce, such as computers, food, and land. These economic goods form the basis of exchange in society and determine the fabric of daily life to some extent. In the economy of attention, goods and information items are not scarce, but rather excessive in most cases. For example, according the statistics of Youtube,[1] there are about 74 h of video, newly uploaded every minute. No one could watch all the videos in just one website, let alone the whole Internet. In one word, this is an extreme buyer's market.

Within traditional economics, when we produce goods, we have to consider both the fixed cost and variable cost. The production of an additional piece involves payment of additional cost, which is the marginal cost. In information economy, research and development of new products costs a lot, while reproducing an item of information costs very little. For example, Microsoft spent millions of dollars to finish Windows 8 but once you get a copy of it, the cost to make an additional copy is less than ten dollars. Compared with the fixed cost, the marginal cost is next to zero.

The circulating currency is also different. In the material world, currency is a kind of generally accepted medium for exchange of goods and services. No matter what is the specific form of currency, it reflects one's ability to get the items and this ability

---

[1] http://www.youtube.com/yt/press/statistics.html

varies greatly among people. In the information economy, the currency is changed to users' attention, of which everyone holds nearly the same amount. For instance, we could access the website of U.S. News for its news and reports without even a penny. But when we choose it, we do pay our attention which adds to its reputation.

In spite of being different in some ways from traditional economics, attention economics must also deal with three basic questions: What to produce? How to produce? And for whom do we produce? Individual preferences and technology are still the keys to answer these questions.

3 Datasets and Their Basic Analysis

In this section, we introduce the dataset, and give some basic characteristics of it. Then we consider the reason for the uneven distribution of famous followees in different categories.

Our dataset was published by Tencent Weibo for KDD Cup 2012.[2] Tencent Weibo was launched in April 2010, and is currently one of the largest microblogging providers in China. As a major platform of SNS, it has 425 million registered users, 67 million daily users, and 40 million new messages each day. The dataset is a sampled snapshot of Tencent Weibo, including user profiles, social graph, famous entities, famous followee adoption history, and so on. In our analysis, we only use these three datasets: social graph, famous entities, and adoption history.

Social graph: contains all the following information at the sample time of the selected users, who were the most "active" ones during the sampling period.

Famous entities: include all the information of famous entities. A famous entity is a special user in Tencent Weibo to be recommended to other users. Typically, well-known celebrities, organizations and groups are selected to be the famous entities. The famous entities and their categories were chosen and assigned by Tencent Inc.

Adoption history: indicates that records of users' new adoption of famous items in the sampling period. This dataset contains both rejections and acceptances records.

In the remaining part of this section, we will introduce and analyze the first two datasets. The adoption history dataset will be discussed when it is used in Sect. 4.

3.1 Social Graph Dataset

First, we give a brief description of the social graph dataset. There are 1,944,589 users, including 1,892,059 followers, 920,110 followees in the dataset. Because

---

[2] http://www.kddcup2012.org/c/kddcup2012-track1/data

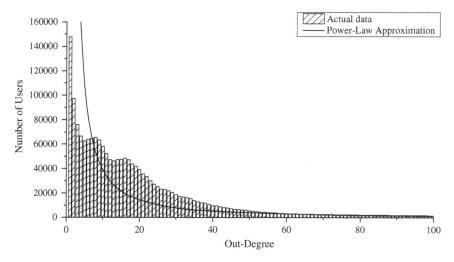

Fig. 3 The Out-Degree Distribution of Followers

this is a sampled snapshot, the dataset is asymmetrical. With 50,655,143 social link records for followers, the average out-degree is 26.77, and for followees, the average in-degree is 55.05. The distributions are partly shown in Figs. 3 and 4. Similar to the results in previous research [9, 24], we find that both the out-degree and in-degree distributions fit a power-law.

Among 1,892,059 followers, there are 83,474 users following more than 100 followees. They account for only 4.41 % of the population, and are not included in Fig. 3. In total, the minimum, median, 90th percent, and maximum out-degree are 1, 14, 52, and 5,188, respectively. Considering only the data in Fig. 3, the out-degree distribution approximately fits the following power-law distribution with R2 of 0.858:

$$Number\_of\_Users = 10^6 \times Out\_Degree^{-1.415} \qquad (1)$$

Out of these 920,110 followees, 19,538, or about 2.12 % in proportion, are followed by more than 20 other users. These 19,538 followees are not shown in Fig. 4. Overall, the minimum, median, 90th percent, and maximum in-degree are 1, 1, 4, and 456,827, respectively. Additionally, only taking the in-degree being equal to or less than 20 into consideration, the in-degree distribution can be approximately represented as the following power-law equation with R^2 being 0.9899:

$$Number\_of\_Users = 840935 \times In\_Degree^{-2.501} \qquad (2)$$

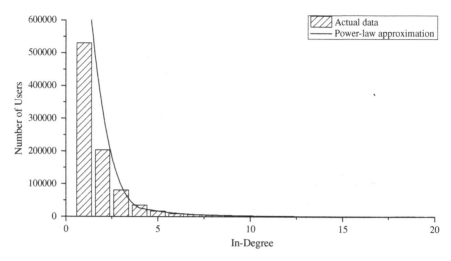

Fig. 4 The In-Degree Distribution of Followees

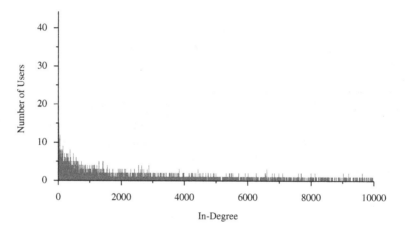

Fig. 5 The in-degree distribution of famous entities

3.2 Famous Entities Dataset

The famous entities dataset includes all famous entities, which are chosen for use in the followee recommendation system, and their category labels.

There are 6,095 famous entities in the dataset. The famous entities and their categories were chosen manually by Tencent Inc. But only 5,796 of them, about 95.09 %, are involved in the social graph dataset. The distribution of these famous entities' in-degree is shown in Fig. 5, only including 4,930 famous entities with equal to or less than 10,000 followers.

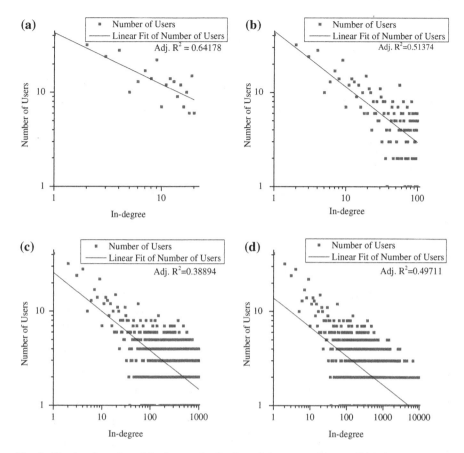

Fig. 6 The log–log plot of in-degree distribution of famous entities. **a** With in-degree ≤20 (5.38 % of all). **b** With in-degree ≤100 (12.03 % of all). **c** With in-degree ≤10³ (44.74 % of all). **d** With in-degree ≤10⁴ (85.06 % of all)

There are 866 famous entities with more than 10,000 followers. These entities account for about 14.94 % of the population, and are not shown in Fig. 5. Totally, there are 44,427,963 social links to these famous entities. Additionally, the minimum, median, average, 90th percent, and maximum in-degree are 1, 1,288, 7,665, 16,509, and 456,827, respectively.

Compared with the in-degree of overall users, which is shown in Fig. 4, the famous entities set has many more followers. Subjectively, these famous entities are well known. Their influence and reputations make them more likely to be identified among millions of users. Additionally, and objectively on the Internet, the famous entities are more likely to be reliable and stable information sources. Consequently, the masses need to follow them to get needed information.

In Figs. 3 and 4, the out-degree of followers somewhat roughly and the in-degree of followees rather well fits power-law distributions. Quite differently in

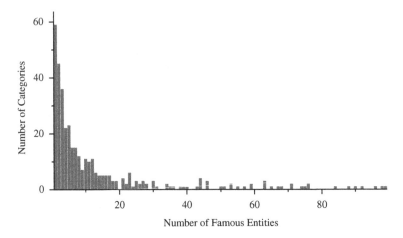

Fig. 7 The distribution of famous entities in each category

Fig. 5, as a whole, the in-degree of famous entities is much more evenly distributed than the preceding two. The log–log plot of their in-degree is shown in Fig. 6 and there is no clear and strong linear correlation found in this figure. It does not fit to power-law in any range.

Even though famous entities generally have many followers, the number of followers of each one varies significantly. The mean value and standard deviation of in-degree for all famous entities are as high as 7,665 and 23,703 respectively. For these with in-degree $\leq 10,000$, the mean value and standard deviation are 1,846 and 2,241, respectively. In other words, some famous entities may not get the same attentions on SNS as in reality. This also implies the importance of microblogging marketing. Without proper dissemination of information and marketing (that is, advertising), it's hard to be a well-known user on SNS, even for a famous entity in real life.

Each famous entity has a hierarchical category label, in form of "a.b.c.d". For example, for Yelp, one popular free application on mobile phones, the category label could be: "science-and-technology.internet.mobile.location-based". These labels are made and assigned to each entity by the staff of Tencent Inc.

In our analysis, we do not care about the hierarchical structure of the categories, and use the full four-level label as a unique category. Two labels, which are not in strict form of "a.b.c.d", are excluded in Fig. 7. As the result, there are 375 categories for these famous entities. In the dataset, famous entities are not evenly distributed in each category, which is shown in Fig. 7.

There are eight categories with more than 100 famous entities in them. They have not been shown in Fig. 7. At the same time, 298 categories hold no more than 20 entities each. The mean and standard deviation of the number of famous entities in each category is about 16.2 and 28.5 respectively. The volume of each category shows significant difference.

(a)   **(b)**

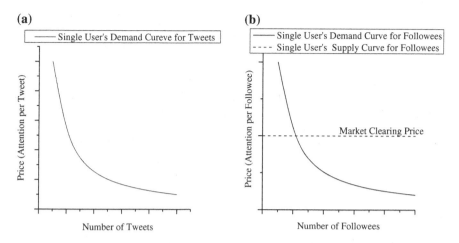

Fig. 8 The demand curve of single user for tweets and followees in one topic. **a** Curves for tweets. **b** Curves for followees

3.3 Using Aggregate Demand-Aggregate Supply Model to Interpret the Uneven Distribution

The uneven distribution of famous entities in different categories reflects the uneven distribution of public's attention on corresponding topics. With attention and information pieces being the exchange goods, we can use the demand–supply model to analyze it [25].

As the carrier and medium of expression, time is an approximate metric of attention [23]. Paying attention implies expenditure of time. For instance, if we follow Justin Bieber on Twitter, we pay attention to show our interest in him and consume his tweets. Simultaneously, we spend time to consume the information provided by him. Thus time is an indirect metric of attention. But it happens sometimes, that we are absentminded while visiting the social media and thus time is an inaccurate measure of attention. Overall, it is reasonable to use time as the metric of attention.

Consider now the demand of a single user. For one specific topic, if some users have no interest completely, the demand curve for tweets or followees would be the y-axis of Fig. 8. In other words, no matter how much attention is needed to purchase and consume the information in the topic, the users would not pay it.

Otherwise, as shown in Fig. 8a, like many ones for the material goods, the demand curve for tweets in any one topic is downward sloping. The demand is determined by preference more than technology. The more one is interested in the topic, the more attention is put into it, and consequently the more time is spent on consuming the related information.

In the short term, the demand curve could be treated as constant, since interests do not change dramatically. But there are exceptions to this rule. When some breaking and important event happens, the public's attention might be attracted on it in a very short time. In daily life, we do not care about earthquakes, because of its rare possibility. But once an earthquake happens, for the public around its epicenter, the demand for earthquake related information will surpass all other kinds of messages in minutes and hours [26, 27].

Given a constant demand for information of one topic, the quantity demanded increases with the reduction of unit cost. For example, if the tweets contain less video than before, information amount decreases, we need less attention to read and think about each message. In this case, the price drops. And we would like to consume more tweets, without paying more attention. The curve could be described by Eq. 3, where attention is a constant.

$$\frac{Attention}{Tweet} \times Number\ of\ Tweets = Attention \approx Time \qquad (3)$$

Theoretically, the equation only holds around the current status. While the price does not change greatly, the substitution effect is so little that we could assume the attention in Eq. 3 approximately remain the same:

$$Attention_{current\ status} \approx Attention_{nearby\ status}$$

But if it goes too far from current status, the substitution effect would change the attention paid to this topic. For example, if the tweets have a much higher price than before, the users would have to reallocate their attentions among different categories to maximize their utility. The details of utility maximization are discussed further in Sect. 5.2.

In practice, the price varies within a small range, unless some form of technology revolution happens. Considering the habits of customers and price strategies of other peers, the reasonable response of each provider is to follow the common price strategy of peers. As a result, information providers would like to maintain the price, and not to change it dramatically. The competition focuses on the quality of tweets. Otherwise, the provider will face the serious risk of loss of customers: a bad price strategy will be a loser even with good information goods.

The demand curve for famous followees is quite similar to the curve for tweets. If we change the x-axis to the number of followees, and the y-axis to a new form of price as attention per followee, the curve would remain the same on the whole. Let the productivity be the same for every followees, we could derive the demand curve for famous followees from Eq. 3.

$$
\begin{aligned}
Attention &= \frac{Attention}{Tweet} \times Number\ of\ Tweets \\
&= \left(\frac{Attention}{Tweet} \times Productivity \right) \times \frac{Number\ of\ Tweets}{Productivity} \\
&= \frac{Attention}{Followee} \times Number\ of\ Followees
\end{aligned}
\qquad (4)
$$

On social media, the demand curve for information provider could describe the users' behaviors more accurately than the curve for tweets because of the consumption pattern. In our daily life as also in some cases in the world of information, the customer needs to find, choose and then consume the goods. For instance, if we want to find an answer to some question on the Internet, we will search the question through search engine, choose some pages in the result set, and finally read them until we find the answer. But after that, we do not care about new information pieces on the viewed websites. In other words, our attention is paid majorly to the specific goods, rather than to the information providers.

In contrast, on social media, the users show a significantly different consumption pattern: subscription. The users have to find out some providers in one interested topic and subscript their tweets. After that, because of the indivisibility of these information sources, the users will receive all the tweets from these followees and have the potential to consume all of them. The attention is allocated to these followees directly.

In a nutshell, on social media, we are not paying our attention to individual tweets. One user, rather than one tweet, is the smallest communication unit. So the demand curve for information provider is better at describing the real social networking service.

The supply curve for followees of a single user is a straight line, which is parallel to the x-axis. The preference of a single user is not powerful enough to influent the followees' producing habit, such as the amount of tweets per day and the information density in tweets. Furthermore, the reading skill of users is constant in short term. Therefore, the price of the supply curve for single user, and the attention needed to consume the tweets of one followee, would not vary and always be the market clearing price. Under such price, as the information always is excessive, there are as many followee candidates as the user wants.

For one specific topic, if we add the demand curves for all users together, we get the aggregate demand curve. Even though the paid attention of users is different, the demand curves are similar in shape. As a result, the aggregate demand curve is approximately the same in shape as that of Fig. 8. The aggregate demand and supply curves, in short term and long term, are shown in Fig. 9.

On market clearing status, the users would like to follow these providers in an equilibrium price. The price is accepted by both followers and followees. Without external forces or changed conditions, no one has the motivation to change the price. As an equilibrium point here, the demand for followees is always equal to supply, as is indicated in the Eq. 5:

$$Aggregate\ Demand = \sum_u \#(Followee_u) = \sum_{u \in users} \#(Follower_u)$$
$$= Aggregate\ Supply \tag{5}$$

where $Followee_u$ means the followees of user u, and $\#(Followee_u)$ means the number of these followees.

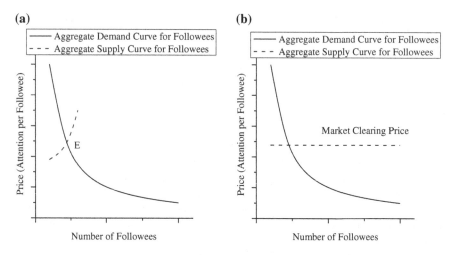

Fig. 9 The aggregate demand-aggregate supply curves for followees in one topic. **a** Short term AD-AS curve. **b** Long term AD-AS curve

In the short term, the aggregate supply curve slopes upward. When the demand of users suddenly increases, the demand curve would move toward the right. The price and supply would both increase. In other words, the users will pay more attention to each followee, and more information providers emerge. For instance, when an earthquake happens, the nearby users would carefully read all information from government originations and news agents. At the same time, some other users, including some persons in the vicinity, would become temporary information sources. As example, more and more breaking news is broadcasted by common users on social media. Due to the low threshold of being information sources, the response of social media market is quite quick.

If the attention change is temporary, only the short term phenomenon would appear. Then everything would go back to the original status with the restoration of attentions. For example, when the Super Bowl takes place, you can see the reports and news everywhere in USA. But during the rest of the year, it is discussed much less. If the attention change is persistent, it leads to a long term modification of behavior. For instance, the development of cloud computing has caused it to become the focus of attention of many researchers. Unlike the temporary event, such attraction stays strong for a long time.

In the long term, at the market clearing price, the elasticity of supply is infinite. That is, there are as many famous followees as the market's aggregate demand wants. On one hand, at the end of the short term drift, all these followees would lower the abnormal price, because it probably would not be accepted by some of their followers. And some temporary information sources turn into qualified long-term followees. On the other hand, with the increase of the supply, the users have more choices. To enhance the diversity of information, users would like to make the price back to normal and follow more sources. After these adjustment, the number of followees increases, while the price approximately remains the same. The situation of decrease in aggregate attention is similar.

Back to the equilibrium price, we will now analyze how the amount of famous followees in one category is determined. Let the $\#(FR)$ be the number of followers, P be the price, and T stand for the threshold, these long-term qualified information sources are willing to produce and provide tweets if and only if they receive enough attention:

$$\#(FR) \times P \geq T$$

$$In\_degree = \#(FR) \geq \frac{T}{P}$$

As widely known, the distribution of both in-degree and out-degree is power law. So we estimate the proportion of famous entities as following:

$$Pr\left(In\_degree \geq \frac{T}{P}\right) = \left(\frac{T}{P}\right)^{-a} \tag{6}$$

For the balance of aggregate supply and demand, it is expressed with the aggregate attention (AA):

$$\frac{AA}{P} = Aggregate\ Suppy = Aggregate\ Demand = \sum_{k=1}^{\infty} \#(\#(FR) = k) \times k \tag{7}$$

where $\#(\#(FR) = k)$ means the number of followees, who have k followers. In terms of the total number of entities (C) in one category, we derive the equation as following to get it.

$$\sum_{k=1}^{\infty} \#(\#(FR) = k) \times k$$

$$= \sum_{k=1}^{\infty} Pr(\#(FR) = k) \times k$$

$$= \sum_{k=1}^{\infty} (Pr(\#(FR) \geq k+1) - Pr(\#(FR) \geq k)) \times C \times k$$

$$= C \times \sum_{k=1}^{\infty} (Pr(\#(FR) \geq k) - Pr(\#(FR) + 1)) \times k \tag{8}$$

$$= C \times (Pr(\#(FR) \geq 1) - Pr(\#(FR) \geq 2) + Pr(\#(FR) \geq 2) \times 2$$
$$- Pr(\#(FR) \geq 3) \times 2 + \ldots)$$

$$= C \times \sum_{k=1}^{\infty} Pr(\#(FR) \geq k)$$

$$= C \times \sum_{k=1}^{\infty} k^{-a} \qquad (a > 1)$$

For the sum, it is hard to get an exact result. But an approximation can be estimated with integration.

$$1 + \int_{k=1}^{\infty} (k+1)^{-a} \leq \sum_{k=1}^{\infty} k^{-a} \leq 1 + \int_{k=1}^{\infty} k^{-a}$$

$$1 + \frac{1}{a-1} \times 2^{-a+1} \leq \sum_{k=1}^{\infty} k^{-a} \leq 1 + \frac{1}{a-1}$$

So we represent the sum as:

$$\sum_{k=1}^{\infty} k^{-a} = 1 + \frac{1}{a-1} \times \varepsilon^{-a+1}, where\ \varepsilon \in (1,2) \tag{9}$$

To combine the Eqs. 7–9, we get the relation between aggregate attention and total number of famous entities:

$$\frac{AA}{P} = C \times (1 + \frac{1}{a-1} \times \varepsilon^{-a+1})$$

$$C = \frac{AA}{P \times (1 + \frac{1}{a-1} \times \varepsilon^{-a+1})} \tag{10}$$

Then, we estimate the number of famous followees ($\#(FFE)$) in each category with Eqs. 6 and 10.

$$\#(FFE) = C \times \Pr\left(In\_degree \geq \frac{T}{P}\right) = \frac{AA}{P \times (1 + \frac{1}{a-1} \times \varepsilon^{-a+1})} \times \left(\frac{T}{P}\right)^{-a}$$

After simplification, the result is:

$$\#(FFE) = \frac{AA \times P^{a-1}}{T^a \times (1 + \frac{1}{a-1} \times \varepsilon^{-a+1})}, where\ a > 1\ and\ \varepsilon \in (1,2) \tag{11}$$

From Eq. 11, the number of famous entities in the topic would change in the same direction of AA's change. If the aggregate attention gets higher, the market needs more information providers. If the situation is converse, less famous entities would survive. And some *"failed"* providers cannot get enough attention as expected, and might switch to other topics or leave the market.

Similarly, the number of famous entities would increase if price rises, and decrease if price drops. When only price increases, and all other parameters stay the same, these followees need fewer followers than before to get enough attention that exceeds the threshold T. As a result, the scope of famous entity candidates is extended, and the total amount increases.

If the threshold for a famous entity to survive changes, the number of famous entities would change toward the opposite direction. For instance, if followees need more attention to act as a famous entity, because of the constant amount of aggregate attention, the number of famous entities would decrease.

Table 1 The brief distribution of top 100 Twitter followees[a] in categories[b]

Category	Gross	Proportion (%)	Example
Singer and Actor	61	61	Justin Bieber (justinbieber); Jim Carrey (JimCarrey)
TV host	10	10	Oprah Winfrey (Oprah); Conan O'Brien (ConanOBrien)
Sport	8	8	SHAQ (SHAQ); FC Barcelona (FCBarcelona)
Technique	7	7	Instagram (instagram); YouTube (YouTube)
Socialite	5	5	Kim Kardashian (KimKardashian); Bill Gates (BillGates)
News	3	3	CNN (CNN); The New York Times (nytimes)
Writer	2	2	Perez Hilton (PerezHilton); Paulo Coelho (paulocoelho)
TV channel	2	2	MTV (MTV); ESPN (espn)
Religion	1	1	Dalai Lama (DalaiLama)
Politics	1	1	Barack Obama (BarackObama)

[a] http://twitaholic.com/top100/followers/ (access on March 31, 2013)
[b] Some celebrities cover many topics. But we only assort them into one category

Generally, in a stable market, the parameter P, T, and a are approximately constant. So the number of famous entities in different topics is not uniform, and strongly determined by the aggregate attention of public. The case of Twitter celebrities, as shown in the Table 1, confirms this rule.

4 Analysis on How Users Adopt the Famous Entities

In this section, we analyze how users follow famous entities based upon the static snapshot of users' social graph and the followee adoption history.

We measure the maximum number of famous followees in one category (*MFFC* for short in the following) for each user. Accordingly, in the Fig. 2b, we could define *MFFC* as:

$$MFFC_a = \max_{C_i \in Categories} \#(C_i \cap FE_a)$$

where $MFFC_a$ means the *MFFC* for user a, and FE_a means the set of followees of a. For example, if user a follows three non-famous entities, two movie stars, and one news agency, its *MFFC* should be two. We assume the category with most followees of each user indicates the user's most interesting field. Thus the *MFFC* measures the users' capacity of followees in one category.

In the dataset of users' social graph, which is described in Sect. 3.1, there are 97,655 followers, about 5.16 % of all, who do not have the social links to famous entities in the dataset, and hence are not involved in the following discussion. Only taking the users with famous followees into account, we obtain the distribution of *MFFC* as shown in Fig. 10.

Deriving from the linear regression result in Fig. 10a, we get the power-law approximation as following in Fig. 10b:

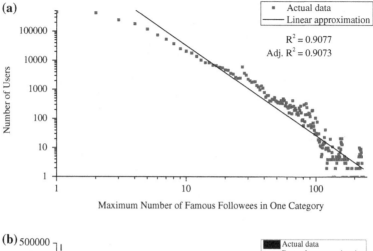

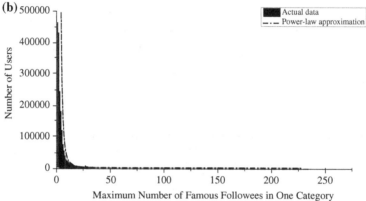

Fig. 10 The Distribution of Maximum Number of Famous Followees in One Category (*MFFC*). **a** In log–log plot in base 10. **b** In original coordinate

$$Number\_of\_Users = 10^{7.601223} \times MFFC^{-3.08936}$$

Overall, with users of *MFFC* being 0, the minimum, median, 90th percent, and maximum *MFFC* are 0, 2, 9, and 234, respectively. The average *MFFC* is about 4.55. To make the power-law property clearer, we make the distribution in smaller ranges in Fig. 11.

Additionally, the percentage cumulative distribution is provided in Fig. 12. Generally, most users do not follow lots of famous entities in one category and this phenomenon may be explained by marginal utility.

With the long-tail property, a small fraction of users have lots of famous followees—up to 234, in one category. Because all the information in these datasets is encoded as random strings and numbers to protect personal privacy and keep fairness in KDD Cup 2012, we cannot make a deeper analysis of this matter here.

(a)

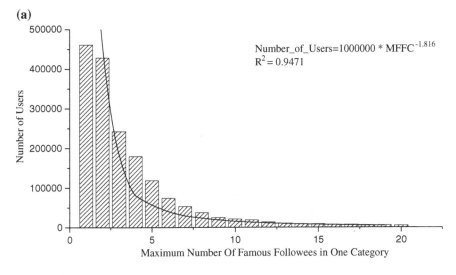

$$Number\_of\_Users = 1000000 * MFFC^{-1.816}$$
$$R^2 = 0.9471$$

(b)

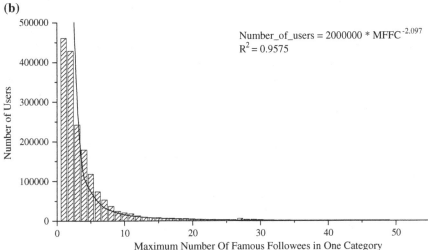

$$Number\_of\_users = 2000000 * MFFC^{-2.097}$$
$$R^2 = 0.9575$$

Fig. 11 The Distribution of *MFFC* with Part of the Users. **a** With *MFFC* ≤ 20 (96.39 % of all users). **b** With *MFFC* ≤ 50 (99.37 % of all users)

We guess that the unusual and excessive adoption of famous entities in one single category may be related with the users' working and living environments. For example, an IT staff might follow more famous entities, in related categories of computer science, than others.

In addition, we check the users' adoption history for famous followees. The users' adoption history contains the users' choice, both rejections and acceptances, to the recommendations from Oct 11 to Nov 11, 2011. Totally there are 73,209,277 records in this dataset. The following two kinds of records are removed and not used:

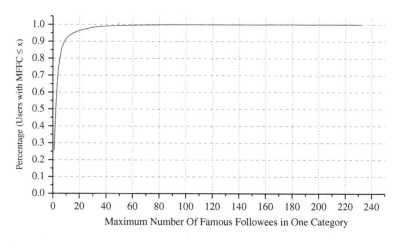

Fig. 12 Percentage cumulative distribution of *MFFC*

(1) the follower in the record does not have its social links information in the social graph dataset; or (2) the followee in the record is not a famous entity in our dataset.

Consequently, there are 62,169,578 (84.92 % of all) valid records in this dataset. Because a user could accept a recommendation to follow one famous entity, then unfollow it, and accept the same recommendation again after some time. There are some repeated records with different timestamps and we did not remove them. For user u, the Adoption Rate (AR) for a specific category C_i is generally defined as following:

$$AR_{u-C_i} = \frac{\#(Acceptances\ in\ C_i\ for\ u)}{\#(Acceptances\ in\ C_i\ for\ u) + \#(Rejections\ in\ C_i\ for\ u)}$$

The average adoption rates for all users are shown in Fig. 13. According to Fig. 10, more than 90 % of users have 9 or less famous followees in the maximum category. As a result, the samples of acceptance rates for the cases, in which the number of famous followees in one category is more than 9, are not sufficient. Thus we combine all these cases into one class as "10+" in Fig. 13.

In particular, the average adoption rate for all categories with n number of famous followees in one category is computed as:

$$\overline{AR}_{\#(FE \cap C_i)=n} = \frac{\sum_{u \in U_n} AR_{u-\bigcup_{\#(FE_u \cap C_j)=n} C_j}}{\#(U_n)}$$

where U_n means the set of users, who have n followees in one or more category.

The users might follow some new entities, and unfollow some old followees in this period. But we only have the adoption history, but do not have the unfollow history. We could not accurately know the number of famous followees in each category for each user at different time in this adoption period. So we use the data in the social graph dataset to statically estimate the number of famous followees for users.

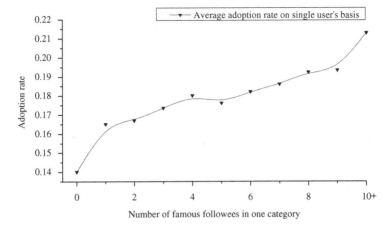

Fig. 13 The distribution of adoption rate

In the beginning, the adoption rate increases rapidly. But with the increase of famous followees in one category, the adoption rate of that category grows much slower and becomes stable.

Assuming that there is no cost for adopting new followees in the interested fields, the users might like to follow as many famous entities as there are to get the most information. Thus the more famous entities are followed in a specific category, the more interest is developed in that realm, and thus it is more likely to adopt new followees in the same field. But in real life, adopting new followees needs more energy and time to digest the additional messages. There is cost associated with adoption.

Overall, for more than 90 % of users, there are 9 or less famous followees in one category. That is, 9 or less information sources in one field are enough to provide sufficient messages with affordable cost. There are also less than 10 % of users who follow more than most of masses. For these users, the value of new messages is much higher than the cost. Thus even though users could get a little additional information by recruiting more followees, they continue to adopt new ones. For example, an analyzer of a company might follow all of its competitors, no matter how many they are. Also, the publicity department of one company is willing to follow and monitor the advertising of all its partners and competitors, no matter how many there are.

In addition, the result in Fig. 13 fits with the general conclusions of Fig. 10. To confirm this, we make an iterative and slightly non-rigorous simulation. Initially, we set the number of users as 1,892,059, which is the same as the total number of followers in the social graph dataset. And the *MFFC* of all users are set to 0 at the beginning. In each iteration round, each user has one opportunity to increase its *MFFC* by one with the average possibility in Fig. 13. Because we are short of samples to evaluate the adoption rate for *MFFC* $\geq$ 10 well, the maximum *MFFC* of users in this simulation is limited to 10.

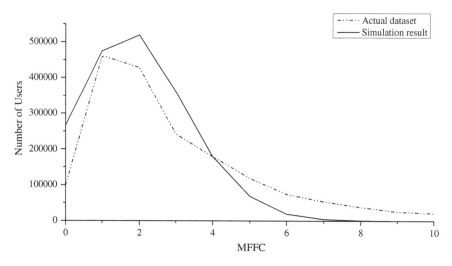

Fig. 14 The simulation result

After 13 rounds, we got a result which is similar to the real situation as shown in Fig. 14. Taking the simplicity of the simulation into account, the simulation matches the real situation very well. In one word, such adoption rates could lead to the power law distribution in Fig. 10.

For the case where *MFFC* is zero, the data in both actual dataset and simulation is not accurate. In the simulation, we did not consider new users, who continually join into the system. And the inaccuracy of real dataset could be an artifact of the sampling method. As we discussed in Sect. 3, the sampled users are the most "active" ones, who are less likely not to adopt any famous entity, compared with the non-selected "inactive" ones. In real life, overall, the users with *MFFC* equal to 0, including the zombie accounts, should be represented by a much greater proportion than it is in our samples datasets.

In terms of the longer and fatter tail in real case, a user with more followees would pay more attention and time on the microblogging service in general and corresponding topic in particular. Thus it is more likely to be exposed to recommendations than others, while our simulation keeps the recommendation chance the same for all users. In other words, even though the difference in adoption rate for one recommendation is not great, taking the amount of receiving recommendations into consideration, the users with more followees have a greater probability to adopt a new followee, than the case in our simple simulation.

Furthermore, in real life, there are other methods in addition to recommendation system for users to choose and adopt new followees. For example, word of mouth and influence of followees in real life play an important and critical role. With these limitations, our simple simulation is unable to accurately fit the real case. But it did show the general shape.

Theoretically, our case is similar with the classic Barabási–Albert model [28]:

1. Expand continuously: in terms of individuals, when they enter the SNS system, they commonly follow some other users in a short time, and then continue to adopt selective followees with a relatively slower pace. In terms of whole system, the existing users continue to choose new followees, and the new users continue to enter the system.
2. Rich get richer: with more famous followees, users are more likely to adopt more. But the increase rate in our case appears slightly different from the Barabási–Albert model.

In sum, the above discussion support that the *MFFC* of users fits to a power law distribution.

5 Explanations on the Adoption from an Attention Economical Perspective

In the previous section, we analyzed how the users adopt famous followees based upon an empirical dataset. In this section, we will analyze and discuss the phenomenon from an attention economical perspective.

5.1 Game Among Followees

One can speak of many naturally defined games within social media [29]. The interactions among followers and followees have inherent conflict and cooperation. On one hand, the followees always want to have more fans and influence, and user loyalty also means a lot for these information providers. On the other hand, the common users want to get information with least attention from the followees. But among the followers, there is little interaction, and nearly no game involved, because the supply is nearly infinite. In the following, we will discuss the game between follower and followee, and among followees.

Generally, the game among followees is a classic public choice problem [30–32]. For example, in the political election, which is the case of the traditional public choice, the voters hold different political views, and the candidates try to articulate a position to maximize the proportion of supporting voters. In the information network, the public's attention has different focus points, and the followees try to produce the information in an optimized subset of the topic to attract as much attention as possible.

For example, when Steven Jobs (co-founder and late CEO of Apple Inc.) and Dennis MacAlistair Ritchie (creator of C programming language and co-creator of UNIX operating system) passed away in October 2011, some information providers within social media had to choose how to allocate their sources to report the events and commemorate these two celebrities. Despite their personal preferences

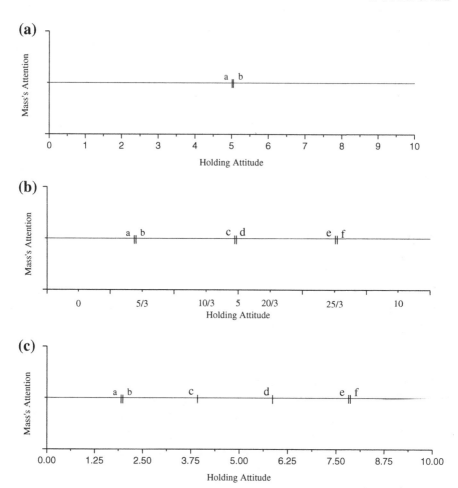

Fig. 15 The Hotelling's model (Example from [33]). **a** The Nash equilibrium with two players. **b** One Nash equilibrium with six players. **c** The other Nash equilibrium with six players

and to maximize the satisfaction of their followers, they had to decide the ratio of the tweets about Jobs to these about Ritchie.

If the public's opinions about ratio are uniformly distributed, as shown in Fig. 15a, and all customers prefer to choose the provider who holds the most similar ratio as wanted, this game problem is simplified into a Hotelling's model. In Fig. 15, the scale of x-axis is from 0 to 10, which represents post tweets about only Jobs and only Ritchie respectively. The middle point 5 means that there are equal post tweets about them.

With only two information providers, say players a and b, as an example of median voter model, the Nash equilibrium would in the middle of the market. Position 0 is strictly dominated by position 2, and position 10 is strictly dominated by position 9. With iterative deletion, the final "best" position for a and b would be the middle of the market. In the middle, they would share the market equally and they cannot get any additional payoff for moving their positions. In this case, by the strong form of the median voter theorem [30], the customers in the middle of the market get the most suitable tweets. The users in the two ends could not get the tweets, which are quite suitable to their demands.

In reality, there are numerously more players in the social media game, such that the median voter model could not be applied here. But it Hotelling's model can still be applied to this case [33]. For three providers, there is no pure strategy Nash equilibrium [34]. For four or more providers, there are pure strategy Nash equilibriums. For some cases, there are two or more unique Nash equilibrium. For example, with six players, there are two Nash equilibriums as shown in Fig. 15b and c. In both situations, no player has motivation to change its attitude.

With more information providers in Hotelling's model, at Nash equilibrium points the followees are dispersedly distributed in the market, rather than stay together in the middle. This distribution increases the diversity of information sources, further segments the market, and provides the customers more targetable and suitable tweets package.

But in practice, the public's attention is definitely not uniformly distributed, unlike the previous discussion. Previous studies [35, 36] have shown that there is the long-tail phenomenon in the public's demand for products. In other words, there are some products, which lots of people like. Other products appear to be attractive to a small fraction of public. But the tail is so long that these niche products account for a considerable market share. Furthermore, every customer is a bit eccentric, such that he/she wants both some popular and niche products [37]. Even though currently there is no direct evidence that the customers for online information show the same behavior as for material products, we reasonably assume they do. As a result, distribution of public's attention among different focus points of one topic would be multi-peak and characterized by asymmetry.

With the uneven distribution of attention, the Nash equilibriums are harder to figure out. But a necessary condition for Nash equilibriums in this case is [33] that no followee's whole market is less than anyone's one-side market. In other words, nobody's market share is greater than twice that of any other's and the potential customers are much more concentrated around the peaks. Thus we could expect the density of followees around the peaks to be much greater than that in the long-tails. The public's demand could be better fulfilled because the followees' strategy is closer to this demand. Our niche preference could not be exactly satisfied sometimes. For example, the public is more familiar with Jobs, so in October 2011, there were many more stories and reports about Jobs, but relatively much less information about Ritchie.

As reality is more complicated than theory, the situation of followees could never be described by a one-dimension model. There are many focus points in a single topic for public and the followees have to choose positions in each dimension. Consequently, the problem becomes that of finding equilibrium in an uneven distributed multi-dimensional space with multi-players. The pure strategy

Table 2 The game between one followee and one follower[a]

		Followee (volume of output)		
		Decrease	Unchanged	Increase
Follower	Follow	±, k	0, 0	±, j
	Unfollow	±, l	±, m	±, n

j > 0>k > l>m > n
[a] The payoff table is measured by change, not gross

Nash equilibrium is hard to find under such circumstances and even does not exist in some cases.

Additionally, the fit between personal preference and followees' attitude is not the only factor to influence the users' choice. For example, if all one's friends show a negative impression of followee c, according to the social balance theory [38, 39], one is less likely to adopt it.

Generally, the game among followees drives the information providers to focus on the public's demand, while not ignoring the niche. As a result, our general interest will be better fulfilled than our niche preference.

5.2 Game Among Followers and Followees

As shown by the previous discussion, the Nash equilibrium in the game among followees determines their market share. But actually the social media is not so "peaceful". There is fierce competition among the players not only for the number of followers, but also for the users' stickiness. The competition could be described with the game among followers and followees.

This is a zero–sum game. Any interaction among single followee and its followers cannot change the total attention from public. And one followee's gain can only come from the loss of others. For example, if one followee increases the quality of its tweets, it attracts more attention from its followers at the cost of attention to other followees.

In practice, the quality of tweets from one followee would not fluctuate greatly in a short time. Thus it could be treated as a constant. This quantity is the core of this game. Generally, the game between a single followee and one of its followers could be simply described by Table 2.

In this game, the payoff of follower is undetermined, except the point (follow, unchanged). At the point (follow, unchanged), the follower continues to follow, and the followee does not change its productivity. They stay the current status. So there is no change about the gross payoffs of both players. Otherwise, no matter which strategy is chosen, the total payoff of follower will either increase or decrease. The fluctuation depends on the interactions among current follower and its other followees.

If there are few similar information sources, the substitution effect is weak. Once one unfollows a followee, the follower will not get enough qualified tweets. The payoff of the follower would be decreased. But if the follower has many information sources on this topic, the excess of tweets would troublesome. So unfollowing an unsuitable followee could ease the information flood and promote the payoff.

When the follower adopts a followee, they stay in the current status, and the payoff of both followee and follower doesn't change if the rate of messages doesn't change. When the followee increases the volume of tweets, if the follower already has too many related tweets, the information flood would become worse and the payoff would decrease. In contrast, if currently the follower does not have sufficient information, the increase of supply could ease the information shortage and add its payoff. Similarly, the payoff of follower would increase or decrease depending on different cases.

In terms of the followee, when a follower chooses to continue to adopt it, if it produces more tweets, the follower has to allocate more attention to digest the items of information. This makes the followee become more important, because its tweets account for greater proportion. As a result, the followee gets more attention and improves its stickiness for users. Conversely, a decrease in productivity would lower the payoff of followee.

When the follower chooses to unfollow, no matter which strategy is selected, the followee will lose one follower completely. But if the productivity is lower, the total cost to make tweets would decrease. And the total lost would be less than the other two cases.

Overall, in this game, there is no strictly dominating strategy for followee. When follower chooses to unfollow, the best response for followee would be to lower the productivity. Otherwise, the best response is to increase the volume of tweets. And because of the uncertainty of the follower's payoff, it is impossible to find any Nash equilibrium.

Within real social media, there are lots of followers and followees. The actual game would be multi-players with infinite strategy space. There is a range of strategies, instead of a few discrete strategies for followees. The strategy space for the followee is infinite. In addition, each individual follower might choose to follow or unfollow according to its specific situation. Therefore the total number of followers might either increase or decrease depends on different reasons.

Generally, when a followee plans to adjust its productivity, it has to analyze market demand and users' stickiness to maximize its payoff as:

$$\max_{PR \in R}(payoff_{PR}) = \max_{PR \in R}(\#(followers_{PR}) * f(price, stickiness) - \cos t_{PR}) \quad (12)$$

where the subscript PR means the productivity level, and the $f(price, stickiness)$ is a function of price and users' stickiness to represent the total payoff from one individual follower.

For followees, the marginal cost of servicing an additional follower is nearly zero. In contrast, in this multi-players game, the production cost is related with the number of tweets. The more tweets are posted, the higher cost is needed. In other words, the marginal cost of producing an additional unit of tweet is not zero.

The Eq. 12 could be used to explain two important phenomena in social media: why there is no followee, who provides everything in the topic; and why there is not a monopoly in social media, even with the marginal cost being zero.

The customer's behavior in social media is different from that in most material transactions. In our daily life, we usually choose the item from a host of candidates, and then consume it. Some merchants provide as many kinds of goods as possible. Nearly every customer could get all wanted items in one stop. This kind of shop, which includes everything (such as Amazon), gives the customers more choices and enhances their shopping experience [35, 36].

But in social media, like Facebook and Twitter, we subscribe from followees, and consume all tweets from them as long as we follow them. If there is one follower, who provide tweets about all information in one topic, few persons could afford to consume all its information pieces. Even though the payoff $f(price, stickiness)$ from one individual follower would be very high, the disadvantage in quantity of followers would lower the total payoff.

As a result, the followees would not produce too many or too few tweets daily. To maximize the number of followers and the total payoff, they would analyze the market demand and the followers' preference to produce the most suitable count of tweets for public. Generally, when the users' stickiness and demand are strong, the elasticity of demand is low, and the followees would like to increase their tweets producing rate. Otherwise, the best response is to keep or lower current productivity.

Furthermore, this statement could explain why there is oligopoly or monopolistic competition on social media, rather than monopoly. Each followee is only willing to produce a limited number of tweets daily. So no followee is able to satisfy all followers' demand. As a result, there are many followees with different focuses in the topic to fulfill the public's demand. If the audience of the topic is only a few, or the topic is very professional and concentrated on few focuses, there would be a few followees to share the market within social media, as oligopoly. More commonly, it is monopolistic competition. Lots of information providers produce tweets on the topic, while each of them has different focus. This scheme guarantees, the follower could get suitable service, and the followees have enough segments of the market to survive.

5.3 Marginal Utility

Everyone has the same time and nearly the same attention. We could use it to work, to rest, or to enjoy the Internet, depends on one's own interest and demand. We assume the persons prefer the most valuable goods and services to them. When we use the social media to seek information, we would like to allocate our attention on different topic in general and different categories of followees in particular so that we can get the maximum satisfaction, like we did in material

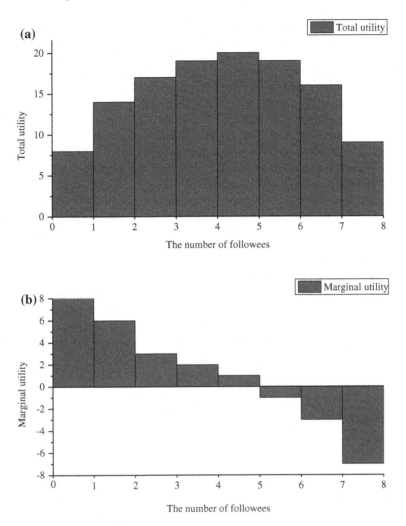

Fig. 16 An example of law of diminishing marginal utility. **a** The relationship between total utility and the number of followees. **b** The relationship between marginal utility and the number of followees

world. The utility is a representation of preferences over different categories of followees, that is, the utility is a measurement of satisfaction of topics [40, 41].

When one user adopts a followee on an interesting topic, it consumes wanted information, is satisfied by the service to some extent, and gets some utility. The Fig. 16 demonstrates an example utility of one user on an individual topic.

At the beginning, the user has not information source at all. When it chooses the first followee, the gain tweets are completely new for it. And as in our assumption, persons prefer the most valuable goods and services to them. The first followee and its tweets would be greatly valuable to the user. So the marginal utility would be large. And the total utility would increase greatly.

With the next adoptions, the marginal utility would be much less, as the law of diminishing marginal utility indicates. On one hand, the user already achieves some validation from the current followees. It is not as eager for the information as at the very beginning. On the other hand, inevitably, there are some overlapped information between the current followees and the new ones. Consuming the overlapped information would be a waste of time and attention, and cannot give the user negative utility. In other words, after adopting some followees, the relative density and absolute number of useful tweets from one additional followee would be less, and marginal utility would decrease.

After the threshold, the total utility turns to decrease, and the marginal utility becomes negative. For example, in Fig. 16, if the user adopts the sixth followee, the marginal utility would -1, and the total utility would decrease from 20 to 19. At the threshold point, the user already gets enough information. Any additional tweets will only annoy the user without any positive affect. In practice, all users would hate to cross the threshold.

If the users have infinite time and attention, the best response to get information would be adopting as many followees as not beyond the threshold value. But actually, the time and attention are the fairest currency. Everyone has the same and limited time and attention. So within social media, we have to consider how to allocate our currency to maximize our utility as:

$$
\max\left(\sum_{C_i \in C}(U_{C_i} - CO_{C_i})\right) = \max\left(\sum_{C_i \in C} U_{C_i} - Total\ cost\right)
$$
$$
= \max\left(\sum_{C_i \in C} U_{C_i}\right) - Total\ cost
$$

(13)

where U_{C_i} means the utility gain from the followees in category C_i, and CO_{C_i} means the corresponding cost. Because the total cost is constant for each individual user, we try to maximize the sum of utility.

Firstly, the marginal utility (MU) from each individual category is a function of multi dependent factors:

$$MU_{Ci} = f(Demand,\ Quantity_L,\ Quality_L,\ Quantity_{-L},\ Quality_{-L},\ Overlap(L, -L))$$

where the $Quantity_L$ and $Quantity_{-L}$ mean the quantity of tweets from the last followee and all the others respectively. Similarly, the $Quantity_L$ and $Quantity_{-L}$ are the measurement of the quality of tweets from the last followee and all the others respectively. The $Overlap(L, -L)$ function is used to estimate the proportion of the last followee's tweets, which is the same or very similar with some tweets from other followees.

To achieve the maximization of total utility, according to the equimarginal principle, we will allocate our attention on each topic, such that the ratio of marginal utility to price is the same for all categories.

$$\frac{MU_{C_1}}{price_{C_1}} = \frac{MU_{C_2}}{price_{C_2}} = \frac{MU_{C_3}}{price_{C_3}} = \cdots = \frac{MU_{C_i}}{price_{C_i}} = \cdots$$

The price of followees in different category is similar in most cases. So the equimarginal principle could be simplified as:

$$MU_{C_1} = MU_{C_2} = MU_{C_3} = \cdots = MU_{C_i} = \cdots = MU \tag{14}$$

At this balance, the users could get the maximum total utility. If the marginal utility for some categories changed, the users will adjust the allocation of attention to re-reach such balance. For example, if the marginal utility of category c is lowered, the user would unfollow someone in the category and withdraw some attention from the category. Then it will reallocate the withdrawn attention to other categories. According to the law of diminishing marginal utility, the marginal utility of category c will increase, and other categories' marginal utility will decrease. As a result, the user would achieve the balance in Eq. 14 again.

If the marginal utility of category one increases, it becomes more valuable for investment. The user would reduce the attention paid on other categories, and put more attention on followees in the category one. The reallocation won't stop until the balance is reached.

The equimarginal principle would explain why one user would follow and unfollow the same entity for several times. If the number of followee is continuous, the users will achieve an exact balance point as in Eq. 14. But the followee is indivisible, and the number of followee is discrete. We can only try to keep the marginal utility being approximately same for all categories. In some cases, the difference in marginal utility is so great that it drives the user to reallocate the attention. But the indivisibility of followees makes it impossible to achieve even an approximate balance. As a result, the user would follow and unfollow the entities frequently.

6 Discussion and Conclusion

In this chapter, we analyze how users adopt famous followees with empirical dataset, and then explain these phenomena from an attention economics perspective.

The used datasets were sampled and provided by Tencent Inc. and they chose the most active users in the sampling period. But the datasets do not provide the precise definition of "active" users. We do not know the standard by which the famous entities were chosen and labeled by the employees of Tencent Inc. Finally, the recommendation algorithm will have some impact on the result in Sect. 5. But at the statistical level, a few outliers cannot affect the general trend. A deeper examination of how different factors affect the results will be studied in the future.

There are still many open issues in this topic. Although we provide confirmation that the maximum number of famous followees in one individual category fits to power law, the factors which affect the upper limit of famous followees in a category for each user are not clear and the model of adopting famous followees is not provided.

In addition, why the in-degree of famous entities varies greatly is unknown. It appears that being well-known in real life does not guarantee success on social networking service. This question requires further research from the perspective of microblogging marketing.

In our analysis, we used some theories and models of material economics. With the aggregate demand-aggregate supply model, we analyzed the capacity of followees in each category. Then game theory was used to investigate the interactions among followees and followers. Finally, utility theory is applied to determine how users choose followees in each individual category. These theories and models are shown to be useful.

In practice, we could use the indifference curve to get the aggregate demand-aggregate supply curves. But within social media, we don't know how to obtain accurate demand and supply curves. This is partly because attention is hard to measure and compare. To investigate the attention market better, we have to understand the users' demand and followees' supply.

With the interactions among followers and followees, the pure strategy Nash equilibrium is difficult to find or does not exist. Under such case, how should the followees determine their market position to maximize their market share? This is an open issue and the answer will be important for social media marketing.

When we measure the utility of users, our analysis combines multiple parameters, such as the user's demand, the quantity and quality of tweets from the last followee and all the others, and the overlap among the tweets. It is easy to figure out the general effect of each factor. But lacking a deep examination, we do not know the exact impact of each factor.

The results of this study could be useful in microblogging marketing. The marketing personnel could discover potential customers better with the results of this chapter. The users with less than 3 followees in the one category do not show significant interest in corresponding field, and have a relative low adoption rate for recommendations. If the microblogging marketers propagate themselves to these users, the efficiency will be low, because of low adoption rate. On the other hand, users following more than 9 entities in the category show great interest, and have a higher adoption rate. But the number of these users is small. And according to their extraordinary interest in their topics, they are not likely to be common users. As a result, on balance, users who follow 3–9 famous entities in the category are the best ones to be targeted for promotion.

Furthermore, analyzing the distribution of users' followees will be helpful in automatic classification of the users. If some users follow many more entities in a single category than most of the masses, they show an extraordinary interest in corresponding field. Such information could be used to find these "uncommon" users.

Finally, in our analyses, we found that many ideas of traditional material economics are also useful in explaining the phenomena of attention economics in social media. These different fields of economics have many differences in details. For example, in traditional economics, the goods are in scarcity. In contrast, in attention economics, the attention of customers is the scarcest thing. But essentially they have the same focuses: what to produce, how to produce, whom to be serviced.

Acknowledgment We appreciate Tencent Inc, the organizers of KDD Cup 2012, for sharing the datasets of microblogging service with the public.

References

1. Borgatti, S.P., Mehra, A., Brass, D.J., Labianca, G.: Network analysis in the social sciences. Science **323**, 892–895 (2009)
2. Acock, A.C., Hurlbert, J.S.: Social network analysis: a structural perspective for family studies. J. Soc. Pers. Relat. **7**, 245–264 (1990)
3. Guo, Y., Chen, J.: A case study: social network and knowledge sharing. 2010 international conference on e-business and e-government (IEEE), pp. 1715–1718 (2010)
4. Cha, M., Mislove, A., Gummadi, K.P.: A measurement-driven analysis of information propagation in the flickr social network. In: Proceedings of the 18th international conference on World wide web—WWW'09, p. 721. ACM Press, New York (2009)
5. Watts, D.J., Strogatz, S.H.: Collective dynamics of "small-world" networks. Nature **393**, 440–442 (1998)
6. Kleinberg, J.: The small-world phenomenon: an algorithmic perspective. In: Proceedings of the thirty-second annual ACM symposium on Theory of computing—STOC'00, pp. 163–170. ACM Press, New York (2000)
7. Facebook Data Team: Anatomy of Facebook, https://www.facebook.com/notes/facebook-data-team/anatomy-of-facebook/10150388519243859
8. Leskovec, J., Horvitz, E.: Planetary-scale views on a large instant-messaging network. In: Proceeding of the 17th international conference on World Wide Web—WWW'08, p. 915. ACM Press, New York (2008)
9. Kwak, H., Lee, C., Park, H., Moon, S.: What is Twitter, a social network or a news media? In: Proceedings of the 19th international conference on World wide web—WWW'10, p. 591. ACM Press, New York (2010)
10. Huberman, B.A., Romero, D.M., Wu, F.: Social networks that matter: Twitter under the microscope (2008)
11. Ebel, H., Mielsch, L.-I., Bornholdt, S.: Scale-free topology of e-mail networks. Phys. Rev. E **66**, 035103R (2002)
12. Watts, D.J.: The "new" science of networks. Annu. Rev. Sociol. **30**, 243–270 (2004)
13. Cha, M., Haddadi, H., Benevenuto, F., Gummadi, K.: Measuring user influence in Twitter: the million follower fallacy. In: Proceeding of the fourth international AAAI conference on weblogs and social media, pp. 10–17 (2010)
14. Ahn, Y.-Y., Han, S., Kwak, H., Moon, S., Jeong, H.: Analysis of topological characteristics of huge online social networking services. In: Proceedings of the 16th international conference on World Wide Web—WWW'07, p. 835. ACM Press, New York (2007)
15. Wilson, M., Nicholas, C.: Topological analysis of an online social network for older adults. In: Proceeding of the 2008 ACM workshop on Search in social media—SSM'08, p. 51. ACM Press, New York (2008)

16. Asur, S., Huberman, B.A.: Predicting the future with social media. 2010 IEEE/WIC/ACM international conference on web intelligence and intelligent agent technology (IEEE), pp. 492–499 (2010)

17. Tumasjan, A., Sprenger, T.O., Sandner, P.G., Welpe, I.M.: Predicting elections with Twitter: what 140 characters reveal about political sentiment. In: Proceedings of the Fourth International AAAI Conference on Weblogs and Social Media, pp. 178–185 (2010)

18. Yu, S., Kak, S.: A survey of prediction using social media, http://arxiv.org/abs/1203.1647

19. Goldhaber, M.H.: The Attention economy and the net. First Monday **2**, 1–27 (1997)

20. Pope, N.: The economics of attention: style and substance in the age of information (review). Technol. Culture **48**, 673–675 (2007)

21. Essig, M., Arnold, U.: Electronic Procurement in Supply Chain Management: An Information Economics-Based Analysis of Electronic Markets. J Supply Chain Manag. **37**, 43–49 (2001)

22. Evans, P.B., Wurster, T.S.: Strategy and the new economics of information. Harvard Bus. Rev. **75**, 70–82 (1997)

23. Erik, B., Joo, H.O.: The attention economy: measuring the value of free digital services on the internet (2013)

24. Mislove, A., Marcon, M., Gummadi, K.P., Druschel, P., Bhattacharjee, B.: Measurement and analysis of online social networks. In: Proceedings of the 7th ACM SIGCOMM conference on Internet measurement—IMC'07, p. 29. ACM Press, New York (2007)

25. Dutt, A.: Aggregate demand-aggregate supply analysis: a history. Hist. Polit. Econ. **34**, 321–363 (2002)

26. Sakaki, T., Okazaki, M., Matsuo, Y.: Earthquake shakes Twitter users: real-time event detection by social sensors. In: Proceedings of the 19th international conference on World wide web—WWW'10, p. 851. ACM Press, New York (2010)

27. Mendoza, M., Poblete, B., Castillo, C.: Twitter under crisis: can we trust what we RT? In: Proceedings of the First Workshop on Social Media Analytics—SOMA'10, pp. 71–79. ACM Press, New York (2010)

28. Barabási, A., Albert, R.: Emergence of scaling in random networks. Science **286**, 1–11 (1999)

29. Hassan, A.S., Rafie, M.A.M.: A survey of Game theory using evolutionary algorithms. 2010 international symposium on information technology (IEEE), pp. 1319–1325 (2010)

30. Congleton, R.D.: The median voter model, http://rdc1.net/forthcoming/medianvt.pdf

31. Holcombe, R.G.: The median voter model in public choice theory. Public Choice **61**, 115–125 (1989)

32. Enelow, J.M., Hinich, M.J.: A general probabilistic spatial theory of election. Public Choice **61**, 101–113 (1989)

33. Eation, B.C., Lipsey, R.G.: The principle of minimum differentiation reconsidered: some new developments in the theory of spatial competition. Rev. Econ. Stud. **42**, 27–49 (1975)

34. Shaked, A.: Existence and computation of mixed strategy Nash equilibrium for 3-firms location problem. J. Ind. Econ. **31**, 93–96 (1982)

35. Chris, A.: The long tail: why the future of business is selling less of more. J. Prod. Innov. Manag. **24**, 274–276 (2007)

36. Erik, B., Yu, J.H., Michael, D.S.: From niches to riches: Anatomy of the long tail. MIT Sloan Manag. Rev. **47**, 67–71 (2006)

37. Goel, S., Broder, A., Gabrilovich, E., Pang, B.: Anatomy of the long tail: ordinary people with extraordinary tastes. In: Proceedings of the third ACM international conference on Web search and data mining—WSDM'10, p. 201. ACM Press, New York (2010)

38. Heider, F.: Attitudes and cognitive organization. J. Psychol. **21**, 107–112 (1946)

39. Cartwright, D., Harary, F.: Structural balance: a generalization of Heider's theory. Psychol. Rev. **63**, 277–293 (1956)

40. Quiggin, J.: A theory of anticipated utility. J. Econ. Behav. Organ. **3**, 323–343 (1982)

41. Starmer, C.: Developments in non-expected utility theory: the hunt for a descriptive theory of choice under risk. J. Econ. Lit. **38**, 332–382 (2000)

A Framework to Investigate the Relationship Between Employee Embeddedness in Enterprise Social Networks and Knowledge Transfer

Janine Viol and Carolin Durst

Abstract Organizations introduce Enterprise Social Networks to support knowledge management and in particular to facilitate knowledge transfer. However, to reap the full benefit of Enterprise Social Networks it is necessary to understand the relations and the interactions between employees within these networks. This book chapter provides a literature-based theoretical framework that enables the analysis of the relationships between an employee's embeddedness in an Enterprise Social Network, their access to social capital, their individual knowledge transfer process and the achieved knowledge transfer in an organization. We develop network-based measures that can be extracted for each framework element using data mining techniques and discuss the relationships among the framework elements. Additionally, suggestions on how to process the network measures using Computational Intelligence methods, e.g., fuzzy logic, are presented. Establishing a strong theoretical groundwork, this book chapter encourages future research crossing the boundaries between information systems, Computational Intelligence, organizational science, and knowledge management.

Keywords Enterprise Social Networks · Organizational social capital · Knowledge-intensive work · Computational intelligence methods · Knowledge transfer

J. Viol (✉) · C. Durst
Institute of Information Systems, Friedrich Alexander University,
Erlangen-Nuremberg, Germany
e-mail: Janine.viol@fau.de

C. Durst
e-mail: Carolin.durst@fau.de
URL: http://www.wi2.uni-erlangen.de/

W. Pedrycz and S.-M. Chen (eds.), *Social Networks: A Framework*
of Computational Intelligence, Studies in Computational Intelligence 526,
DOI: 10.1007/978-3-319-02993-1_12, © Springer International Publishing Switzerland 2014

1 Introduction

The ability to transfer knowledge influences an organization's productivity and has been found to be crucial for its survival [3]. Consequently, organizational structures should be designed in a way to ensure and support an efficient knowledge exchange. However, knowledge transfer and knowledge-intensive work in organizations oftentimes occur within informal networks which are removed from the organizational charts and formalized standard procedures [13]. Facilitating an efficient knowledge transfer, these informal organizational networks are based on social relations characterized by a high degree of trust among the collaborating employees [2]. To ensure an effective and efficient handling of knowledge, it is important to understand and to manage these informal networks. Yet, the social relations and cultural factors fostering informal networks are considered as being "imprecise and difficult to manage" [53, p. 200].

Traditionally, studies of informal social networks have applied social network analysis methods to self-reported data, e.g., collected in surveys or questionnaires. Recently, companies have started to use internal online social networks—similar to Facebook—in order to support different goals, e.g., to improve knowledge sharing and collaboration among employees. Belonging to the category of Enterprise 2.0 applications [48], these Enterprise Social Networks (ESN) are used to connect employees within a company. Well-known examples include "Blue-pages" at IBM or "Harmony" at SAP [64]. In general, Boyd and Ellison [12] define social network sites as "web-based services that allow individuals to (1) construct a public or semi-public profile within a bounded system, (2) articulate a list of other users with whom they share a connection, and (3) view and traverse their list of connections and those made by others within the system". Compared to public and open social networking sites, such as Facebook or the professional network LinkedIn, ESN are closed applications within a company's intranet and include features such as identity management, expert search and relationship management as well as supporting functions like interfaces to other enterprise applications.

With the advent of ESN, it is now possible to extract and analyze rich communication data to investigate social interactions within an organization using the principles and practice of Computational Intelligence for the data collection, analysis, and interpretation.

A number of studies have already addressed the relationship between properties of "offline" organizational networks and knowledge transfer, e.g. [2, 3, 60, 62]. Yet, to the authors' best knowledge these relationships have not yet been investigated in an online, that is, ESN, context. For instance, interactions occurring in ESN could reflect the existing formal and informal relationships in an organization and/or add a completely new communication layer to the organizational network.

This book chapter presents a theoretical framework which is based on the theory of social capital. It enables the analysis of an employee's embeddedness in an ESN, his or her knowledge transfer process and the overall knowledge transfer achieved in an organization. In particular, we offer starting points to investigate the following research questions:

- How are employees embedded in ESN compared to the formal organizational structure?
- How can social capital be identified and measured in ESN using the principles and practice of Computational Intelligence and how is social capital associated with the knowledge transfer process of individual employees?
- How can we use these insights to enhance the overall organizational knowledge transfer?

Applying methods of Computational Intelligence, e.g., fuzzy logic, this book chapter adds to the literature by introducing a novel framework that facilitates an understanding of interactions and especially informal relationships and social capital in ESN. Companies introduce and use ESN with the objectives of improving the access to information and knowledge, increasing productivity and team performance [9], improving collaboration [63] and reducing costs, for instance. The variety of these objectives may indicate that the goals of the introduction of ESN are oftentimes not clear and insufficiently targeted. For practitioners, this book chapter provides insights into how the knowledge transferring behavior of employees in ESN can be analyzed and how these insights can be used to influence and to reinforce knowledge transfer by designing the formal organizational structure accordingly.

This book chapter is structured as follows: Providing the theoretical background for this research, Sect. 2 defines the central constructs knowledge transfer, organizational social networks and social capital and reviews the relationships between these constructs. Section 3 shows how the constructs can be measured using social network analysis and Computational Intelligence methods. Integrating the previously presented information, Sect. 4 presents the theoretical framework, the corresponding measurement items and indicators, and suggestions on how to use Computational Intelligence methods to process the measurement items to aggregate values for the framework constructs. Concluding this book chapter, the last section provides implications for theory and practice as well as directions for future research.

2 Theoretical Background

2.1 Knowledge Transfer

According to Argote et al. [4] knowledge transfer involves the sharing, interpreting, combining and storing of information. In organizations knowledge transfer describes the process through which one unit (e.g., individual, group, department, division) is affected by the experience of another [4], for instance

	Informal knowledge	Formal knowledge
Tacit knowledge	• Learning how to motivate a specific colleague by observing his or her befriended colleagues	• Learning new skills, e. g. negotiating strategies, by observation
Explicit knowledge	• Getting hints from colleagues to better manage a task	• Receiving information from business reports or process charts

Fig. 1 Classification of transferred knowledge

through learning from each other [20]. The transfer of knowledge causes costs, especially for the source of knowledge, which could be the individual offering the knowledge, for instance. This individual needs to make an effort and invest time in order support others in understanding his or her knowledge [60]. Therefore, it is more likely that knowledge transfer occurs and is successful when it is easy and does not require high investments in terms of time and effort.

With regard to the transferred knowledge, explicit and tacit knowledge as well as formal and informal knowledge can be distinguished. Explicit or codified knowledge can be transmitted in formal and systematic language, e.g., through documents such as reports or guidelines [52]. On the contrary, tacit knowledge is acquired through personal experience and is difficult to formalize and to communicate [52]. In the context of organizations, formal knowledge refers to institutionalized and approved knowledge and is part of standard procedures whereas informal knowledge is unapproved and shared within a community [45]. Figure 1 gives examples for different types of knowledge transferred in organizations.

An example for the transfer of informal and explicit knowledge is getting advice from colleagues on how to perform a certain task, e.g., by applying a special technique. Learning how to deal with a specific colleague, e.g., how to motivate him or her, by observation is an example for the transfer of informal and tacit knowledge. Examples related to the transfer of formal and tacit knowledge concern the learning of new skills or the improvement of existing skills, such as getting to know new negotiating strategies by observing one's supervisor at a meeting. Formal and explicit knowledge is contained in official company documents, such as an organization manual.

According to Argote et al. [4] and Retzer et al. [62], the transfer of formal and informal knowledge in organizations occurs in many different ways using mechanisms like training, personnel movement, communities of interest, manuals and reports, patents or relationships with other companies or customers. These mechanisms influence knowledge transfer at different organizational levels: At the individual level, knowledge transfer is e.g., influenced by training. At the group level, factors influencing knowledge transfer include moving staff from one group to another. At the department level, the available communication tools and technologies [4], the firm's strategy or the organizational form [54] may influence knowledge transfer.

Of particular interest in the context of this research is the influence of social relations on knowledge transfer. Hansen [29] found that weak social relations facilitate a cost-effective search for new information whereas strong social relations enable the transfer of complex information and tacit knowledge. Argote and Ingram [3] claimed the need for more research on properties of organizational social networks that facilitate (or impede) knowledge transfer. Additional communication channels within enterprises, like ESN, may also have an influence on intra-organizational knowledge transfer [9]. The analysis of knowledge transfer in ESN may facilitate a better understanding of the informal communication flow within enterprises [45].

2.2 Social Networks in Organizations and Knowledge Transfer

A social network "consists of a finite set or sets of actors and the relation or relations defined on them" [72, p. 20]. Actors are described as social entities, e.g., individual, corporate, or collective social units. The transfer of immaterial or material resources between them happens along relational ties. Most commonly, ties are classified into the categories of strong and weak ties. The tie strength is determined by the amount of time, the emotional intensity, the intimacy, and the reciprocal services which characterize the tie [26]. According to Granovetter [26], strong ties are intimate bonds between family members or close friends which are maintained regularly and permanently. Being of informal nature, they tend to be concentrated in particular groups and occur between network members with a shared social identity. Contrary, weak ties emerge as non-intimate bonds between acquaintances. Maintained infrequently and inconsistently, weak ties may be formal contacts and are more likely to link members of different small groups [65]. Actors and their actions within social networks are considered as interdependent and the structure of social networks may provide opportunities for or place constraints on individuals and their actions [72, p. 4].

Taking up the social network perspective, organizations can be considered as a web of formal and informal linkages between employees. While formal social networks are prescribed by the management and represent the organizational structure of a corporation, informal social networks are emergent and connect groups of individuals across the formal structures in an organization [2]. The knowledge that flows in organizational networks can be split into two classes: non-working domain and working domain [69]. Relating the two types of

	Informal social networks	Formal social networks
Working domain	• Communities of practice	• Business unit
Non-working domain	• Running group	• Company excursion

Fig. 2 Classification of organizational social networks

organizational networks and the two types of knowledge transfer occurring in these networks, Fig. 2 shows examples for the different combinations.

An example for a given organizational structure is the business unit, which is formal and in which work-related knowledge is transferred. Also transferring work-related knowledge but of rather informal nature are communities of practice, where employees with similar working or research interests group together to collaborate and share information and experiences [2]. Furthermore, non-working domain knowledge may be transferred in formal networks, e.g., between the participants of a company excursion in or informal networks, e.g., between employees participating in a running group [69].

Due to their particular importance for knowledge transfer and knowledge-intensive work [2], this research focuses on informal social network structures. Knowledge transfer in informal networks oftentimes occurs through activities related to seeking and giving advice. According to Cross et al. [18], individuals help each other in five ways: These include the provision of solutions, meta-knowledge, problem reformulation, validation, and legitimization. Solutions directly help the person seeking information to solve a problem whereas obtained meta-knowledge comprises pointers to individuals with expertise or relevant documents. The reformulation of a problem helps the advice seeker to view a problem from different angles and by providing validation an individual's plan or solution is affirmed. Finally, legitimization refers to a respected person having reviewed an individual's solution [18].

A lot of researchers investigated organizational social networks in order to understand the dynamics, e.g., knowledge-sharing behavior or knowledge transfer within the organizational structure.

Takahashi [68] examined the network of R&D project teams within a global software company in order to investigate how network characteristics vary across the different R&D teams and how these network characteristics influence the performance of the R&D teams. The analysis showed a distinction between

Fig. 3 Enterprise Social Networks and knowledge transfer

knowledge transfer within project teams and outside the project team. Whereas the knowledge transfer within project teams was found to be essential for the execution of the R&D project, external knowledge transfer was more relevant to achieve practical research impact [68]. The fact that the only project with research impact had low values for network density, network centralization, and distance-based cohesion indicates the necessity of external knowledge transfer.

Hansen [29] studied the impact of tie strength and type of knowledge on knowledge transfer. He found strong ties to be more efficient in transferring tacit knowledge than weak ties. Both strong and weak ties are feasible to transfer explicit knowledge.

Investigating information and knowledge transfer in a sample of 16 German regional innovation networks with almost 300 firms and research organizations, Fritsch and Kauffeld-Monz [24] found that strong ties are more beneficial for the exchange of knowledge and information than weak ties. Moreover, their results suggest that broker positions in the network tend to be associated with social returns, e.g., the enhancement of the extent of information transferred to network partners, rather than with private benefits.

The research of Reagans and McEvily [60] considers how different features of informal networks affect knowledge transfer. The results show that measures concerning the network as a whole, such as social cohesion and network range foster knowledge transfer, over and above the effect of the strength of the tie between two people.

Based on evidence of studies focusing on the relationship between offline organizational social networks and knowledge transfer, we propose that ESN and knowledge transfer are similarly associated (Fig. 3). However, additional factors may influence the relationship between ESN and knowledge transfer as shown by the placeholder in Fig. 3. To understand the dynamics within ESN in terms of how ESN—as an additional communication channel—may influence an employee's knowledge transfer process, we take a closer look at the theory of social capital in the next section.

2.3 Social Capital in Organizations and Knowledge Transfer

A social network's patterns of relationships define the structure in which different types of resources, i.e. social support, and social capital are embedded. Lin [43] defines social capital as "resources embedded in one's social network, resources that can be accessed or mobilized through ties in the network". Formal and weak social ties are more likely to provide informational support, whereas informal and strong social ties are associated with the generation of emotional support [65]. With regard to the different forms of social capital, Putnam et al. [58] distinguish between bonding and bridging social capital. Bonding social capital emerges in networks of people who show a high degree of homogeneity, expressed in e.g., similar interests and characteristics, and who are connected by strong ties. Bridging social capital, however, is based on networks with weak ties. Connecting heterogeneous people across social groups, it provides access to sets of non-redundant resources. It has been argued that weak social ties provide better access to informational and instrumental resources, e.g., job-related information. Yet, not all bridging links are equally useful [42]. In the case of job seeking, for instance, a weak tie contact may be particularly valuable if the contact person is in a more influential and powerful position than the job seeker [6].

Employees embedded in formal and informal company networks have access to various kinds of resources forming their social capital. Informal social networks have been found to be especially rich in social capital and more efficient than traditional corporate structures in enabling knowledge transfer, e.g. by [17, 38, 40, 55].

In this research, we concentrate on the analysis of an employee's embeddedness in an ESN, his or her knowledge transfer process and the overall knowledge transfer achieved in an organization using the concept of social capital. Summarizing the body of literature on the influence of electronic social networks on social capital, Ellison et al. [21] identified four major directions:

1. Online social networks encourage both bonding and bridging social capital.
2. Online social networks allow people to generate new social capital, for example by triggering interaction with previously unknown people.
3. The use of online social networks strengthens people's offline social network and supports the individual in maintaining existing relationships, for example by enhancing traditional means of communication.
4. Maintaining a large network of weak ties, and especially bridging weak ties, has been shown to result in positive outcomes stemming from the increased access to a diverse set of resources.

Social network sites have been suggested to have a specifically positive impact on the formation and enhancement of bridging social capital by [22, 67], for example by allowing contact maintenance with colleagues of different business units. Moreover, using social network sites may be particularly beneficial for individuals who are "network disadvantaged", i.e. members of social networks poorer than others in social resources. Studying social capital in an organizational context, Steinfield et al. [66] found that the use of an internal social network site allows individuals to locate useful information, draw on resources and make contributions to the network. According to their study internal social network sites do not only support knowledge management processes but also provide greater social capital benefits for employees in otherwise "network disadvantaged"

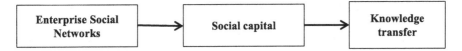

Fig. 4 Enterprise Social Networks, social capital and knowledge transfer

positions, especially newer and younger employees and those employees in a hierarchically lower position.

In our theoretical framework we suggest that bonding and bridging social capital can be used to explain the relationship between an employee's embeddedness in an ESN and the overall knowledge transfer achieved in an organization (Fig. 4).

3 Measurement

3.1 Knowledge Transfer

A differentiation can be made between the explicit and implicit measurement of knowledge transfer. Explicitly, knowledge transfer is measured based on questionnaire or survey data. Implicitly, knowledge transfer can be identified by observation or qualitative text analysis, e.g., by analyzing documents. Measuring knowledge transfer, different studies focus on different aspects, e.g., on the occurrence of knowledge transfer, the persons transferring knowledge, the quantity or quality of the transferred knowledge or the effects of knowledge transfer.

Fritsch and Kauffeld-Monz [24] analyzed the occurrence and quantity of knowledge transfer in innovation networks using questionnaires. Differentiating between knowledge transfer and knowledge absorption, knowledge transfer is measured by asking the questionnaire participant whether a network partner benefitted from assistance provided by him or her. Conversely, knowledge absorption is identified by measuring the extent of assistance received from network partners.

Items related to seeking and receiving information, knowledge, assistance and advice are commonly used to identify knowledge flows between members of an organization as well as knowledge sources and sinks, e.g. in [16, 24, 40, 61]. In a study by Chan and Liebowitz [16], members of an organization were asked questions like "Who are the top three people you would ask for each of the following areas: general advice, management and leadership advice, institutional knowledge, subject matter expertise, and technical/procedural knowledge?" within a so-called knowledge audit survey. The information provided in this survey facilitates the identification of knowledge sources and sinks within the organization and is visualized in a knowledge map showing the knowledge flows between different members of the organization.

Going one step further, Helms and Buijsrogge [31] used a knowledge network analysis approach to analyze knowledge transfer between employees using the concept of velocity and viscosity of the knowledge transfer. Knowledge velocity is defined as the speed of knowledge transfer in an organization and knowledge viscosity describes the richness of the transferred knowledge. The approach of knowledge network analysis includes (1) the identification of knowledge areas in the organization, (2) the identification of the main actors for each area and their levels of expertise and (3) the participation of all members of the organization in a questionnaire to identify the knowledge flows. The results of this analysis can be used to create a push graph or a pull graph [31]. The push graph identifies the different levels of expertise of the network members and shows the viscosity of the knowledge transferred between members of the organization. Thus, it reveals the distribution of expertise within an organization and whether the level of expertise is likely to increase over time. The pull graph, however, focuses on the velocity of the organizational knowledge transfer. It visualizes the awareness and access to knowledge of the members of the organization when they need it independent of different levels of expertise.

Some studies focus on the effect of knowledge transfer. Assuming that knowledge transfer leads to changes in the knowledge or the performance of the recipient units, Argote and Ingram [3] suggest measuring knowledge transfer explicitly, by assessing individual knowledge change through questionnaires or implicitly by observing changes in performance. Other implicit approaches to identify and measure knowledge transfer rely on the analysis of company documents. Analyzing the effect of inter-firm knowledge transfers within strategic alliances, Mowery et al. [51] measured the change in the partner firms' technological capabilities by using the citation patterns of partner firms' patent portfolios. That way, they were able to measure the extent of overlap between the partners' technological resources as an indicator for knowledge transfer within the alliance.

With regard to online social media, Huang and DeSanctis [35] coded and analyzed the entries in a professional online forum. Searching for text strings such as "Where can I find", "Do you know" or "Does anybody have" they categorized the forum entries as information seeking, information providing, explicit knowledge sharing and tacit knowledge sharing.

This research focuses on the analysis of ESN in order to support the knowledge transfer process of individual employees and thus, to improve the overall knowledge transfer in an organization. Some of the presented approaches to identify and measure knowledge transfer in organizations can be translated to the (partly) automated identification and measurement of individual knowledge transfer processes and knowledge transfer in an ESN. Hereby, we need to focus on the transfer of explicit knowledge of formal or informal nature as the transfer of tacit knowledge can be hardly observed in ESN. This knowledge transfer may occur in formal or informal networks in the working or non-working domain. The application of the knowledge network analysis approach by [31] seems to be particularly relevant. In a first and manual step, the different knowledge areas within an organization need to be identified, e.g., by considering the different functions in the organization. Next, different levels of expertise of employees could be identified by looking at their positions and skill sets available on the profile pages in the ESN. This information could be extracted using data mining techniques. Thirdly, the occurrence of knowledge transfer or the knowledge flows can be identified by analyzing the communication between employees, e.g., the exchange of wall posts and comments. Using text mining techniques, wall posts of employees could be scanned for text strings such as "Does anyone know" or "Does anybody have" in order to identify the seeking and giving of advice in the ESN. Simultaneously, the knowledge viscosity is measured taking into account the roles and levels of expertise of the employees [31]. Creating a directed organizational advice network based on e.g., exchanged wall posts in the ESN, experts or knowledge sharers are identified by calculating and comparing the frequency of being asked for advice across different employees.

The posting of documents, factual information or links could be another indicator of knowledge transfer, particularly when this information is shared by another employee. If the ESN provides the possibility to evaluate the information posted by an employee, e.g., through a rating system, the usefulness or quality of the transferred knowledge could also be identified. The sharing of posts of another employee may also be an indicator for useful information.

3.2 Network-Based Social Capital

This study considers social capital as inherent property of social networks and employs social network concepts and methodology to measure individual social capital as requested by scholars, (e.g. in [11, 39, 44, 50]). Social capital measures may be based on the structure of social networks, the presence of specific alters, and the diversity of social resources, for instance. As a well-established research area, social network research offers readily developed guidelines for operationalization and tools for data collection. The name generator, position generator and resource generator are three survey instruments used for the network-based measurement of individual social capital in offline contexts [70]:

1. The *name generator* estimates the network size and identifies the respondent's social network structure and content. In a first step, the respondent mentions names of persons she or he knows (name generator). Secondly, the respondent provides information about the relationships to and attributes of the previously named people. In a last step, respondents assess alter–alter ties.
2. The *position generator* is based on the assumption that occupations of network members reflect social resources and that highly ranked occupations correspond with high returns. The interviewee is given a list of 10–30 different occupations and states whether she or he "knows" anyone having this occupation. Subsequently, the interviewee provides information on the relationships to people in these positions (family members, friends, and acquaintances). The position

generator measures the highest accessed prestige, range in accessed prestige as well as the number of different positions accessed.

3. When using the *resource generator*, the respondent's access to social resources is checked against a list of useful and concrete resources. Afterwards, the respondent provides information on relationships to people having these resources (family members, friends, and acquaintances).

The advantages and disadvantages of the three techniques to measure individual social capital according to [32, 70] are listed in Table 1.

As to the measurement of bridging and bonding social capital, Williams [73] developed the Internet Social Capital Scales (ISCS) which can be applied to online as well as offline contexts. The measurement tool consists of a bridging and a bonding social capital subscale based on the criteria for the two social capital types [59]. Thus, bridging social capital measures e.g., consider if someone's social network is outward-looking and whether a person interacts with a broad range of people. Contrary, bonding social capital measures focus on social capital generated by strong ties, e.g., emotional support. William's ISCS was found to be valid and is the predominantly used survey tool in studies examining the association between social networking sites and social capital, e.g. in [14, 22, 66].

This research aims to collect full personal networks to assess the effects of social capital emerging from ESN structures on knowledge transfer. The analysis of personal networks is comparative and looks at varieties in e.g., size, quality and shape of different personal networks [34]. Sociometric network measures such as density and centrality are designed for whole networks and normally used in structural analyses. Yet, there are no mathematical or statistical reasons prohibiting the use of those sociometric measures to personal network data [49]. Enabling the collection of (full) personal networks, the data collection on ESN is less subject to logistic restraints than traditional offline strategies (e.g., the name generator). The application of sociometric measures of social capital to personal networks of ESN users appears reasonable as these networks can be assumed to be sufficiently large. Moreover, they can include many types of network members (e.g., colleagues, friends, and acquaintances) as well as a high number of weak and indirect ties that may influence the network structure and ego's behavior. The following sections provide an overview of structural, positional, and functional measures of social networks and their relation to social capital.

Structural Measures can be applied to egocentric network data. They focus on how actors in a network are connected to each other and provide interesting information for this research in that they enable the analysis of how structural network properties, e.g., density, influence the flow of resources in the network. Based on [11, 39], Table 2 provides an overview of selected egocentric measures and their relation to social capital.

Just as bridges, the measure network constraint is drawn from Burt's [15] theory of structural holes. Varying with respect to size, density and hierarchy of the network, network constraint will be high in small networks with actors who are closely tied to each other (density) or tied to one central contact (hierarchy).

Table 1 Measurement tools for individual social capital

Technique	Advantages	Disadvantages
Name generator	• Useful to identify single alters and their attributes • Enables study of individual network structure, relationship-specific attributes and relationship multiplexity	• High costs of data collection • Lengthy and repetitive interviews • Tendency to identify mainly strong ties
Position generator	• Enables estimation of potential pool of resources accessible to respondent • More respondent-friendly than name generator • Can be broadly applied • Constructed from a firm theoretical basis • Standardized construction of indicators • Collection of hierarchical positions	• Emphasis on instrumental use • Validity/reliability problem
Resource generator	• Respondent-friendly • Actual measurement of social resources	• Validity/reliability problem • Network size and density are not measured

Table 2 Egocentric structural measures of social capital

Measure	Description	Relation to social capital
Network size	Number of alters ego is directly connected to	Positive More relationships increase the change of accessing resources
Egocentric network density	Extent to which alters are connected to each other	Ambiguous A densely connected network may limit the access to non-redundant sets of resources
Network constraint	Extent to which ego is constrained by alters	Negative A highly constrained actor will have fewer opportunities for action

Table 3 Sociometric positional measures of social capital

Measure	Description	Relation to social capital
Degree centrality	Degree to which an actor is directly connected to most others in the network	Positive Access to network resources is more readily available when the actor is directly connected to others
Betweenness centrality	Degree to which an actor lies on the shortest path connecting others in the network	Positive Actors with a high betweenness centrality connect others who would otherwise not be connected and create new opportunities for exchange of resources
Closeness centrality	Degree to which an actor is close to other actors in the network	Negative The greater the distance to other actors, the less the chance of accessing resources
Eigenvector centrality	Degree to which an actor is connected to well-connected others in a network	Positive Being connected to well-connected others increases the chance of accessing resources
Bridges	Bridges link unconnected groups of others in the network	Positive Actors acting as bridges facilitate access to new and possibly diverse resources

Contrary, large networks with disconnected ties will have a low network constraint. According to Burt [15], more constrained networks span fewer structural holes, are associated with more network closure and generate less social capital.

Positional Measures reflect the position of actors in a network and focus on either the centrality of an actor or an actor's bridging location in the network. Requiring alter–alter ties between all actors in the network, they can be applied to whole networks only [39]. Table 3 lists sociometric positional measures and their relation to social capital as suggested in [11, 39].

Given that weak ties may enable a cost-effective search for new information [29], the analysis of actors providing bridges between several subsections of the overall network may be of particular interest in this research. According to Burt's

[15] theory of structural holes, bridge connections generate social capital in that they link otherwise disconnected groups which may result in better access to more and diverse information.

Functional Measures focus on the network resources provided by specific ties [39], functional measures are of particular interest in this research which seeks to analyze which ties may influence an individual's knowledge transfer process. Drawing on the work of Borgatti [11], Lakon et al. [39] and the PRI [57], Table 4 gives an overview of relevant egocentric functional measures and their relation to social capital.

The network heterogeneity and compositional quality of a personal ESN can be measured in a relatively straightforward way provided that information on the attributes of network members is available. However, assessing the strength of ties may be difficult in an ESN. According to Granovetter [26, p. 1362], tie strength is a "(probably linear) combination of the amount of time, the emotional intensity, the intimacy (mutual confiding), and the reciprocal services which characterize the tie". Multiplexity, which refers to network ties based on more than one relationship context (e.g., as a colleague and friend) and thus represents overlaps in one's social network, is suggested to be an additional measure of tie strength [26]. Seeking to construct valid measures of tie strength, Marsden and Campbell [46] used survey data on friendship ties and found that time spent together in a relationship as well as the depth of the relationship are two distinct elements of tie strength. Measures of closeness or intensity appeared to be the best indicators of tie strength whereas the frequency and duration of contact as well as kinship were less related to the concept. The findings of Marsden and Campbell [46] are supported by Mathews et al. [47] who studied university students and concluded intimacy to be the best predictor of tie strength. The principle of "proximity", that is, the similarity between two individuals, is considered as another tie strength indicator. Applying fuzzy sets theory and operations, Hassan et al. [30] determined the similarity of individuals to simulate the chances of two individuals to become friends. This approach could also be used for a data set collected in a virtual community, such as data from an ESN.

Several studies sought to quantitatively determine tie strength in virtual communities. Having obtained quantitative tie-strength data using their Virtual Tie-Strength Scale (VTS-Scale) questionnaire in a web forum, Petrczi et al. [56] found that closeness, intimacy, mutual confiding, multiplexity and shared interests are equally important indicators of tie strength in virtual groups and offline networks. Gilbert and Karahalios [25] constructed a predictive model capable to distinguish between strong and weak ties with over 85 % accuracy. Using friendship data from Facebook users, the authors found that intimacy, intensity and duration are the most relevant dimensions to determine tie strength. Xiang et al. [75] presented an unsupervised approach based on profile similarity and interaction activity to model relationship strength in online social networks. Assuming that homophily strongly correlates with tie strength, the authors analyzed counts of common networks, groups and friends for pairs of Facebook users to determine their profile and connection similarity. Moreover, user interaction features were computed based on

Table 4 Egocentric functional measures of social capital

Measure	Description	Relation to social capital
Network heterogeneity	Diversity of alters' backgrounds regarding relevant dimensions	Positive A heterogeneous network gives access to diverse sets of resources
Compositional quality	Number of alters who possess characteristics that are of interest to ego	Positive
Tie strength	Tie strength enables measuring the resources provided by specific ties	Strong ties are likely to generate social support, e.g., emotional and instrumental support. They create bonding relations which may results in the exchange of resources requiring a significant emotional, financial, or time investment Weak ties or ties that serve as bridges with other networks provide access to diverse non-redundant resources

wall postings and picture taggings. The combination of profile and interaction information summarized in a relationships graph performed well in determining meaningful relationships. Arnaboldi et al. [5] found that the recency of contact between individuals has the highest relevance in the prediction of tie strength. The combination of this measure with e.g., indices about the social similarity between people may lead to more accurate predictions. Other indicators of meaningful links in online social networks include picture friends to identify reciprocal or asymmetric friendships [41], wall posts and photo comments as suggested by [71, 74]. Due to the vagueness of the concepts influencing tie strength, such as "intimacy", Fazeen et al. [23] used a fuzzy logic approach to estimate tie strength. To determine the keenness with which a tweeter on Twitter is followed, they considered the parameters "Reply Message Percentage", "Common Follower Percentage", and "Normalized Mean Reply Delay" as input set of their system. Each of these three input parameters is represented by five fuzzy sets: very low, low, medium, high, and very high. These fuzzy inputs are then processed by the rule base and a fuzzy output for the tie strength is generated. The application of this fuzzy logic approach is advantageous as it is capable to process input parameters that are members of different categories.

Due to the similar functionalities of online social networks or Twitter and ESN, the discussed tie strength indicators and the use of a fuzzy logic approach can be assumed to be equally applicable in ESN.

4 Theoretical Framework

Integrating the previously presented information, the proposed framework (Fig. 5) shows how different variables may influence the knowledge transfer process of an employee and the achieved knowledge transfer.

Adapted from Durst et al. [19], the framework differentiates between the sources (determinants) and effects of social capital and uses indicators and measurement tools drawn from the field of social network analysis [57]. Computational intelligence methods, such as fuzzy logic and text mining in combination with machine learning, may be used to process these indicators. Social networks are considered as a precondition of social capital and place where social capital emerges [65]. Knowledge transfer processes and the achieved knowledge transfer are recognized as the effects of different social capital types. Including three areas of network characteristics—*network structure*, *network ties* and *network members*—the framework enables the analysis of both network locations and resources embedded in the network [42].

The framework element *network structure* includes measurement items such as network size, density, cliques, brokerage and constraint as described in more detail in Table 5. Using methods from the field of social network analysis, these structural properties are calculated mainly to identify the degree of *network openness/ closure*. Processing the structural properties, the application of fuzzy logic seems

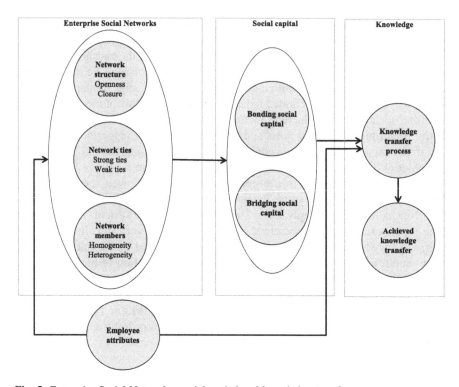

Fig. 5 Enterprise Social Networks, social capital and knowledge transfer

appropriate to determine whether a network is open or closed. The network structure may be connected with the *strong* and *weak ties* and the *network homogeneity/heterogeneity*, too. For instance, high network density may be a predictor of a rather closed and homogeneous network with strong ties between members.

Strong and *weak ties* are determined by considering the ESN's relational properties. The framework element *network ties* examines e.g., the network's composition, tie multiplexity, duration of a contact and communication mode as well as the intimacy and intensity of a contact. A detailed description of the recommended measurement items for the framework element network ties is given in Table 6. Looking at the network composition enables the identification of e.g., team members, colleagues in general or friends in an employee's ESN and provides information on the context and role in which persons have got to know each other. It is also connected to the determination of multiplex ties which are based on knowing a person in more than one relationship context, for instance, colleagues who are good friends at the same time. Multiplexity is an indicator of tie strength [26], yet, the intimacy, intensity and duration of a contact are the best variables to predict tie strength in ESN [25]. The frequency and different types of activities happening along the ties in an ESN can be used to distinguish actively used ties

Table 5 Network structure—description of suggested measurement items and indicators in ESN

Measurement items	Description	Indicator in ESN	References
Size	Number of alters that an ego is directly connected to	• Number of colleagues that an ego is directly connected to on the ESN	[1, 8, 15, 57]
Density	The percentage of ties that exist in a network out of all possible ties	• The ratio of the number of colleagues connected to ego compared to all possible ties in the observed employee's network on the ESN	[1, 8, 15, 49, 57]
Cliques	Sets of alters who are all directly tied to each other	• Sets of colleagues in the observed employee's network who are all directly tied to each other	[28, 49]
Components	Sets of alters who are connected to one another directly or indirectly	• Sets of colleagues in the observed employee's network who are connected to one another directly or indirectly	[28, 49]
Brokerage	Number of pairs not directly connected	• Number of pairs of colleagues in the observed employee's network not directly connected	[28]
Effective size	Number of alters that ego has, minus the average number of ties that each alter has to other alters	• Number of colleagues in the observed employee's network, minus the average number of ties that each colleague has to other colleagues	[11, 28]
Constraint	The extent to which ego's connections are to others who are connected to one another	• The extent to which the connections of the observed employee are to others who are connected to one another	[11, 15, 28]

Table 6 Network ties—description of suggested measurement items and indicators in ESN

Measurement items	Description	Indicator in ESN	References
Network composition	Different units in within the network members	• Types of colleagues an employee is connected to on the ESN, e.g., team members, project group members or members of the same department	[1]
Multiplexity	Multiplex ties are based on knowing a person in two or more relationship contexts. Multiplexity reflects overlap between an individuals social networks	• Number of different relationship contexts a tie is based on, e.g., a team member may also be member of the running group the observed employee participates in	[8]
Duration	Length of time the ego knows an alter	• Date of the first interaction with a colleague the observed employee is connected to on the ESN	[8]
Frequency	How often the network members have contact with each other	• Frequency of exchanged wall posts • Frequency of exchanged comments • Frequency of other forms of communication available on the ESN	[1, 8, 57]
Communication mode	Method or range of methods that individuals use in keeping in touch with others in their networks	• Wall posts • Comments • Other forms of communication available on the ESN	[1]
Intensity	Strength and nature of a relationship in terms of emotional investment	• Content and importance of matters that are discussed in, e.g., wall posts or comments • Length of the exchanged wall posts or comments (number of characters)	[1, 57]
Reciprocity	The extent to which exchanges or transactions are even or reciprocal	• Comparison of the length of the exchanged wall posts or comments (number of characters) between an employee and a connected colleague	[8]
Spatial proximity	Members who maintain face-to-face relationships on a regular basis	• Common tags in uploaded photos • Participation in events published on the ESN	[57]

Table 7 Exemplary rules of the fuzzy rules base

Number	Rule
1.	IF common_colleagues IS VL AND shared_projects IS VL AND replies IS VL AND mean_reply_time IS VL THEN tie_strength IS VL
2.	IF common_colleagues IS VL AND shared_projects IS VL AND replies IS VH AND mean_reply_time IS VL THEN tie_strength IS L
3.	IF common_colleagues IS VL AND shared_projects IS L AND replies IS VH AND mean_reply_time IS VL THEN tie_strength IS M
4.	IF common_colleagues IS VL AND shared_projects IS VH AND replies IS VH AND mean_reply_time IS M THEN tie_strength IS H
5.	IF common_colleagues IS M AND shared_projects IS VH AND replies IS VH AND mean_reply_time IS H THEN tie_strength IS VH

from inactive ties [74]. In a next step, actively used ties can be classified as strong or weak ties. Quantitative procedures, such as text mining, may be feasible to detect specific words in wall posts indicating closer relationships between employees [33]. Moreover, communication relationships could be classified by using a text mining approach to compare the wall post of an employee with previous postings of other employees [10]. Similarities between postings detected using this approach may point to a closer relationship or a certain degree of homophily between employees.

The final estimation of the tie strength between two employees could be done using a fuzzy logic approach. Adapting the approach of Fazeen et al. [23] to the context of ESN, the four parameters "Common Colleagues Percentage", "Shared Project Groups Percentage", "Reply Message Percentage", and "Normalized Mean Reply Delay" are regarded as indicators of tie strength in ESN and used as inputs for the fuzzy system. Each of these five inputs is assigned to five fuzzy sets: very low (VL), low (L), medium (M), high (H), and very high (VH). Having processed the fuzzy inputs using the rule base, the fuzzy logic generates a fuzzy output for the tie strength. Table 7 shows five exemplary rules among the total of 625 rules that were created in cooperation with a domain expert.

The framework element *network members* refers to other employees or colleagues and employee is connected to on the ESN and considers the same measurement items as the framework element *employee attributes* described in the next section. For each member of the observed employee's ESN, sociodemographic variables, e.g., age, gender and department affiliation as well as his or her interests and activities, need to be identified, as shown in Table 8. The degree of similarity among the network members with respect to different attributes determines the *network's homogeneity/heterogeneity* as suggested in [1, 57].

The framework element *employee attributes* includes sociodemographic measurement items [57] such as a person's age, gender and department affiliation as well as his or her interests, activities and intensity of ESN usage. Employee attributes are suggested to influence the ESN characteristics, activities related to knowledge transfer processes (and, consequently, achieved knowledge transfer). The ESN characteristics could be influenced by the employee's frequency and intensity of ESN usage and by the employee's job position, e.g., by the fact if he or

Table 8 Network members and employee attributes—description of suggested measurement items and indicators in ESN

Measurement items	Description	Indicator in ESN	References
• Age • Gender • Profession • Departmental affiliation • Activities • Interests • Intensity of ESN usage	Dimensions of member attributes or employee attributes	• Information available on an employee's profile page, e.g., name, department, skill profile etc	[8] [1]

Table 9 Knowledge transfer process—description of suggested measurement items and indicators in ESN

Measurement items	Description	Indicator in ESN	References
• Knowledge seeking • Knowledge providing	An employee's activities related to asking and providing knowledge	• Number of wall posts with text strings indicating seeking advice, such as "Does anyone know…" or "Does anybody have…" • Number of wall posts including answers to knowledge requests	[35]

she needs to regularly maintain relationships with people in different departments or branches of the company. An employee's activities related to knowledge transfer processes may be influenced by different personality traits, such as his or her openness or extraversion [7].

With regard to the emerging forms of social capital, the framework makes a distinction between bonding and bridging social capital. As previously discussed, *bonding social capital* occurs in homogeneous groups of people sharing similar interests or characteristics whereas *bridging social capital* is associated with networks with weak ties between heterogeneous people. Classifying bonding or bridging social capital in an employee's ESN can be done by processing the previously described ESN characteristics using fuzzy logic. A high degree of network closure, high reliance of the individual on strong tie contacts and high degree of network homogeneity are suggested to predict bonding social capital [65]. On the contrary, open networks increase bridging social capital by allowing individuals to establish new links which may provide access to more diverse resources and knowledge [57].

The framework element *knowledge transfer process* examines the observed employee's activity in knowledge transfer processes, that is, activities related to the seeking and providing of knowledge, operationalized as advice as shown in

Table 10 Achieved knowledge transfer—description of suggested measurement items and indicators in ESN

Measurement items	Description	Indicator in ESN	References
• Number of information requests • Number of responses	The ratio of information seeking to information providing	• Ratio of the number of wall posts indicating advice seeking to the number of wall posts indicating providing advice	[35]

Table 9. The seeking or providing of knowledge could be identified by scanning wall posts for text strings like "Does anyone know" or "Does anybody have". These operations may be automated using methods like text mining and machine learning. The employee's embeddedness in the ESN may enable or constrain activities related to knowledge transfer as indicated by the arrows between *bonding or bridging social capital* and *knowledge transfer process*. For instance, Han and Hovav [27] suggest that bonding social capital positively affects knowledge sharing and project performance.

The framework element *achieved knowledge transfer* is the dependent variable and identifies the aggregated knowledge transfer occurring in the observed sample of the ESN, e.g., a team or department within an organization. It is determined by identifying the overall number of information requests and responses and by comparing the ratio of information seeking to information providing [35] (Table 10).

5 Conclusions

Based on social capital theory, this book chapter introduced a theoretical framework to analyze the relationship between an employee's embeddedness in an ESN, his or her knowledge transfer process and the overall knowledge transfer achieved in an organization. Knowledge transfer having been identified as an important enabler for competitive capabilities and advantages, this framework enables researchers and practitioners to investigate knowledge transfer processes within ESN using social network analysis and Computational Intelligence methods.

Companies already using ESN can identify key individuals, e.g., key knowledge sharers, in ESN, and see whether key individuals in the ESN are those who would be expected in key positions and—if that is not the case—take corrective actions, e.g., provide coaching sessions to help these individuals better fulfill their roles [2]. Given that the advantages of using ESN are still unclear, companies could benefit from the results of this study by applying data mining techniques in their ESN to find out how to influence and enhance knowledge transfer in their organizations, e.g., by learning from informal networks detected in the ESN in order to (re)design the formal organizational structure accordingly.

This research offers great potential for future research. Elaborating on this book chapter, future studies could consider in more detail human-centric aspects of social networks, such as the measurement of social capital or the determination of strong and weak ties in ESN. The principles and practice of Computational Intelligence can be expected to play a crucial role in the analysis, design, and interpretation of the functioning of ESN. The classification of bonding or bridging social capital in ESN by using fuzzy logic is only one use case. Furthermore, the analysis of dynamics in ESN, such as the diffusion of information and knowledge and the identification of influential employees using simulations and e.g., the concept of swarm intelligence [36] or neuro-fuzzy systems [37] may be interesting topics to be addressed in further studies.

References

1. ABS: Measuring social capital: An Australian framework and indicators. Australian Bureau of Statistics, Canberra (2004)
2. Allen, J., James, A.D., Gamlen, P.: Formal versus informal knowledge networks in R&D: A case study using social network analysis. R&D Manag. **37**(3), 179–196 (2007)
3. Argote, L., Ingram, P.: Knowledge transfer: a basis for competitive advantage in firms. Organ. Behav. Hum. Decis. Process. **82**(1), 150–169 (2000)
4. Argote, L., Ingram, P., Levine, J.M., Moreland, R.L.: Knowledge transfer in organizations: learning from the experience of others. Organ. Behav. Hum. Decis. Process. **82**(1), 1–8 (2000)
5. Arnaboldi, V., Guazzini, A., Passarella, A.: Egocentric online social networks: Analysis of key features and prediction of tie strength in Facebook. Comput. Commun. **36**, 1130–1144 (2013)
6. Avenarius, C.B.: Starke und Schwache Beziehungen. In: Stegbauer, C., Häußling, R. (eds.) Handbuch Netzwerkforschung, pp. 99–111. VS Verlag für Sozialwissenschaften, Wiesbaden (2010)
7. Barrick, M.R., Mount, M.K., Judge, T.A.: Personality and performance at the beginning of the new millennium: what do we know and where do we go next? Int. J. Sel. Assess. **9**(1&2), 9–30 (2001)
8. Berkman, L.F., Glass, T., Brissette, I., Seeman, T.E.: From social integration to health: Durkheim in the new millennium. Soc. Sci. Med. (1982) **51**(6), 843–857 (2000)
9. BITKOM: Einsatz und Potenziale von Social Business für ITK-Unternehmen. Tech. rep., Bundesverband Informationswirtschaft, Telekommunikation und neue Medien e. V., Berlin (2013)
10. Bodendorf, F., Kaiser, C.: Detecting opinion leaders and trends in online social networks. In: Proceedings of the 2nd workshop on social web search and mining. pp. 65–68. Hong Kong (2009)
11. Borgatti, S.P., Jones, C., Everett, M.: Network measures of social capital. Connections **21**(2), 27–36 (1998)
12. Boyd, D.M., Ellison, N.: Social network sites: definition, history, and scholarship. J. Comput.-Mediat. Commun. **13**(1), 210–230 (2007)

13. Brown, J.S., Duguid, P.: Structure and spontaneity: Knowledge and organization. In: Nonaka, I., Teece, D.J. (eds.) Managing Industrial Knowledge: Creation, Transfer and Utilization, pp. 44–67. SAGE Publications, London (2001)
14. Burke, M., Marlow, C., Lento, T.M.: Social network activity and social well-being. In: Proceedings of the 28th international conference on Human factors in computing systems-CHI'10 p. 1909 (2010)
15. Burt, R.S.: Structural holes versus network closure as social capital. In: Lin, N.,Cook, K.S., Burt, R. (eds.) Social capital: Theory and research, pp. 31–56. Transaction Publishers, New Brunswick (2001)
16. Chan, K., Liebowitz, J.: The synergy of social network analysis and knowledge mapping: a case study. Int. J. Manag. Decis. Mak. **7**(1), 19 (2006)
17. Chang, H.H., Chuang, S.S.: Social capital and individual motivations on knowledge sharing: Participant involvement as a moderator. Inf. Manag. **48**(1), 9–18 (2011) (Enterprise Social Networks and Knowledge Transfer 27)
18. Cross, R., Borgatti, S.P., Parker, A.: Beyond answers: dimensions of the advice network. Soc. Netw. **23**(3), 215–235 (2001)
19. Durst, C., Viol, J., Wickramasinghe, N.: Online social networks, social capital and health-related behaviors: a state-of-the-art analysis. Commun. Assoc. Inf. Syst. **32**(1), Article 5 (2013)
20. Easterby-Smith, M., Lyles, M.A., Tsang, E.W.K.: Inter-organizational knowledge transfer: current themes and future prospects. J. Manage. Stud. **45**(4), 677–690 (2008)
21. Ellison, N., Lampe, C., Steinfield, C., Vitak, J.: With a little help from my friends: Social network sites and social capital. In: Papacharissi, Z. (ed.) A networked self: Identity, community and culture on social network sites, Chap. 6, pp. 124–145. Routledge, New York (2010)
22. Ellison, N., Steinfield, C., Lampe, C.: The benefits of facebook friends: social capital and college students use of online social network sites. J. Comput.-Mediat. Commun. **12**(4), 1143–1168 (2007)
23. Fazeen, M., Dantu, R., Guturu, P.: Identification of leaders, lurkers, associates and spammers in a social network: Context-dependent and context-independent approaches. Soc. Netw. Anal. Min. **1**(3), 241–254 (2011)
24. Fritsch, M., Kauffeld-Monz, M.: The impact of network structure on knowledge transfer: an application of social network analysis in the context of regional innovation networks. Ann. Reg. Sci. **44**(1), 21–38 (2008)
25. Gilbert, E., Karahalios, K.: Predicting tie strength with social media. In: Proceedings of the 27th international conference on Human factors in computing systems CHI'09, pp. 211–220 (2009)
26. Granovetter, M.S.: The strength of weak ties. Am. J. Sociol. **78**(6), 1360–1380 (1973)
27. Han, J., Hovav, A.: To bridge or to bond? Diverse social connections in an IS project team. Int. J. Proj. Manag. **31**(3), 378–390 (2012). doi:10.1016/j.ijproman.2012.09.001
28. Hanneman, R.A., Riddle, M.: Introduction to Social Network Methods. University of California, Riverside (2005)
29. Hansen, M.T.: The search-transfer problem: the role of weak ties in sharing knowledge across organization subunits. Adm. Sci. Q. **44**(1), 82 (1999)
30. Hassan, S., Salgado, M., Pavón, J.: Friendship dynamics: modelling social relationships through a fuzzy agent-based simulation. Discret Dyn Nat Soc. **2011**, 1–19 (2011)
31. Helms, R., Buijsrogge, K.: Knowledge network analysis: A technique to analyze knowledge management bottlenecks in organizations. In: 16th international workshop on database and expert systems applications (DEXA'05), pp. 410–414. IEEE (2005)
32. Hennig, M.: Soziales Kapital und seine Funktionsweise. In: Stegbauer, C., Häußling, R. (eds.) Handbuch Netzwerkforschung, pp. 177–189. VS Verlag für Sozialwissenschaften, Wiesbaden (2010)

33. Hine, C.: Virtual ethnography: Modes, varieties, affordances. In: Fielding, N., Lee, R.M., Blank, G. (eds.) The SAGE handbook of online research methods, Chap. 14, pp. 257–270. SAGE, London (2008)
34. Hogan, B.: Analysing social networks via the internet. In: Blank, G., Lee, R.M., Fielding, N. (eds.) The SAGE handbook of online research methods. Sage Pubn Inc, London (2008)
35. Huang, S., DeSanctis, G.: Mobilizing informational social capital in cyber space: Online social network structural properties and knowledge sharing. In: ICIS 2005 proceedings, p. 18 (2005)
36. Kaiser, C., Kröckel, J., Bodendorf, F.: Ant-based simulation of opinion spreading in online social networks. In: 2010 IEEE/WIC/ACM international conference on web intelligence and intelligent agent technology, pp. 537–540 (Aug 2010)
37. Kaiser, C., Schlick, S., Bodendorf, F.: Warning system for online market research identifying critical situations in online opinion formation. Knowl.-Based Syst. 24(6), 824–836 (2011)
38. Labianca, G.: Group social capital and group effectiveness: the role of informal socializing ties. Acad. Manag. J. 47(6), 860–875 (2004)
39. Lakon, C., Godette, D., Hipp, J.: Network-based approaches for measuring social capital. In: Kawachi, I., Subramanian, S., Kim, D. (eds.) Social Capital and Health, pp. 63–81. Springer, New York (2008)
40. Levin, D.Z., Cross, R.: The strength of weak ties you can trust: the mediating role of trust in effective knowledge transfer. Manage. Sci. 50(11), 1477–1490 (2004)
41. Lewis, K., Kaufman, J., Gonzalez, M., Wimmer, A., Christakis, N.A.: Tastes, ties, and time: a new social network dataset using Facebook.com. Soc. Netw. 30(4), 330–342 (2008)
42. Lin, N.: Building a network theory of social capital. In: Lin, N., Cook, K.S., Burt, R. (eds.) Social capital: Theory and research, Chap. 1, pp. 3–29. Sociology and economics, Transaction Publishers, New Brunswick (2001)
43. Lin, N.: Social Capital: A Theory of Social Structure and Action. Cambridge University Press, Cambridge (2001)
44. Lin, N., Cook, K.S., Burt, R.S.: Social Capital: Theory and Research. Sociology and economics, Transaction Publishers, New Brunswick (2001)
45. Maier, R.: Knowledge management systems: Information and communication technologies for knowledge management. Springer-Verlag New York Incorporated, New York (2004)
46. Marsden, P.V., Campbell, K.E.: Measuring tie strength. Soc. Forces 63(2), 482–501 (1984)
47. Mathews, K.M., White, M.C., Long, R.G., Soper, B., Von Bergen, C.W.: Association of indicators and predictors of tie strength. Psychol. Rep. 83(3), 1459–1469 (1998)
48. McAfee, A.P.: Enterprise 2.0: the Dawn of emergent collaboration. MIT Sloan Manag. Rev. 47(3), 21–28 (2006)
49. McCarty, C.: Measuring structure in personal networks. J. Soc. Struct. 3(1), 1 (2002)
50. Moore, S., Shiell, A., Hawe, P., Haines, V.A.: The privileging of communitarian ideas: citation practices and the translation of social capital into public health research. Am. J. Public Health 95(8), 1330–1337 (2005)
51. Mowery, D.C., Oxley, J.E., Silverman, B.S.: Strategic alliances and interfirm knowledge transfer. Strateg. Manag. J. 17, 77–91 (1996)
52. Nonaka, I.: A dynamic theory of organizational knowledge creation. Organ. Sci. 5(1), 14–37 (1994)
53. O'Reilly, C., Tushman, M.: Using culture for strategic advantage: Promoting innovation through social control. In: Tushman, M., Anderson, P. (eds.) Managing Strategic Innovation and Change: A Collection of Readings, pp. 200–216. Oxford University Press, Oxford (1997) (Enterprise Social Networks and Knowledge Transfer 29)
54. Osterloh, M., Frey, B.S.: Motivation, knowledge transfer, and organizational forms. Organ. Sci. 11(5), 538–550 (2000)
55. Parker, A., Borgatti, S.P., Cross, R.: Making invisible work visible: using social network analysis to support strategic collaboration. Calif. Manag. Rev. 44(2), 25–47 (2002)
56. Petróczi, A., Nepusz, T., Bazsó, F.: Measuring tie-strength in virtual social networks. Connections 27(2), 39–52 (2007)

57. PRI: Measurement of social capital: reference document for public policy research, development, and evaluation. http://www.horizons.gc.ca/doclib/Measurement_E.pdf (2005). Accessed 22 Sept 2011
58. Putnam, R.D.: Bowling Alone. Simon & Schuster, New York (2001)
59. Putnam, R.D., Leonardi, R., Nanetti, R.Y.: Making Democracy Work: Civic Traditions in Modern Italy. Princeton University Press, Princeton, New Jersy (1994)
60. Reagans, R., McEvily, B.: Network structure and knowledge transfer: the effects of cohesion and range. Adm. Sci. Q. **48**(2), 240 (2003)
61. Retzer, S.: Inter-organisational knowledge transfer among research and development organisations: Implications for information and communication technology support. Ph.D. Thesis, Victoria University of Wellington (2010)
62. Retzer, S., Yoong, P., Hooper, V.: Inter-organisational knowledge transfer in social networks: A definition of intermediate ties. Inf. Syst. Frontiers **14**(2), 343–361 (2010)
63. Richter, A., Bullinger, A.C.: Enterprise 2.0-Gegenwart und Zukunft. In: MKWI 2010, pp. 741–753 (2010)
64. Richter, A., Koch, M.: The enterprise 2.0 story in Germany so far. Paper presented at the Workshop "What to expect from Enterprise 3.0: Adapting Web 2.0 to Corporate Reality" in International Conference on Computer-Supported Collaborative Work (2008)
65. Rostila, M.: A resource-based theory of social capital for health research: can it help us bridge the individual and collective facets of the concept?. Soc. Theor. Health **9**(2), 109–129 (2011). doi:10.1057/sth.2011.4
66. Steinfield, C., DiMicco, J., Ellison, N.: Bowling online: Social networking and social capital within the organization. In: Proceedings of the Fourth Communities and Technologies Conference, pp. 245–254 (2009)
67. Steinfield, C., Ellison, N., Lampe, C.: Social capital, self-esteem, and use of online social network sites: a longitudinal analysis. J. Appl. Dev. Psychol. **29**(6), 434–445 (2008)
68. Takahashi, M.: R&D network-case study and social network Analysis. In: ICIS 2012 Proceedings, pp. 1–19 (2012)
69. Toni, A.F.D., Nonino, F.: The key roles in the informal organization: a network analysis perspective. Learn. Organ. **17**(1), 86–103 (2010)
70. Van Der Gaag, M., Webber, M.: Measurement of individual social capital. In: Kawachi, I., Subramanian, S., Kim, D. (eds.) Social Capital and Health, pp. 29–49. Springer, New York (2008)
71. Viswanath, B., Mislove, A., Cha, M., Gummadi, K.P.: On the evolution of user interaction in Facebook. In: Proceedings of the 2nd ACM workshop on online social networks-WOSN'09, p. 37 (2009)
72. Wasserman, S., Faust, K.: Social Network Analysis: Methods and Applications. Structural Analysis in the Social Sciences. Cambridge University Press, Cambridge (1994)
73. Williams, D.: On and off the net: scales for social capital in an online era. J. Comput.-Mediat. Commun. **11**(2), 593–628 (2006)
74. Wilson, C., Boe, B., Sala, A., Puttaswamy, K.P., Zhao, B.Y.: User interactions in social networks and their implications. In: Proceedings of the Fourth ACM European Conference on Computer Systems-EuroSys'09, p. 205 (2009)
75. Xiang, R., Neville, J., Rogati, M.: Modeling relationship strength in online social networks. In: Proceedings of the 19th International Conference on World Wide Web-WWW'10, p. 981 (2010)

A Novel Approach on Behavior of Sleepy Lizards Based on K-Nearest Neighbor Algorithm

Lin-Lin Tang, Jeng-Shyang Pan, XiaoLv Guo, Shu-Chuan Chu and John F. Roddick

Abstract The K-Nearest Neighbor algorithm is one of the commonly used methods for classification in machine learning and computational intelligence. A new research method and its improvement for the sleepy lizards based on the K-Nearest Neighbor algorithm and the traditional social network algorithms are proposed in this chapter. The famous paired living habit of sleepy lizards is verified based on our proposed algorithm. In addition, some common population characteristics of the lizards are also introduced by using the traditional social net work algorithms. Good performance of the experimental results shows efficiency of the new research method.

Keywords Social network analysis (SNA) · K-Nearest neighbor (KNN) algorithm · Sleep lizard · Computational intelligence

1 Introduction

The social network analysis research work which is evolved from psychology, sociology, anthropology and mathematics rises in the 1930s. It is initially used for studying the real relationship among people in society. And it has been continuously developed, especially in the 1930s to the late 1970s; Harvard University and the University of Manchester have made a significant contribution to social

L.-L. Tang (✉) · J.-S. Pan · X. Guo
Harbin Institute of Technology Shenzhen Graduate School, Xili,
Shenzhen 518055, Nanshan, China
e-mail: linlintang2009@gmail.com

S.-C. Chu · J. F. Roddick
School of Computer Science, Engineering and Mathematics,
Flinders University of South Australia, GPO Box 2100
Adelaide 5001, South Australia

W. Pedrycz and S.-M. Chen (eds.), *Social Networks: A Framework*
of Computational Intelligence, Studies in Computational Intelligence 526,
DOI: 10.1007/978-3-319-02993-1_13, © Springer International Publishing Switzerland 2014

network analysis [1]. The Harvard researchers drew algebraic models of groups together and then by making use of the set theory and the multidimensional scaling, they established the concepts such as the strength and distance of connections [2]. In recent years, with the rapid development of computer technology, the application of social network analysis is increasingly being used and has already been extended to the information science.

The basic idea of social network is to establish and analysize the relationship among the members of the social group for finding some potential relation among them which will be used in some special research of social group. There are several classical research methods in the social network analysis research area, such as the Sociogram algorithm, the school of Manchester algorithms and the Structure Hole theory. Each of them has gained a lot of research results which give a great push to the development of the social network analysis.

In fact, the K-Nearest Neighbor (KNN) algorithm [3] which was introduced by Cover and Hart in 1968 is often used for classifying. The basic idea of KNN is to find the K nearest members in the training set. Then the classification of the given member X can be determined by category of these nearest members. The common rule is that X has the same category of the most of the K nearest members. The most commonly used distance measuring criterion is the classical Euclidean distance. It is also used in our chapter. Actually this useful tool has already been widely used in computational intelligence for its important place in classification [4]. In fact, part of the work has already been proposed by authors in Ref. [5] and the new improve work will also introduced here.

The whole chapter is organized as follows. Some related knowledge such as the basic knowledge of the social network will be introduced in Scct. 2. And the simple introduction of KNN algorithm will also be proposed in Sect. 3. Our testing sample and the proposed KNN based algorithms and the simulation results will be shown in Sect. 4. The conclusion is shown in Sect. 5.

2 Related knowledge

2.1 Social Network Analysis

Actually, the Social Network Analysis (SNA) can be described as a research study for structures which consist of at least two social entities (usually more) and the links among them. There are many kinds of relationships which are used for linking, such as the kinship relations, social roles, actions, affective, material exchanges and common behaviors. The graph method and the matrix method are usually used for expressing those structures. The Fig. 1 gives the general idea of such graphs.

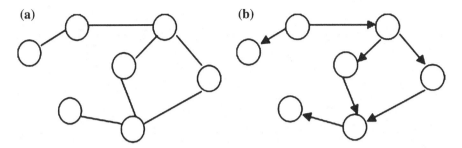

Fig. 1 **a** Non-directive graph. **b** Directive graph

Table 1 Adjacency matrix example

	a	b	c	d	e	f	g
(1) Non-directive matrix							
a	–	1	0	1	0	0	0
b	1	–	1	0	0	1	0
c	0	1	–	0	0	0	1
d	1	0	0	–	0	0	0
e	0	0	0	0	–	1	0
f	0	1	0	0	1	–	1
g	0	0	1	0	0	1	–
(2) Directive matrix							
a	–	0	1	0	0	0	1
b	0	–	0	0	1	0	0
c	0	0	–	1	1	1	0
d	0	0	0	–	0	0	0
e	0	0	0	0	–	0	0
f	0	0	0	1	0	–	0
g	0	0	0	0	0	0	–

Another important method for SNA is the matrix method. The most commonly used matrixes are the adjacency matrix, incidence matrix, distance matrix and the valued matrix. One of the most commonly used and simplest is the adjacency matrix. The adjacency matrix is a $n \times n$ matrix for a given graph G with n nodes. If there is a link between node i and node j, the value in the place (i, j) will be set 1. It is similar for the graph method and the matrix method in classification. In fact, the Matrix method can also be divided into the non-directive and the directive matrix methods based on the different relationship between the different models as shown in Table 1.

Generally speaking, the non-directive matrix is a symmetric matrix and the directive matrix is an asymmetric matrix. And there are some other matrix methods such as the adjacent matrix with weight. The weight value reflects the different correlation intensity. One example is shown in the Table 2.

Table 2 Adjacency matrix example

	a	b	c	d	e	f	g
a	–	0	5	1	0	0	0
b	0	–	5	5	22	13	0
c	5	5	–	0	0	0	3
d	1	5	0	–	0	0	0
e	0	22	0	0	–	9	0
f	0	13	0	0	9	–	7
g	0	0	3	0	0	7	–

There are many useful definitions for the SNA.

Degree: It is the adjacent number of the actor. Geodesics are the shortest path between the two nodes.

Distance: It is the length of the Geodesics.

Diameter: It is the length of the longest Geodesics.

Connected graph: It is a graph that each pair of the nodes in it can be connected.

And the Disconnected graph's definition is on the contrary.

Size: It is the number of a group.

Density: It is the tightness of among the different nodes.

Cohesion: It is the strength between the actors in the group.

Centrality: It gives the location information of the actors.

The most useful two definitions which are used for analyzing the group behavior character are the degree centrality analysis and the middle centrality analysis. The definition can be shown in the following formulas.

$$C_{AD-non}(i) = d(n_i) = \sum_j x_{ij} = \sum_i x_{ji} \tag{1}$$

$$C_{AD-dir}(i) = d(n_i) = \sum_j x_{ij} + \sum_i x_{ji} \tag{2}$$

Here, the above two formulas (1) and (2) represent the absolute non-directive and directive point degree centrality respectively. And $x_{i,j}$ is the path number from other points to the point n_i. $x_{j,i}$ is the path number from other points to the point n_i. In fact, the relative point degree centrality is the standardization of the absolute point degree centrality. And the relative non-directive and the directive point degree centrality can be shown in the following formulas.

$$C_{RD-non}(i) = \frac{d(n_i)}{n-1} \tag{3}$$

$$C_{RD-dir}(i) = \frac{d(n_i)}{2(n-1)} \tag{4}$$

Here, n is the point number in whole graph.

Another useful definition is the point degree central potential. The point degree centrality concerns on the point itself. And the point degree central potential is used to describe the tightness and the central tendency which can be shown as below. C_{Dmax} and C_{Di} are the largest point degree centrality and any other point degree centrality respectively.

$$C_D = \frac{\sum_{i=1}^{n}(C_{Dmax} - C_{Di})}{\max\left[\sum_{i=1}^{n}(C_{Dmax} - C_{Di})\right]} \tag{5}$$

It is the same as the above point degree centrality; the point degree central potential also has two different definitions for the absolute and the relative which are given in the following formula (6) and (7).

$$AD = \frac{\sum_{i=1}^{n}(C_{ADmax} - C_{AD}(i))}{(n-1)(n-2)} \tag{6}$$

$$C_{RD} = \frac{\sum_{i=1}^{n}(C_{RDmax} - C_{RD}(i))}{(n-2)} \tag{7}$$

The betweenness centrality is another indictor for the actor point. Its definition is shown in the following formula (8–10).

$$C_{AB}(i) = \sum_{j}^{n}\sum_{k}^{n} b_{jk}(i), \; j \neq k \neq i, \quad j < k \tag{8}$$

$$C_{RB}(i) = \frac{2C_{AB}(i)}{(n-1)(n-2)} \tag{9}$$

$$b_{jk}(i) = \frac{g_{jk}(i)}{g_{jk}} \tag{10}$$

Here, g_{jk} is the geodesics number between j and k. $g_{jk}(i)$ is the number of geodesics of the third point i which is on the geodesics between j and k. And $b_{jk}(i)$ can been seen as the probability of the point i for controlling the point k and j. n is the point number of the group.

According to the point and centre definition, there is another definition named the central potential shown as below.

$$C_B = \frac{\sum_{i=1}^{n}(C_{ABmax} - C_{AB}(i))}{(n-1)^2(n-2)} = \frac{\sum_{i=1}^{n}(C_{RBmax} - C_{RB}(i))}{n-1} \tag{11}$$

Fig. 2 KNN algorithm graph

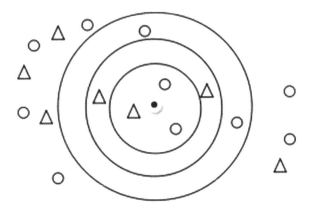

Generally speaking, the commonly used research method for the SNA can be described as follows.

(1) To confirm the research object.
(2) To collect the data through the questionnaire or other methods depends on your own needs.
(3) To establish the corresponding relationship matrix by making use of the relation among the research object.
(4) To deal with or analysize the funded relationship matrix by using the current SNA software or other methods.
(5) To give a conclusion for the research results.

2.2 KNN Algorithm

The KNN algorithm is introduced by the famous Nearest Neighbor (NN) algorithm which is designed to find the nearest one point of the observed object. The main idea for the KNN algorithm is to find the K nearest points and it is a classification algorithm based on learning and analogy.

The Euclidean distance shown in the following formula (12) is often used in KNN. Here, a and c_t are two vectors.

$$d(a, c_t) = \sqrt{\sum_{i=1}^{n} (a_i - c_{ti})^2} \tag{12}$$

The basic idea of KNN algorithm is shown in the Fig. 2. If the K value in the KNN algorithm equals to 3, then the testing point will have the same classification with the round points. And if the K value is 5, then the testing point will belong to the triangle set.

In fact, there are a lot of different improvements for the traditional KNN algorithm, such as the Wavelet based K-nearest neighbor Partial Distance Search

(WKPDS) algorithm which was proposed by Hwang and Wen [6], Equal-average Nearest Neighbor Search (ENNS) algorithm which was introduced in reference [7] by Pan in 2004, Equal-average Equal-norm Nearest Neighbor code word Search (EENNS) algorithm which was shown in reference [8] by Qiao and the Equal-average Equal-variance Equal-norm Nearest Neighbor Search (EEENNS) algorithm given in reference [9] by Lu. In this chapter, we use the ENNS algorithm in our experiment. The basic theory in this algorithm can be described as follows:

Let the sum of the vector a as the following formula (13) shows.

$$S_a = \sum_{i=1}^{N} a_i \tag{13}$$

As we all know that $N \cdot D(A, C_j) \geq (S_A - S_{C_j})^2$. Here N is the dimension of the vector space. A and C_j and the wavelet coefficients vector for a and c_j. If there is following inequality (14) is satisfied, then there is the formation of the formula (15).

$$|S_A - S_{C_j}| \geq \sqrt{N \cdot D(A, C_i)} \tag{14}$$

$$D(A, C_j) \geq D(A, C_i) \tag{15}$$

And if the formula (16) is satisfied, then the inequality (17) will also be satisfied.

$$|S_A - S_{C_j}| \geq \sqrt{\frac{N}{2} D(A, C_i)} \tag{16}$$

$$D(A, C_j) \geq D(A, C_i) \tag{17}$$

The steps of such an algorithm are proposed as follows:

Step 1.   Apply the wavelet transform [10] onto the training set to get the wavelet coefficient vector. And so does the testing vector. Then, the transformed vector should be sorted by comparing the approach coefficients from small to large.

Step 2.   Compare the testing vector with the training vectors and select the point which has the least absolute value between them as the initially center vector. Then we choose K-vectors with the center vector that has the least absolute difference value between the testing vector and it. And the initial distance is chosen as the largest distance between our testing vector and these training K-vectors shown in the following formula (18).

$$D_{\min} = D(A, C_i) \tag{18}$$

Step 3.   For the vectors before the K-vector block, check the inequality (19) shown below to judge if it should be in the K-vector block. And for the vectors behind the K-vector block, check the inequality (20). If they follow the rules respectively, then the K-vectors are the aim group. The searching process comes to an end.

$$S_{c_j} \leq S_a - \sqrt{\frac{N}{2^n}D_{\min}} \qquad (19)$$

$$S_{c_j} \geq S_a + \sqrt{\frac{N}{2^n}D_{\min}} \qquad (20)$$

Step 4.   If the vectors don't follow the rules above respectively, then the distance between the testing vector and them should be compared with the initial largest distance $D_{\min}$ to decide if the vector should be changed into the K-vector block. The comparison algorithm is chosen as the WKPDS algorithm. Then the steps 3 and 4 are repeated until all the vectors in the training set have been checked.

In fact, the KNN algorithm has a lot of advantages for solving some unknown and uniform distribution data and it is also to be applied. And on the other hand, it also has some disadvantages for many reasons. All these properties have been concluded shown below.

Advantages:

(1) The algorithm is simple and easy to be applied.
(2) The training process is simple.
(3) There is no need to give a description for the rule because the training sample itself is the rule. And the noise data is permitted.
(4) It can avoid number imbalance problem of the samples because there is little relation for the non-neighbors.

Disadvantages:

(1) It is difficult to choose a perfect K value.
    Different choice of K value will give a great effect for the final result. And there is no such a good method to choose a perfect K value for one time.
(2) The speed for classification is slow.
    The KNN algorithm is one cased-based lazy learning method which means that every testing sample will be calculated the distance between it and every other sample in the training set. The time complexity is $(m * n)$ where m is the sample dimensions and n is the number of training samples.
(3) It is easily affected by the irrelevant properties of the samples.
    The equal weight values of all the different classification property will result the probably large error for some properties' irrelative.
(4) The result is easily affected by the training samples' distribution.
    For the KNN algorithm is one classification method based on the traditional statistics, the result is mainly dependent on the distribution of the training samples. If the distribution of the training samples is totally non-uniform, the effect of KNN algorithm will be poor.

2.3 KNN Algorithm's Improvement

The basic idea for improving the traditional KNN algorithm can be shown in the following.

(1) To improve the classification speed of the KNN algorithm.
(2) To improve the classification accuracy by improving the distance formulas or the weighted values.
(3) To improve the choice of K value.

In fact, the low efficiency of the traditional KNN algorithm is the distance or similarity calculation between every testing sample with any one sample in the training sample set. And if too many properties of the samples, when calculating the distance between the samples, its time complexity will of course increase, and the algorithm speed will be slow. But if the comparison samples' former several property distance can judgment out if the sample is one of the k neighbors, and if there is one method which can classify if the sample is one of the K neighbors and need not calculate the distance between the samples which are waiting for test and the training samples. Then the efficiency of the KNN algorithm will be improved. Based on the above ideas, several improved methods for the KNN algorithm can be introduced in the following.

(1) Haar wavelet transform

Firstly, the definition of the Haar wavelet transform is proposed and the basic idea for Haar wavelet transform is shown in the following formula (21). Here, A is gotten from the vector $a = \{a_1, a_2, \ldots, a_N\}$ and N is the dimension value. $A_{S_{j,n}}$ and $A_{d_{j,n}}$ are the approximate and the detailed coefficients for the vector a after jth transform respectively.

$$\begin{cases} A_{S_{j,n}} = \frac{1}{\sqrt{2}} \left[A_{S_{j-1,2n}} + A_{S_{j-1,2n+1}} \right] \\ A_{d_{j,n}} = \frac{1}{\sqrt{2}} \left[A_{S_{j-1,2n}} - A_{S_{j-1,2n+1}} \right] \end{cases} \tag{21}$$

When $2^{k-1} < N \leq 2^k$ and after k times wavelet transform, the vector x can be totally decomposed and the only approximation coefficient will be $C_a = \frac{N \cdot M_a}{\sqrt{2^k}}$ and $M_a = \frac{1}{N} \sum_{i=1}^{N} a_i$. The approximation coefficients and the detailed coefficients construct the wavelet coefficients vector. For the Haar wavelet transform is an orthogonal transformation, so for any two vectors a and b, we all have b and $\|a\| = \|A\|$. Here, A and B are the wavelet coefficients of the a and b, $d(a, b)$ is the Euclid distance between a and b. So, it is the same for finding k neighbors in the original domain and in the wavelet domain.

(2) WKPDS algorithm

A KNN method based on the part distance research in wavelet transform was proposed by Hwang and Wen named Partial Distance Search (WKPDS) its main process is shown below.

Suppose $D_{\min} = D(A, C_t) = \sum_{i=1}^{n} (A_i - C_{ti})^2$ is the max distance among the k neighbors of the sample a to be tested. Here, A and C_t are the wavelet coefficients of the vector a and c_t.

In fact, the main idea of the WKPDS algorithm is first to transform the training sample c_j and the testing sample a with Haar wavelet transform and then to get the wavelet coefficients vectors C_j and A. Then secondly to calculate the former h dimensions' square distance between A and C_j. It can be described as $\sum_{i=1}^{h} (A_i - C_{ji})^2$.

If it is satisfied with the condition $\sum_{i=1}^{h} (A_i - C_{ji})^2 \geq D_{\min}$, $1 \leq h \leq n$ then we can have $D(A, C_j) \geq D(A, C_t)$. It mains that c_j is not one of the k neighbors of the testing sample a for the distance between them is larger than the max distance between the testing sample a and the current k neighbors. If when $h = n$, the above inequality is still not set up, then c_j can instead c_t of to be one of the current k neighbors and to calculate the $D_{\min}$ value again. Use this method until to find the k neighbors of a.

Although, there are many advantages for this method, there are still some thing can be done to improve the efficiency of such a KNN algorithm.

(3) ENN algorithm

According to the comparison between the testing sample and the all training samples, one new algorithm named "Equal-Average Nearest Neighbor Search, ENNS" was proposed by Jeng-Shyang Pan in 2005.

Suppose the sum of all the elements for vector x is $S_a = \sum_{i=1}^{N} a_i$. It is easy for us to prove the following inequality. Here, A and C_j are the wavelet coefficients of vector a and c_j.

$$N \cdot D(A, C_j) \geq (S_a - S_{c_j})^2 \tag{22}$$

If $|S_a - S_{c_j}| \geq \sqrt{N \cdot D(A, C_i)}$ then we have $D(A, C_j) \geq D(A, C_i)$. Considering the mean value formula and the approximation formula for coefficients, then we can have $D(A, C_j) \geq D(A, C_i)$.

The main idea of the ENNS algorithm is firstly to transform the training samples with Haar wavelet, then sort from smallest to largest based on the value of the approximation coefficients. Secondly, transform the testing sample a with the Haar wavelet and find the number of the sample which has the minim distance between their approximation coefficients. Then the chosen sample can be defined as the centre and initialize k neighbor samples. The process can be shown in the following.

1. Suppose p is the number of the sample with minim distance and S is the number of the training samples. The sorting number is j_i, $i = 1, 2, \ldots, k$ after the initializing the k neighbor samples.
2. If $p - \lfloor k/2 \rfloor < 1$, then $j_i = i$.

3. If $p - \lfloor k/2 \rfloor > S - k + 1$, then $j_i = S - k + i$.
4. Otherwise, then $j_i = p - \lfloor k/2 \rfloor + i$.

In fact, let the $D_{\min} = D(A, C_i)$ be the square value of the maxim distance between the testing sample a and the current k neighbors and if it satisfies $S_{c_j} \leq S_a - \sqrt{\frac{N}{2^n} D_{\min}}$ or $S_{c_j} \geq S_a + \sqrt{\frac{N}{2^n} D_{\min}}$, then we can have $D(A, C_j) \geq D(A, C_i)$. Then it is easy to judge that c_j is not belong to the k neighbor samples and there is no need to be compared again.

For the training samples have been sorted from small to large, so if the approximation coefficient with number j satisfies j or $S_{c_j} \geq S_a + \sqrt{\frac{N}{2^n} D_{\min}}$, then the numbers before j or after j satisfy these inequalities, too. It means that the sample before or after it is no possible to be one of the k neighbors of the testing sample a. And if the approximation coefficient can not satisfy either of the two inequalities, we can use the WKPDS algorithm to judge the relation between the distance between c_j and a and $\sqrt{D_{\min}}$.

(4) EENNS algorithm

After the ENNS algorithm, Yulong Qiao proposed another new KNN improve algorithm named the Equal-Average Equal-Variance Nearest Neighbor Search (EENNS). It is an improve for the above ENNS algorithm and the efficiency is obviously improved. Its main process is shown in the following.

1. Let the variance of the vector a be $V_a = \sqrt{\sum_{i=1}^{n} (a_i - M_a)^2}$. It is easy to prove that $D(A, C_j) \geq (V_A - V_{C_j})^2$. Here, the A and C_j are the wavelet coefficients of the vectors a and c_j.
2. If it is satisfied with $(V_A - V_{C_j})^2 \geq D(A, C_i) \Leftrightarrow |V_A - V_{C_j}| \geq \sqrt{D(A, C_i)}$, then we have $D(A, C_j) \geq D(A, C_i)$.

In fact, the main idea of EENNS algorithm is to some thing similar to the ENNS algorithm. After the initializing step, we can judge the training sample coefficient S_{c_j} satisfy the inequality $S_{c_j} \leq S_a - \sqrt{\frac{N}{2^n} D_{\min}}$ (or $S_{c_j} \geq S_a + \sqrt{\frac{N}{2^n} D_{\min}}$) or not. If it is satisfied, then the sample before c_j (or after c_j) is impossible to be one of the k neighbors of the testing vector a. Otherwise, we can continue the size judgment of the variance value V_{C_j} of the training sample c_j's wavelet coefficient C_j. If it is satisfied with the inequality $V_{C_j} \leq V_A - \sqrt{D_{\min}}$ or $V_{C_j} \geq V_A + \sqrt{D_{\min}}$, then we have c_j is impossible one of the k neighbors of the testing vector a. So much more other computing time can be saved. If the above judge methods do not work, then we can make use of the WKPDS algorithm to judge the size relation between the distance of c_j and a and $\sqrt{D_{\min}}$.

Table 3 Part of the data after process

Easting	Northing	Month	Day	Hours	Minute
991	976	9	20	17	30
974	970	9	21	9	30
977	977	9	21	9	40
971	971	9	21	9	50
978	973	9	21	11	40
986	989	9	21	12	40
982	978	9	21	13	20

(5) EEENNS algorithm

An improvement work was proposed by Z. M. Lu and S. H. Sun based on the EENNS algorithm named the Equal-Average Equal-Variance Equal-Norm Nearest Neighbor Search (EEENNS). Its main process can be shown as below.

1. Suppose the norm value of vector a is $\|a\| = \sqrt{\sum_{n=1}^{k} a_n^2}$.
2. Let $norm_{max} = \|A\| + \sqrt{D_{min}}$, $norm_{min} = \|A\| - \sqrt{D_{min}}$.
3. If the wavelet coefficient norm value of the vector c_j satisfy the inequality $\|C_j\| \geq norm_{max}$ or $\|C_j\| \leq norm_{min}$, then we have $D(A, C_j) \geq D_{min}$.

The main idea of EEENNS is to use the same initial step and then to judge whether $S_{c_j} \leq S_a - \sqrt{\frac{N}{2^n} D_{min}}$ or $S_{c_j} \geq S_a + \sqrt{\frac{N}{2^n} D_{min}}$, $V_{C_j} \leq V_A - \sqrt{D_{min}}$ or $V_{C_j} \geq V_A + \sqrt{D_{min}}$, $\|C_j\| \geq norm_{max}$ or $\|C_j\| \leq norm_{min}$. If any one of these inequalities is not satisfied, then the c_j is impossible one of the k neighbors of a.

3 Introducing and Preprocessing for Our Sleepy Lizard Samples

3.1 Basic Sample Information Introduction

Professor Stephan T. Leu and his colleagues in Flinders University got the living data about the 55 sleepy lizards that live in one area near the Bundey Bore Station after three months observation from Sept. 15th to Dec. 15th 2009 and they have also done much work based on the observed data [11, 12]. There are three prime contents in this record: body temperature in every 2 min, step number in 2 min and the location information in every 10 min. We take the location information as an example which is shown in the Table 3.

The first two columns in the Fig. 3 are the X-coordinate values and the Y-coordinate values after some reprocess respectively. Considering the living

Fig. 3 The time-location
graph for lizards

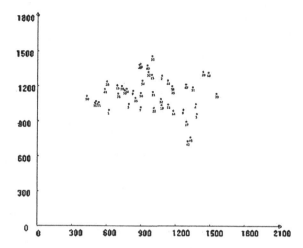

property of sleepy lizards and the accuracy of our equipments, 14 m are chosen for
the threshold to judge the encounter area for two sleepy lizards. If we take the
encounter matter as a basic information transmission relationship, then the communication relation between different sleepy lizards can be got.

3.2 The Activities Scope and Time Analysis of the Sleepy Lizards

When people observe a group of animals, the commonly beginning is the active
space and active time. Based on the long-time observes, the scientists can easily
give a report for the living law of one kind animals.

Firstly, quantization preprocess is done on these data for the location information which is recorded by the Universal Transverse Mercator (UTM) is so large
that something must be done to translate them into smaller ones. The processed
data is shown in the Table 3. Here, the Easting and Northing is position coordinates and the unit is m.

3.2.1 The Activities Scope Analysis

The location information which is got by the GPS is recorded every 10 min. So if
there is no record in one 10 min, we can believe that the observed lizard does not
move in this 10 min and the location information is the same with the above one.

In order to give a clear analysis for this group of lizards, we can give a detail
graph of the every time location for the whole group as shown in the Fig. 3. It is
the location information at 15:30 on 17th September.

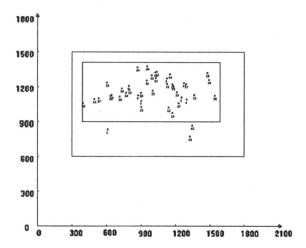

Fig. 4 The activities scope graph for lizards

And the activities scope graph also can be got by the same way. One example is shown in the Fig. 4. It is the activities scope information at 8:50 on 15th September.

Generally speaking the lizards are such one kind of animals who do not like move from time to time based the above analysis graphs. This research result is consistent with the well known lizard life characteristics.

3.2.2 The Activities Time Analysis

In order to get a more clear life characteristic, the moving number for lizard group is recorded in this chapter and the activities time during which the moving number is above some threshold is also recorded. And one of the example graphs is shown as below.

Here, take the Fig. 5a for an example. The horizontal axis represents hours and minutes in the day, the vertical axis represents month and day (from September 15 to December 15). If the group's number of lizard is active in more than a certain percentage at some point, correspond to this moment in the day and the time between 10 min before time to draw a line, it means that in 10 min more than the proportion of lizard has moved.

3.3 The Lizard Society Analysis Based on the SNA

It has been known that the average habit area for one lizard is 4 m². So we can suppose the two lizards have communicated one time when the distance between them is within 2 m in our chapter. And based on the analysis of Stephan T. Leu, the average error for the GPS is 6 m. Then we can generally believe the two lizards have communicated one time when the distance between them within 14 m.

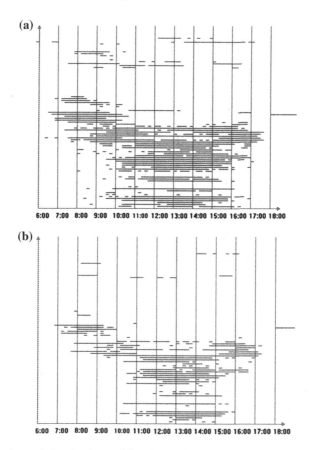

Fig. 5 Time-rule graph for the observed lizard group. **a** The over half activity segments graph. **b** The over two-thirds activity segments graph

3.3.1 To Fund a Network Model

In order to get more information from the action of lizards, we use the communication relation between lizards as the basic relationship to fund the relation matrix. And the communication times are taken as the strength.

Based on the Table 1, we can get a communication matrix for those lizards in the Fig. 6.

3.3.2 The Analysis for the Active Time Rule

For a more deep study of the active time rule of lizards, we use the following methods during our research:

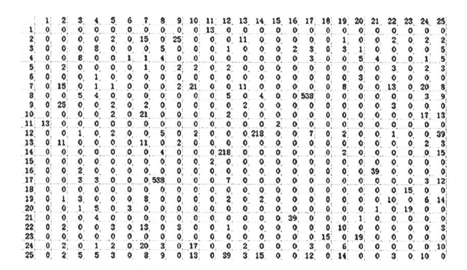

	1	2	3	4	5	6	7	8	9	10	11	12	13	14	15	16	17	18	19	20	21	22	23	24	25
1	0	0	0	0	0	0	0	0	0	0	0	13	0	0	0	0	0	0	0	0	0	0	0	0	0
2	0	0	0	0	2	0	15	0	25	0	0	0	11	0	0	0	0	0	1	0	0	2	0	2	2
3	0	0	0	8	0	0	0	5	0	0	0	1	0	0	0	2	3	0	3	1	0	0	0	0	5
4	0	0	8	0	0	1	1	4	0	0	0	0	0	0	0	3	0	0	5	4	0	0	0	1	5
5	0	2	0	0	0	1	0	2	2	0	2	0	0	0	0	0	0	0	0	0	3	0	0	2	3
6	0	0	0	1	0	0	0	0	0	0	0	0	0	0	0	0	0	0	0	3	0	0	0	0	0
7	0	15	0	1	1	0	0	0	2	21	0	0	11	0	0	0	0	0	8	0	0	13	0	20	8
8	0	0	5	4	0	0	0	0	0	0	5	0	4	0	0	538	0	0	0	0	0	0	0	3	9
9	0	25	0	0	2	0	2	0	0	0	0	2	0	0	0	0	0	0	0	0	3	0	0	0	0
10	0	0	0	0	2	0	21	0	0	0	0	2	0	0	0	0	0	0	0	0	0	0	0	17	13
11	13	0	0	0	0	0	0	0	0	0	0	0	0	2	0	0	0	0	0	0	0	0	0	0	0
12	0	0	1	0	2	0	0	5	0	2	0	0	0	218	0	0	7	0	2	0	0	1	0	0	39
13	0	11	0	0	0	0	11	0	2	0	0	0	0	0	0	0	0	0	0	0	0	0	0	2	3
14	0	0	0	0	0	0	0	4	0	0	0	218	0	0	0	0	0	2	0	0	0	0	0	0	15
15	0	0	0	0	0	0	0	0	0	0	2	0	0	0	0	0	0	0	0	0	0	0	0	0	0
16	0	0	2	0	0	0	0	0	0	0	0	0	0	0	0	0	0	0	0	39	0	0	0	0	0
17	0	0	3	3	0	0	0	538	0	0	0	7	0	0	0	0	0	0	0	0	0	0	0	3	12
18	0	0	0	0	0	0	0	0	0	0	0	0	0	0	0	0	0	0	0	0	0	15	0	0	0
19	0	1	3	0	0	0	8	0	0	0	0	2	0	2	0	0	0	0	0	0	10	0	0	6	14
20	0	0	1	5	0	3	0	0	0	0	0	0	0	0	0	0	0	0	1	0	19	0	0	0	0
21	0	0	0	0	0	0	0	0	0	0	0	0	0	39	0	0	0	1	0	0	0	0	0	0	0
22	0	2	0	0	3	0	13	0	3	0	0	1	0	0	0	0	0	0	10	0	0	0	0	0	3
23	0	0	0	0	0	0	0	0	0	0	0	0	0	0	0	0	15	0	19	0	0	0	0	0	0
24	0	2	0	1	2	0	20	3	0	17	0	0	2	0	0	3	0	6	0	0	0	0	0	0	10
25	0	2	5	5	3	0	8	9	0	13	0	39	3	15	0	0	12	0	14	0	0	3	0	10	0

Fig. 6 Part of the communication matrix between sleepy lizards for our samples

(1) To divide the one-day time into three periods (0:00–8:00, 8:00–16:00, 16:00–24:00). Based on the analysis for communication in every period, we can find the most communication has been completed during 8:00–16:00.

(2) To divide the whole observing time from the 15th Sept. to 15th Dec. into two periods as the first two months and the last month. Based on the whole statistic data for communication, we can find the most communication has occurred in the first two months.

The Fig. 7 gives the communication matrix between 8:00 and 16:00 in the first two months. As we can see that it has little difference from the whole observed data in three months.

Based on the above analysis, we can draw a conclusion for such a kind of animal: its main active time is from 8:00 and 16:00 in the first two months. And this conclusion also proved that the well known living properties of lizards.

3.3.3 The Centrality Analysis

After the global analysis, we pay some attention to the individuals. We will give some research results to see the effects of the important role for the whole group.

The classical degree centrality analysis and the betweenness centrality analysis graphs are shown in the Fig. 8.

The mean purpose of such graphs is to find the different social status and different resource controlling ability for every sleepy lizard in the samples.

	1	2	3	4	5	6	7	8	9	10	11	12	13	14	15	16	17	18	19	20	21	22	23	24	25
1	0	0	0	0	0	0	0	0	0	0	13	0	0	0	0	0	0	0	0	0	0	0	0	0	0
2	0	0	0	0	0	0	11	0	17	0	0	0	11	0	0	0	0	0	1	0	0	2	0	1	0
3	0	0	0	4	0	0	0	0	5	0	0	0	1	0	0	0	2	3	0	3	1	0	0	0	5
4	0	0	4	0	0	0	1	1	4	0	0	0	0	0	0	0	0	3	0	0	4	4	0	1	5
5	0	0	0	0	0	0	0	0	1	0	0	2	0	0	0	0	0	0	0	0	0	2	0	0	0
6	0	0	0	1	0	0	0	0	0	0	0	0	0	0	0	0	0	0	0	2	0	0	0	0	0
7	0	11	0	1	0	0	0	0	1	15	0	0	8	0	0	0	0	0	8	0	0	6	0	13	7
8	0	0	5	4	0	0	0	0	0	0	0	5	0	3	0	0	491	0	0	0	0	0	0	3	9
9	0	17	0	0	1	0	1	0	0	0	0	0	2	0	0	0	0	0	0	0	0	3	0	0	0
10	0	0	0	0	0	0	15	0	0	0	0	0	0	0	0	0	0	0	0	0	0	0	0	10	5
11	13	0	0	0	0	0	0	0	0	0	0	0	0	0	2	0	0	0	0	0	0	0	0	0	35
12	0	0	1	0	2	0	0	5	0	0	0	0	199	0	0	0	7	0	2	0	0	0	0	0	35
13	0	11	0	0	0	0	8	0	2	0	0	0	0	0	0	0	0	0	0	0	0	0	0	2	1
14	0	0	0	0	0	0	0	3	0	0	0	199	0	0	0	0	0	0	2	0	0	0	0	0	14
15	0	0	0	0	0	0	0	0	0	0	0	2	0	0	0	0	0	0	0	0	0	0	0	0	0
16	0	0	2	0	0	0	0	0	0	0	0	0	0	0	0	0	0	0	0	28	0	0	0	0	0
17	0	0	3	3	0	0	0	491	0	0	0	7	0	0	0	0	0	0	0	0	0	0	0	2	11
18	0	0	0	0	0	0	0	0	0	0	0	0	0	0	0	0	0	0	0	0	0	0	0	0	0
19	0	1	3	0	0	0	8	0	0	0	0	2	0	2	0	0	0	0	0	0	0	8	0	4	10
20	0	0	1	4	0	2	0	0	0	0	0	0	0	0	0	0	0	0	0	0	1	0	16	0	0
21	0	0	0	4	0	0	0	0	0	0	0	0	0	0	0	28	0	0	0	1	0	0	0	0	0
22	0	2	0	0	2	0	6	0	3	0	0	0	0	0	0	0	0	0	8	0	0	0	0	0	3
23	0	0	0	0	0	0	0	0	0	0	0	0	0	0	0	0	0	7	0	16	0	0	0	0	0
24	0	1	0	1	0	0	13	3	0	10	0	0	2	0	0	0	2	0	4	0	0	0	0	0	7
25	0	0	5	5	0	0	7	9	0	5	0	35	1	14	0	0	11	0	10	0	0	3	0	7	0

Fig. 7 The first two months' communication matrix

Here, the UCINET (a windows software product) will translate the communication strength to 1 when do the pair communication betweenness centrality analysis. The computation formula is shown as below.

$$C_{AB}(i) = \sum_{j}^{n} \sum_{k}^{n} b_{jk}(i), \ j \neq k \neq i \quad \text{and } j < k, \ b_{jk}(i) = \frac{g_{jk}(i)}{g_{jk}} \tag{23}$$

As we can see from the Fig. 8, communication strength for every sleepy lizard and places they take during the communication are shown clearly. For example, though the 39th sleepy lizard's degree centrality is low which means communication strength is weak, the betweenness centrality is high which means the place it takes is very important.

3.4 Paired Living Analysis

The famous paired living property of sleepy lizards can also be studied by a relation graph based on the communication definition mentioned above which is shown in the Fig. 9a. And when the threshold value is large enough, the paired living character can also be verified in Fig. 9b.

As we can see from the Fig. 9, the famous paired living character can be verified from the communication relationship.

Further more, when the threshold value is 60, the communication graph can be shown in the Fig. 10.

As we can see from the Fig. 10, all the left relation is almost the one to one except for the lonely points when the threshold value is large enough.

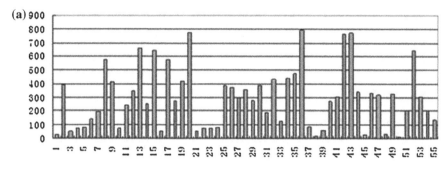

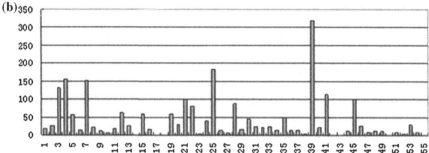

Fig. 8 **a** The degree centrality analysis graph and **b** the betweenness centrality analysis graph

4 Our Proposed Method

4.1 One Behavior Research Method Based on the KNN Algorithm

We introduce the location information based KNN method to study the intimacy and even to verify the paired living character in this chapter. The reason for us to do so is based on the pre-analysis of this kind of animal whose motion range is relative fixed.

The steps of our proposed method basically follow the ENNS algorithm mentioned above. And the location information which is a two demotions vector is used for one sleepy lizard sample. The experimental result example for a sample 42nd sleepy lizard's K nearest friends can be shown in the Fig. 11.

As we can see that the 43rd is the most 'nearest' for the 42nd sample. And the 3-nearest frequency simulation results for the 43rd shown in Fig. 12 can help us to verify the paired living character.

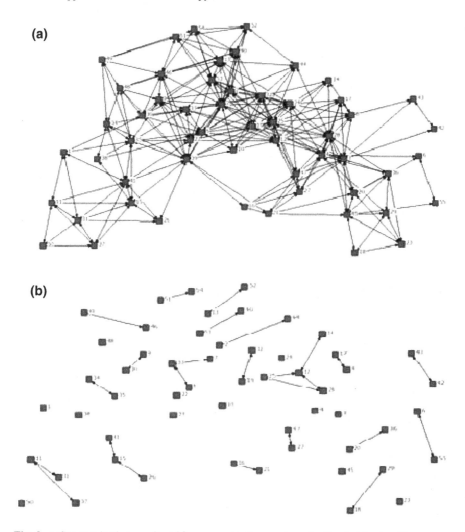

Fig. 9 a Communication graph and **b** communication graph under threshold value 30

4.2 Intimacy Based Algorithm Analysis of Sleepy Lizard Living Property

From the group point of view, the KNN based algorithm analyzes the recent distance K sample individual for the tested sample one time. But the distance of KNN is relative. KNN based algorithm does not take into consideration of the actual distance to the analysis result influence. So this chapter also introduces a kind of intimate degree based algorithm for the sleepy lizard living analysis method.

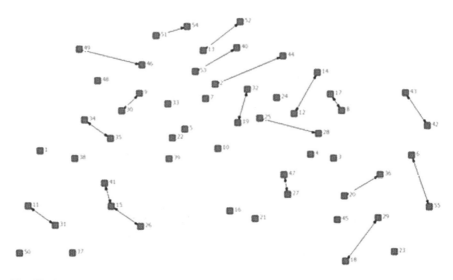

Fig. 10 Communication graph under threshold value 30

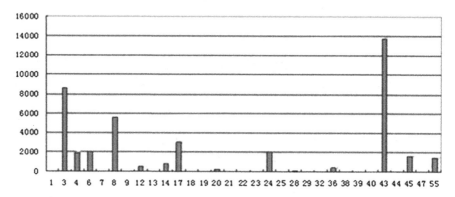

Fig. 11 3-nearest frequency for the 42nd sample based on KNN algorithm

4.3 Intimate Degree Based Algorithm

This chapter proposes the intimate degree based algorithm on the basis of actual distance to analyze the intimacy between sleepy lizards. The closer of the actual distance, the greater the intimate degree value, and vice versa.

The intimate degree algorithm procedure can be shown as follows:

Step 1.   To initialize begin to end time for the algorithm analysis and to set the tested sample. To initialize each intimate degree value between tested samples to the rest of sample of the whole time. And set current time to the starting time of the algorithm analysis.

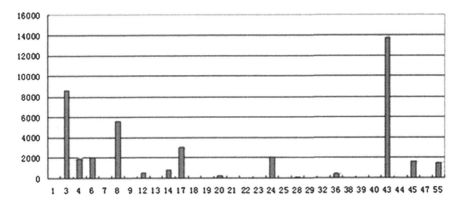

Fig. 12 3-nearest frequency for the 43rd sample based on KNN algorithm

Step 2.   If the current time is larger the end time, go to Step 6; if not, go to Step 3;
Step 3.   To compute the intimate degree values among the entire instances at current time. And for each line of intimate degree is normalized.
Step 4.   To update intimate degree between tested samples to the rest sample during the whole time and normalize them.
Step 5.   Update the current time. And then go to Step 2.
Step 6.   Output the analysis result.

4.3.1 Analysis Result of Intimate Degree Algorithm

Consider a sleepy lizard (such as the 43rd sample) as the observed object. Analyze its intimate degree with the others in this small group.

Suppose the adjustable parameter σ to be 6,000, the algorithm analysis results can be shown as below.

The experimental result shows that the 43rd and the 42nd sleepy lizard have the maximum intimate degree value. The relationship between them is also the most intimate. Consider the 42nd sample as the training sample. After the intimate analysis, we can get the result.

The experimental result shows that the 42nd and the 43rd sleepy lizard have the maximum intimate degree value, which is also far greater than other value. It is to say that their relationship is the most intimate one. As we can see from the Figs. 13 and 14, the 42nd and the 43rd sleepy lizards only have the most intimate relationship. Because of the arbitrary of choosing research object, we can come to the conclusion that sleepy lizards have the paired living property.

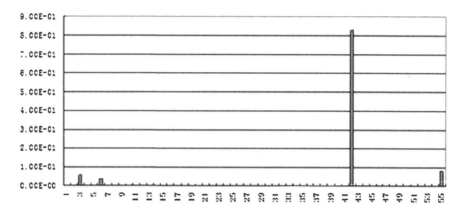

Fig. 13 Intimate relationship between the 43rd sample and other sleepy lizards

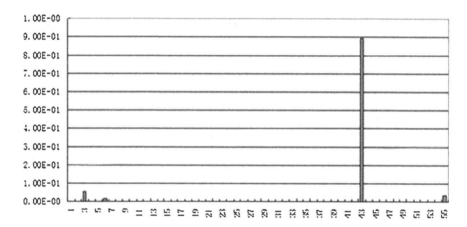

Fig. 14 Intimate relationship between the 42nd sample and other sleepy lizards

4.4 KFN Based Sleepy Lizard Affinity-Non-affinity Relationship Analysis

4.4.1 KFN Algorithm Description

The so called KFN algorithm is the abbreviation of the K-Farthest Neighbors algorithm, as the name suggests, this algorithm is to study the farthest K samples from the tested sample focus on the training samples. On the contrary that KNN is to find the nearest K samples.

It has been many years for the appearance of KNN algorithm, which apply in every field including text classification, cluster analysis, and so on. While the KFN algorithm rarely known by people. The KFN algorithm has slowly draw people's attention for its ability to solve the real problems in applications.

As known to us all, the nearest algorithm proposed firstly to find the nearest point to one testing point; it is used to analyze the relationship between the two so as to classify the category. In order to avoid the limitation and the low accuracy problems, the KNN algorithm is proposed.

The KFN algorithm is the opposite of the KNN algorithm, a product of reverse thinking of KNN algorithm. The two algorithms have some similarity and some differences.

The similarities are:

(1) The construction manner of the training instance sets are the same.
(2) The same way to set the initial K value.
(3) There is no need extra data to describe the rule, training instance is their rule.
(4) Both of the two methods are based on the distance or similarity as a measure of the affinity and non-affinity relationship between samples.

The differences are:

(1) The KNN algorithm is to find the K nearest samples of the testing sample and to judge the tested sample category based on the K nearest samples' categories. Its purpose is to classify or cluster. While the KFN algorithm is to determine the sample activities scope or research ranges through finding the K farthest samples.
(2) Due to the different purpose of the two algorithms, the selection of K value will bring them different influence. For the KNN algorithm, selection of K value decides the accuracy of classification. For the KFN algorithm, selection of K value has an accuracy effect on the activities scope.

The analysis process of KFN algorithm is similar to the KNN algorithm, its research steps can be summarized as the follows:

(1) To construct the training instance set T.
(2) To select the K value.
(3) To find the K farthest samples to the testing sample.
(4) To analyze the Practical significance based on the experiment result.

There are many real applications of the KFN algorithm. For example, for some a supermarket, consider the consumers of this supermarket as the research objects, by using the KFN algorithm to analyze the K persons live in the farthest from the supermarket who consume in it. Then we can probably get the business scale of the supermarket.

4.4.2 KFN Algorithm Experiment Results

The same as the intimate degree based algorithm to analyze the intimacy between sleepy lizards based on the KNN algorithm, a method based on KFN algorithm named the KFN Based Sleepy Lizard Affinity-Non-affinity Relationship Analysis method is also proposed here.

Some experiment results are shown in the Fig. 15.

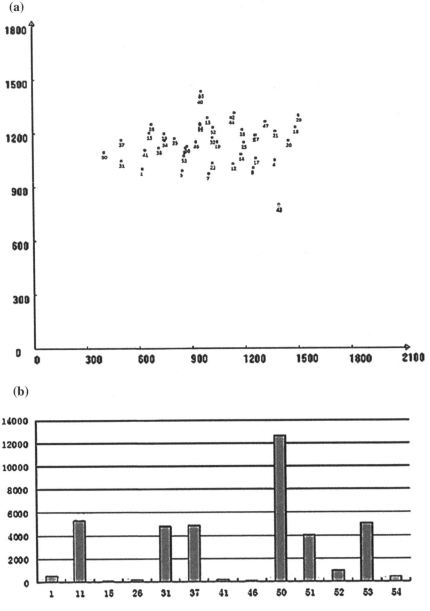

Fig. 15 KFN algorithm experiment results for our lizard group. **a** The KFN graph at one time. **b** 3 Farthest neighbors for the 43rd sample

5 Conclusion

This chapter proposes a new research and its improvement for a small sleepy lizard group. By introducing the traditional KNN algorithm, a good performance for the relationship between the samples can be given. Even the famous paired living character also can be verified. To make full use of the data base and find more new living habits of such a small group sleepy lizards is our future work.

Acknowledgments The authors would like to thank the reviewers for providing very helpful comments and suggestions. The authors would also like to thank for the support from the project named Research on Multiple Description Coding Frames with Watermarking Techniques in Wavelet Domain which belongs to the NSFC (National Natural Science Foundation of China) with the Grant number 61202456.

References

1. Granovetter, M.: Getting a Job: A Study of Contacts and Careers. Harvard University Press, Cambridge (1974)
2. Granovertter, M.: The strength of weak ties. Am. J. Sociology **78**(6), 1360–1380 (1973)
3. Kuncheva, L.: Fitness function in editing KNN reference set by genetic algorithms. Pattern Recognit. **30**(6), 1041–1049 (1997)
4. Berton, L., de Andrade Lopes, A.: Informativity-based graph : exploring mutual KNN and labeled vertices for semi-supervised learning. In: Fourth International Conference on Computational Aspects of Social Networks (CASoN), pp. 1120–1123, August 2012
5. Guo, X., Chu, S.-C., Tang, L.-L., Roddick, J. F., Pan, J.-S.: A research on behavior of sleepy lizards based on KNN algorithm. In: IEEE the 4th Asian Conference on Intelligent Information and Database System (ACIIDS-2012), Taiwan, pp. 109–118, March 2012
6. Hwang, W.J., Wen, K.W.: Fast KNN Classification Algorithm Based on Partial Distance Search. IEEE Trans. Comput. **24**(7), 750–753 (1975)
7. Pan, J.S., Lu, Z.L., Sheng, H.S.: An efficient encoding algorithm for vector quantization based on subvector technique. IEEE Trans. Image Process. **12**(3), 265–270 (2003)
8. Pan, J.S., Qiao, Y.L., Sun, S.H., A fast k nearest neighbors classification algorithm. IEICE Trans. Fundam. **E87-A**(4), 961–963 (2004)
9. Lu, Z.M., Sun, S.H.: Equal-average equal-variance equal-norm nearest neighbor search algorithm for vector quantization. IEICE Trans. Inf. Syst. **86**(3), 660–663 (2003)
10. Hassana, G.K., Shaker, K.A., Zou, B.J.: Optimal approach for texture analysis and classification based on wavelet transform and neural network. J. Inf. Hiding Multimedia Sig. Process. **2**(1), 33–40 (2011)
11. Stephan, T.L., Bashford, J., Kappeler, P.M., Bull, C.M.: Association networks reveal social organization in the sleepy lizard. Anim. Behav. **79**(1), 217–225 (2010)
12. Stephan, T.L., Kappeler, P.M., Bull, C.M.: Refuge sharing network predicts ectoparasite load in a lizard. Behav. Ecol. Sociobiol. **64**(9), 152–160 (2010)

An Evaluation Fuzzy Model and Its Application for Knowledge-Based Social Network

Hyeongon Wi and Mooyoung Jung

Abstract Knowledge-based organizations (KBOs) such as universities, research institutes, and research centers at businesses and industries manage their projects in pursuit of the goals of their organizations. It is well understood that collaboration becomes one of the most important factors for successful completion of these projects. In performing a project jointly, it is important for project team members to know who has the required knowledge. Thus it is imperative to assess what and how much the employees know. Using a knowledge-based social network and its basic approach, a new method is proposed to analyze the knowledge and collaboration collectively possessed by the employees of a KBO. This study first deal with the methodologies of evaluating knowledge and collaboration possessed both by individuals and the organization. Since the quantitative evaluation is essential in developing a series of evaluating methods, the measures of knowledge and collaboration are derived. Then, the knowledge network types of KBO and the network roles of KBO members are discussed. Four types of knowledge-based social network and four roles of network members are also discussed respectively. An evaluation fuzzy model is proposed to test the feasibility of the knowledge-based social network and its measures. A case study is used to demonstrate effectiveness of the proposed model.

Keywords Knowledge-based social network · Knowledge competence · Collaboration competence · Knowledge evaluation fuzzy model · Familiarity evaluation fuzzy model

H. Wi
Technology Commercialization Group, Research Institute of Industrial Science and Technology (RIST), Hyoja San 32, Pohang-si, Pohang 790-330 Republic of Korea
e-mail: whg@rist.re.kr

M. Jung (✉)
School of Technology Management, Ulsan National Institute of Science and Technology (UNIST), UNIST-gil 50, Ulsan, Ulsan 689-798 Republic of Korea
e-mail: myjung@unist.ac.kr

W. Pedrycz and S.-M. Chen (eds.), *Social Networks: A Framework of Computational Intelligence*, Studies in Computational Intelligence 526, DOI: 10.1007/978-3-319-02993-1_14, © Springer International Publishing Switzerland 2014

1 Introduction

Knowledge-based organizations (KBOs) such as research institutes, universities and various research centers in businesses and industries manage their projects in pursuit of the goals of their organizations. Research institutes and universities apply to the government or to the industries and corporations to undertake their projects whereas the research centers belong to corporations are engaged in developing products or processes that would promote business opportunities for their corporations. The performance of these projects is evaluated at the time of their completion. The project performance depends on how the project goals have been accomplished in three aspects: quality, time and budget [1].

Individual capabilities of the project team members who carry out the project are comprised of the factors of knowledge and collaboration competence, time availability and cost which are mobilized to meet the requirements of the project's quality, time and budget. These factors vary depending on how the project team members are selected [2]. Ultimately, the quality of project performance is influenced by project team formation [3]. For these reasons, the core of project management is shifting from technologies involved in the project to human resources [4].

Pettersen [5] has categorized human resource management into eight areas in terms of project management—human resource planning, reception of necessary human resource, selection of project team members, job analysis, remuneration, education and training, performance assessment, and career planning. Out of these eight areas, human resource planning is perhaps the most important area in accomplishing the mid and long term goals of KBOs. Human resources equipped with specialized knowledge in the related areas are required for KBOs to search business opportunities in the market and plan and carry out projects which can produce the products in demand or develop processes necessary to implement envisioned business opportunities. It is the task of the human resource management department to make mid and long term plans to supply such human resources whenever they are called for.

The human resource management department must construct an IT foundation with a variety of systems which would make it possible to create a synergy effect by sharing not only the knowledge possessed by individuals but also by sharing the collective knowledge possessed by the organization. Knowledge sharing in an organization is very important because no matter how greatly talented are the people employed and assigned at the right time to participate in a project, good project performance cannot be expected if there is no collaboration among the project team members. For this reason, the human resource management department must conduct a quantitative evaluation of the knowledge competence of each individual as well as the knowledge competence collectively possessed by the organization. In addition, a quantitative evaluation must also be conducted on the collaboration competence of each individual.

Fig. 1 Conceptual
framework for knowledge-
based social network

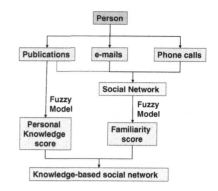

In performing a project jointly, it is important for project team members to know who has the required knowledge [6]. Thus, the knowledge network, which consists of webs of personal relationship, is essential in performing group activities effectively, and interpersonal awareness of others' knowledge is needed [7]. The four types of knowledge network—hobby network, professional learning network, best practice network, and business opportunity network—are divided by two axis of benefit level and the amount of managerial support [8]. The human resource management department must determine what stage the KBO's current knowledge network has reached in implementing its mid and long term strategies and suggest appropriate actions to be taken at each stage. Furthermore, the human resource management department must evaluate the knowledge and collaboration compe- tence of members of the knowledge network and define appropriate roles and assign them to individual members in a way that prevent overlapping and ineffi- cient human resource assignment and distribution. Thus quantitative evaluation is essential in developing a series of evaluating methods as well as appropriate actions to be taken. This study defines types of the knowledge networks of a KBO and roles of the network members of a KBO and presents methods of evaluating the knowledge competence and collaboration competence possessed both by individuals and the organization.

To evaluate the knowledge and collaboration level of an organization quantita- tively, the fuzzy models presented in the previous study [9] are used. The conceptual framework of the knowledge-based social network was proposed as shown in Fig. 1. Personal knowledge score is rated as a score obtained through the knowledge evaluation fuzzy model using keywords comparison. A social network is formed using the co-authors of publications, e-mails and phone calls. And then, the famil- iarity score among the personnel is evaluated through the familiarity fuzzy model, using the number of co-authors, average intervals of publications, the number of e-mails, the number of phone calls, and the sum of times of each phone call.

The rest of this study is organized as follows. Previously proposed measures for evaluating the employees on the basis of knowledge are presented and the liter- ature related to the evaluation of organizations is surveyed in Sect. 2. New mea- sures for the knowledge and the collaboration are summarized in Sect. 3. Section 4

Fig. 2 Conceptual diagram
of the knowledge-based
social network

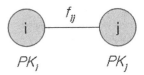

presents four types of the knowledge network and four roles of network members of a KBO. A case study and a prototype for testing the feasibility of the knowledge-based social network and its measures are presented in Sects. 5, and 6 presents conclusions.

2 Background

2.1 Definition of the Knowledge-Based Social Network

The knowledge-based social network presented in the previous study [9] is a social network that is comprised of PK_i, personal knowledge and f_{ij}, familiarity amongst personnel as shown in Fig. 2.

The knowledge competence score is evaluated using the personal knowledge score and the familiarity score. The knowledge competence of the person i can be shown in Eq. (1).

$$KC_i = PK_i + KSN_i \tag{1}$$

where
KC_i: knowledge competence of person i
PK_i: personal knowledge of person i
KSN_i: knowledge from social network of person i.
The following Eq. (2) is for the case in Fig. 2:

$$KC_i = PK_i + \sum_{j=1}^{N} f_{ij} * PK_j \tag{2}$$

where
f_{ij}: familiarity score between persons i and j ($0 \leq f_{ij} \leq 1$)
N: the number of personnel.

2.2 Evaluation of an Organization

Although knowledge management has been a major research object since the late 1980s, few methods have been proposed for the evaluation of a KBO. Assessing a KBO is difficult because many investments are made in intangible assets that are hard to measure. Two conventional methods have mostly been used: the net present value (NPV) and the comparative valuation using financial multiples. The NPV method uses an appropriate discount rate to discount the cash flows generated by a proposed project [10]. Comparative valuation using financial multiples uses Tobin's Q ratio equals the market value/asset value [11]. A positive Q ratio can be attributed to the intangible part of intellectual capital not captured by the traditional accounting systems [12].

However, this method is unable to distinguish the value of nonintellectual capital from the value of intellectual capital. Knowledge assets is defined as a sum of net R&D and net training [13]. Although this approach is simple, it tends to underestimate the knowledge assets of the organization as a whole.

There are two ways to measure the financial value of knowledge asset. One way is to measure the overall value of the knowledge of organization [14, 15], and the other method is to measure the value of separate knowledge asset [13, 16]. The former is not appropriate for measuring low level knowledge assets [16] whereas the latter is a method of measuring the activities related to application of knowledge.

Wilkins et al. [13] have defined the value of knowledge asset as the sum of the cost-based value and the added value, summed over all the relevant processes in which it is a resource. Dekker and de Hoog [16] defined the value of knowledge item as profit from the product ascribed to the knowledge item. The problem with these two methods is that they measure the value of knowledge derived only from the process that is being considered. As a result, neither method considers additional values derived from other processes.

Another knowledge measurement model calculates the period expense incurred during the service life of human capital through recognition and distribution of the cost of human capital, deducting it from the valuation account of the human capital [17]. This method differs from all the methods discussed above in that it approaches the human capital that possesses the knowledge. However, it does not measure knowledge competence of the entire organization. Furthermore, the traditional evaluation methods are incapable of reflecting future opportunities and risks. To make up for these shortcomings, a new evaluation method using the Real Options theory has been developed [18]. However, it should be noted that this method does not properly evaluate a KBO at the knowledge level as required by the management of the organization in project management.

2.3 Evaluation of the Knowledge and the Collaboration Collectively

Most studies in the literature evaluate organizations in terms of finance. Although financial evaluation is an important factor from the viewpoint of KBO, the evaluation of the knowledge level is as important as financial evaluation.

If it is possible to evaluate the knowledge level of a KBO, it would be enabled the management department or the human resource management department not only to establish mid and long term strategies but also analyze the effects of introducing various resources such as human resource, budget and equipment by evaluating different knowledge levels required at different stages of strategy implementations. Evaluation of the current knowledge level of the organization makes it possible to analyze the knowledge network type and develop actions to be taken for the success of future projects at the organizational level. Quantitative evaluation of collaboration levels of the members of the organization is also very important. By taking into account the knowledge levels as well as the collaboration levels of employees, the human resource management department will be able to analyze the network role required by the knowledge network type of the organization and input appropriate human resource well-qualified for the network role. If mid and long term strategies are carried out by evaluating the knowledge levels as well as the collaboration levels, it would be possible to achieve a greater success for the organization.

3 Measures for the Knowledge-Based Social Network

To measure the knowledge and collaboration level of an organization, new measures of knowledge-based social network are presented. To analyze a knowledge-based social network, personal competence is divided into knowledge competence and collaboration competence. Knowledge competence is divided into personal knowledge and knowledge from the social network, and collaboration competence is divided into density that represents communication, degree centrality that represents coordination, and closeness centrality that represents cooperation [19].

3.1 Evaluation of Knowledge Competence

The knowledge competence score of a KBO shown in Eq. (3) is the sum of personal knowledge scores and knowledge scores from knowledge-based social network. The formula is shown in detail in the previous study of Wi et al. [9].

$$OKC = OPK + OKSN$$
$$= \sum_{i=1}^{N} PK_i + \sum_{i=1}^{N} \sum_{j=1}^{N} (t_{ij} * f_{ij} * PK_j) \tag{3}$$

where
   OKC: knowledge competence score of a KBO
   OPK: sum of personal knowledge scores of employees of a KBO

OKSN: sum of knowledge scores from knowledge-based social network of a KBO

PK_i: personal knowledge score of person i

t_{ij}: the existence of link from persons i and j (1:exists, 0:does not exist)

f_{ij}: familiarity score between persons i and j

N: the number of personnel.

3.2 Evaluation of Collaboration Competence

To analyze the knowledge-based social network, density, degree centrality, and closeness centrality as the measures showing the degree of communication, coordination and cooperation, these three aspects of collaboration can be used [20]. The definitions used for the traditional social network analysis are as follows [21, 22]. The density of a social network is the quotient of the connections that exists among the network actors by the total of connections that could possibly exist in the network. Degree centrality is a measure which counts the number of ties a person has. Closeness centrality is the sum of all the smallest ties of the indirect path that must be taken by a person to go to other persons within the network. Closeness becomes smaller as a person is nearer to the center. For a person located in the boundary of the network, closeness becomes greater because he or she has to go through many persons to contact other persons. Since the existing measures are calculated based only on the number of ties, the measure should be adjusted to utilize the personal knowledge score and familiarity score.

The density shown in Eq. (4) is the quotient of the sum of knowledge scores from the knowledge-based social network by the sum of personal knowledge score and knowledge scores from the knowledge-based social network. The degree of communication of a person in a specific sector can be measured by the density. Likewise, the sector featuring strong points or weak points can be identified in terms of communication by comparing the density of each sector. When a KBO focuses on researches on the sector carrying relatively high density, significant results can be obtained [23].

$$density_i = \frac{\sum_{j=1}^{N} (t_{ij} * f_{ij} * PK_j)}{PK_i + \sum_{j=1}^{N} (t_{ij} * f_{ij} * PK_j)} \tag{4}$$

where

$density_i$ = person i's density of knowledge-based social network

PK_i : personal knowledge score of person i

t_{ij} : the existence of link from persons i and j (1:exists, 0:does not exist)

f_{ij} : familiarity score between persons i and j

N : the number of personnel.

Degree centrality shown in Eq. (5) is a measure which the sum of personal knowledge score and knowledge scores from knowledge-based social network a person has. The larger this value is, the better the coordination that can be expected. Degree centrality correlates with innovation output in exploratory and service communities [24].

$$degree\ centrality_i = \sum_{j=1}^{N} (t_{ij} * f_{ij} * PK_j) \tag{5}$$

where

degree centrality$_i$ = person i's degree centrality of knowledge-based social network

PK_i: personal knowledge score of person i

t_{ij}: the existence of link from persons i and j (1:exists, 0:does not exist)

f_{ij}: familiarity score between persons i and j

N: the number of personnel.

Closeness centrality shown in Eq. (6) refers to the sum of inverse of maximum of knowledge competence scores corresponding to the indirect path that must be taken by a person to go to other persons within the network. The larger this value is, the better the cooperation that can be expected. The presence of more central persons in the research team has a positive influence on quality [25]. One person represents direct and indirect influences on other persons. Using the closeness centrality in combination with the degree centrality, the direct influences and indirect influences on other persons within the knowledge-based social network of a specific person can be identified.

$$closeness\ centrality_i = \sum_{j=1}^{N} d_{ij} \tag{6}$$

where

closeness centrality$_i$ = person i's closeness centrality of knowledge-based social network

d_{ij} = inverse of maximum of knowledge competence scores in an indirect path from person i to person j.

The collaboration competence score of a KBO shown in Eq. (7) is the sum of personal density scores, degree centrality scores and closeness centrality scores from knowledge-based social network.

$$OCC = \sum_{i=1}^{N} density_i + \sum_{i=1}^{N} degree\ centrality_i + \sum_{i=1}^{N} closeness\ centrality_i \tag{7}$$

where

OCC: collaboration competence score of a KBO

density$_i$: personal density score of person i

degree centrality$_i$: personal degree centrality score of person i *closeness centrality$_i$*: personal closeness centrality score of person i

N: the number of personnel.

Using the density, the degree centrality and the closeness centrality, persons who can collaborate well within the knowledge-based social network can be selected for a project. Persons with core competency in a specific sector can be selected using the density, degree centrality and closeness centrality; a person who has three high measures and is carrying out research on the sector should be selected as the project manager for high project performance. Persons with three low measures are given opportunities to make contact with other persons by creating a natural atmosphere and providing education needed for communication.

4 Types of the Knowledge-Based Social Network and Roles of Network Members

The types of knowledge-based social network and roles of network members are discussed below using the measures for the knowledge-based social network defined in Sect. 3.

4.1 Types of Network

Four types of knowledge network have been introduced—hobby network, professional learning network, best practice network and business opportunity network—based on X-axis and Y-axis of benefit level (individual-organization) and the amount of managerial support (supported-self-managed) [8]. Hobby networks are formed on the basis of individual interests and are usually small-sized groups. Professional learning networks extend beyond hobbies by building an individual knowledge base. If their importance is recognized, they will receive the management's support. Best practice networks are essentially the institutionalized forms of knowledge sharing in organizations. They are characterized by multidirectionality: each member and each unit, in principle, learn from all the others. Business opportunity networks are business-driven, entrepreneurial networks, which are potentially the most innovative and attractive from the growth perspective.

Büchel and Raub [8]'s four network types are applied in a modified form in this study, and the sum of the personal knowledge scores of a specific sector of the KBO and the sum of knowledge scores from the knowledge-based social network are indicated on the X-axis and Y-axis, respectively, as shown in Fig. 3.

Each network is categorized as follows. The maximum of X-axis was derived by multiplying the target personal knowledge score per person calculated from successful past projects by the number of personnel of the sector to manage.

Fig. 3 Four types of
knowledge-based social
network

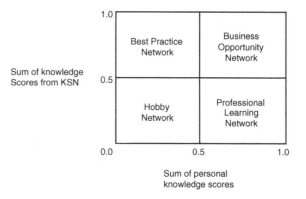

Similarly, the maximum of Y-axis is derived by multiplying the target knowledge score from knowledge-based social network derived from successful past projects by the number of personnel of the sector to manage.

The target personal knowledge score and target knowledge score from knowledge-based social network differ depending on the characteristics or scale of the applied organization. When the maximum of X-axis and Y-axis are derived, each axis is divided into maximum value for normalization. X-axis and Y-axis are categorized into two each, $0 \sim 0.49$ and $0.5 \sim 1.0$.

The Hobby network is a network smaller than 0.5 in both personal knowledge and knowledge from knowledge-based social network, and this network type is usually found in the beginning stage of a mid or long term project. The professional learning network is a network larger than 0.5 in personal knowledge and smaller than 0.5 in knowledge from knowledge-based social network. This network is found in the mid stage of the project in progress, and it is a type of network in which a small number of experts share knowledge amongst themselves, like a community of practice. Publication productivity is significantly linked to professional learning network [26]. Best practice network is a network smaller than 0.5 in personal knowledge and larger than 0.5 in knowledge from knowledge-based social network. In this network type, knowledge is shared forcefully in an organization, and Quality Control (QC) [27] and Total Quality Management (TQM) [28] are good examples. Business opportunity network is a network larger than 0.5 in both personal knowledge and knowledge from knowledge-based social network. In this network type, members share knowledge efficiently either forcefully or voluntarily, and the Six Sigma [29] is a good example.

The current status of the network in a specific sector of the KBO can be determined using the knowledge from the knowledge-based social network representing the degree of collaboration with the personal knowledge of personnel. The human resource management department must establish and implement strategic human resource plans covering a specific sector of the KBO. At this time, if the status of the network covering a specific sector of the current KBO can be recognized using the above mentioned measures, we can learn how to react under the current status. Suppose that the network for a specific sector of the current KBO is the hobby network, and that the organization has included the sector in the future growth directions of the organization. In such a case, we must develop the hobby network into a business opportunity network; this in turn necessitates inviting experts from external sources in the sector. Internally, we should promote familiarity by increasing the communication opportunities between personnel [30]. When experts are employed or when the communication opportunities are increased between personnel, the knowledge-based social network is reorganized by updating the data related to publications. The sums of personal knowledge scores and the knowledge scores from the reorganized knowledge-based social network are indicated on the X-axis and Y-axis, respectively, to compare them with the values prior to implementation of employment or communication opportunities. This way, the influences of a specific program on the KBO's knowledge-based social network can be calculated quantitatively. In addition, the implementation of various programs of the human resource management department and the results of their implementation can be traced and managed, and the development of a knowledge-based social network covering a specific sector of the KBO can be monitored.

4.2 Roles of Network

Knowledge-based social networks require a set of differentiated roles to develop over time. Patterns of four typical roles are observed in the most effective networks [8]. The network coordinator plays a pivotal role in most networks. The coordinator assesses the health of the network on a regular basis and acts as a catalyst connecting the network members. The coordinator is assisted by a support structure. The support's functions may include organizing and posting information generated by the network members. Highly effective networks rely in most cases on one or more editors to validate the content of the network. A sponsor plays the role of allowing the effective networks to benefit from top management support. The sponsor maintains contact with it, largely through the coordinator, reviews the network activities, contributes to keep them aligned with business strategy, and makes sure that an appropriate support is available when needed.

Although Büchel and Raub [8] have defined the network roles, they did not specify how to evaluate the abilities of members and assign them to appropriate groups. To find persons qualified to take on such network roles, we demonstrate the personal knowledge score for each person on the X-axis and the collaboration competence which is the sum of density, degree centrality and closeness centrality on the Y-axis as shown in Fig. 4.

The maximum of X-axis is determined by calculating the target personal knowledge score per person from successful past projects. Similarly, the maximum of Y-axis is determined by deriving the target collaboration competence per person from successful past projects.

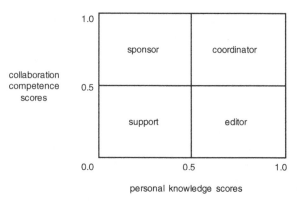

Fig. 4 Four roles of network members

The target personal knowledge score and the target collaboration competence vary in accordance with the characteristics or the scale of an applied organization. When the maximum value of X-axis and Y-axis are determined, each axis is divided by the maximum value for normalization. X-axis and Y-axis are categorized into two each, 0 ~ 0.49 and 0.5 ~ 1.0.

Support group is a group whose personal knowledge is smaller than 0.5 and whose collaboration competence is also smaller than 0.5. The members belonging to this group are suited for simple repetitive work, and technical assistants in KBOs belong to this group. People with personal knowledge larger than 0.5 and the collaboration competence smaller than 0.5 belong to the Editor group. Experts, specialists, and senior researchers in KBOs belong to the Editor group. Sponsor group is made up of people whose personal knowledge is smaller than 0.5 and whose collaboration competence is larger than 0.5. Department managers or executives usually belong to this group. Coordinator group is made up of people whose personal knowledge is larger than 0.5 and whose collaboration competence is also larger than 0.5. The members of this group are usually project managers. In other words, they have full knowledge of the project and collaborate well with team members. Persons plotted in the relevant area on the graph are those who are qualified for the network roles. Any changes in the person's network roles can be managed by plotting the trend changes in the personal knowledge score and collaboration competence over time.

5 A Case Study

This section presents a simple example of the knowledge-based social network and compares the traditional social network and the knowledge-based social network that considers familiarity and personal knowledge. The prototype is developed to handle an actual case of the type of knowledge-based social network and the role of personnel.

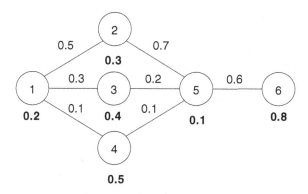

Fig. 5 An example of knowledge-based social network

Table 1 The result of measures of knowledge-based social network

		Traditional SNA	Knowledge-based social network
Density	Person 1	0.20	0.74
	Person 2	0.13	0.68
	Person 3	0.13	0.37
	Person 4	0.13	0.18
	Person 5	0.26	0.79
	Person 6	0.06	0.28
Degree centrality	Person 1	3	0.32
	Person 2	2	0.17
	Person 3	2	0.08
	Person 4	2	0.03
	Person 5	4	0.82
	Person 6	1	0.06
Closeness centrality	Person 1	5	1.86
	Person 2	6	1.99
	Person 3	6	2.55
	Person 4	6	7.60
	Person 5	2	3.57
	Person 6	9	1.10

5.1 An Example

A simple knowledge-based social network is configured as shown in Fig. 5 to
calculate the density, degree centrality and closeness centrality presented in
Sect. 3. The personal knowledge scores and familiarity scores are indicated on
nodes and links. Table 1 shows the calculation results of the density, degree
centrality, and closeness centrality of the knowledge-based social network.

In terms of the density through which we can check the degree of communication, Person 5 showed the highest values under the traditional Social Network Analysis (SNA), followed by Person 1. Person 5 again showed the highest values under knowledge-based social network, followed by Person 1. In the case of degree centrality, which indicates direct influences on coordination, Person 5 had the highest value under the traditional SNA followed by Person 1. Person 5 again showed the highest values under knowledge-based social network, followed by Person 1. In terms of the closeness centrality through which we can check the degree of cooperation, Person 5 showed the lowest values under the traditional SNA, followed by Person 1. Person 4 again showed the highest values under knowledge-based social network, followed by Person 5. In the traditional SNA, lower closeness centrality is closer to the center of the network, and higher closeness centrality is closer to the center of the network under knowledge-based social network. Since only the number of ties was considered under the traditional SNA, it is difficult to reflect all the characteristics of the persons forming the network.

5.2 A Prototype

A prototype of the evaluation method of knowledge-based social network was developed to validate the proposed approach. A research institute is selected as the basis of data collection for the prototype. The publications of 45 researchers including articles, project reports, and patents published from 2001–2006 were selected. The tools used in the development of the prototype are VC++ 6.0 [31], Xfuzzy 3 [32], and LabWindows/CVI 6.0 [33]. Xfuzzy 3, a modeling tool for fuzzy model, was used to model the fuzzy inference system for evaluating the personal knowledge and familiarity. The VC++ was used to implement the calculation of measures for knowledge-based social network. The LabWindows/CVI was used to design the user interface.

First, the knowledge-based social network was constructed with co-author information of articles, project reports, and patents. The number of co-authors times and the average year of the intervals of publications are the inputs of familiarity evaluation on publications. The familiarity among the researchers is calculated using the familiarity evaluation fuzzy model. When the required keywords of specific sector are "strip casting, magnesium alloy, friction stir, arc welding, and water cooling", personal knowledge scores of 45 researchers for each keyword of a publication are evaluated by the fuzzy inference system. The summation of the evaluated knowledge scores for publications is the personal knowledge score of a researcher for a keyword. In this case, five personal knowledge scores are evaluated for each researcher. As shown in Fig. 6, the developed prototype displays personal knowledge scores, knowledge scores from knowledge-based social network, density, degree centrality, and closeness centrality of 45 researchers. Clicking No. right below the screen displays the

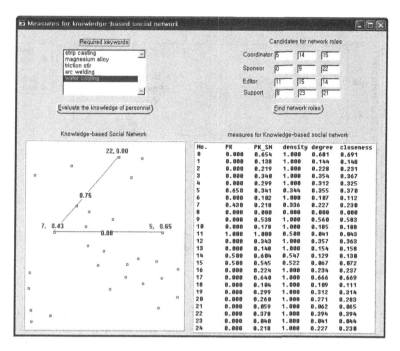

Fig. 6 The result of measures for knowledge-based social network

researcher's knowledge-based social network on the same screen as shown in the Fig. 6 left below. Specifically, the personal knowledge scores and familiarity scores of each researcher may be viewed.

5.3 Result and Discussion

The research institute cited in the prototype has conducted Project A since 2003, succeeding in commercializing the result of Project A in 2007 by investing 19 million USD until 2006. The target personal knowledge score and the target knowledge score from knowledge-based social network were derived from the projects that had been conducted from 1998–2002, and the scores were multiplied by 45, which is the number of personnel managed in the sector, to establish the maximum personal knowledge score and the maximum knowledge score from the knowledge-based social network. For the 10 keywords extracted from Project A, the normalized personal knowledge scores and the knowledge scores from the knowledge-based social network of 45 researchers involved in Project A from 2003 up to 2006 are depicted in Fig. 7. Thus, the organization's knowledge-based social network type evolved from the hobby network through the professional learning network ultimately to the business opportunity network.

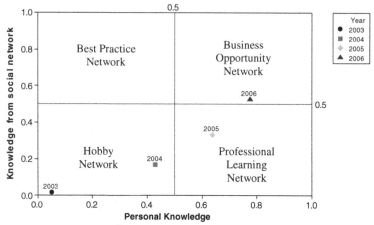

Fig. 7 Change in network type over time

The analysis of the network type revealed that the network development track is mainly attributed to the Six Sigma. Six Sigma was introduced into the research institute, and the results of the five stages—Define, Measure, Analyze, Improve, and Control—of Six Sigma on all projects were registered in the client-server system to be shared by all the researchers. Since 2005 when the enterprise portal was incorporated in the existing client-server system, smooth knowledge sharing has been in progress. The cost-effect analysis on the policies of the research institute can be performed with the development stage of the network type. For an ordinary enterprise, the organization's network type is expected to evolve into a business opportunity network from the hobby network after going through the best practice network. Small and medium sized enterprises are sharing and spreading knowledge through QC or TQM, and Six Sigma that has been introduced in large enterprises has grown to business opportunity networks in many cases.

The target personal knowledge score and the target collaboration competence were extracted from the projects that had been completed between 1998 and 2002, and we set up the maximum personal knowledge score and the collaboration competence. Figure 8 displays the normalized personal knowledge scores and the collaboration competence of the 45 researchers who had been involved in Project A between 2003 and 2006 in relation to the 10 keywords, respectively. In terms of the researchers' network roles, a comparison between the 2003 and 2006 figures suggests that the number of coordinators increased from 2 in 2003 to 6 in 2006 when commercialization of the result of Project A was nearly completed, and that the knowledge of supporters expanded as well.

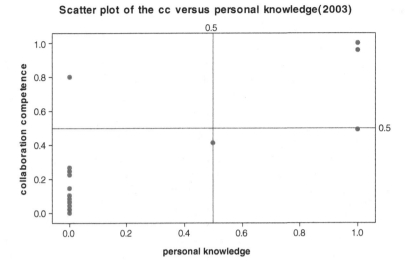

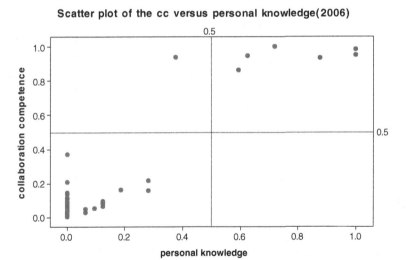

Fig. 8 Changes in the network role of researchers over time

6 Conclusion

In this study, a framework was established with the related measures for evaluating the network type of knowledge-based social network and the network roles in a specific sector of a KBO. Excluding the conventional qualitative evaluations deliberately, the knowledge of a personnel was evaluated using a fuzzy model on the articles, reports, and books published by the employees. In addition, the knowledge competence of the KBO was evaluated in terms of the knowledge from knowledge-based social network and also in terms of personal knowledge. Based on these evaluations, knowledge possessed by a specific sector of the KBO was quantified. As the result of a case study indicates, the personal knowledge and knowledge from knowledge-based social network of researchers has increased from 2003 to 2006. The measures of the knowledge-based social network were developed to analyze the network roles of the personnel of the KBO. The density, degree centrality and closeness centrality represented the level of the communication, coordination and cooperation. In addition, candidates suitable for network roles were selected by the density, degree centrality and closeness centrality, which are the measures that can compare the knowledge and collaboration of personnel in the KBO. By monitoring the measures over time, the degree of knowledge and the degree of collaboration in a specific sector in the organization were checked. After the Project A had been performed, for one year, the personal knowledge and knowledge from knowledge-based social network were in low level. Such evaluation and analysis of a KBO in terms of knowledge and collaboration helps the human resource management department make mid- and long-term strategic decisions. Since the enterprise portal and Six Sigma were introduced into the research institute for the systematic knowledge sharing, the level of knowledge and collaboration has increased. The result of the implemented decisions can also be monitored by observing the measures over time.

Acknowledgments This work was supported by the 2008 Research Fund (No. 1.080008.01) of UNIST, Ulsan 689–798, Korea.

References

1. Pheng, L.S., Chuan, Q.T.: Environmental factors and work performance of project managers in the construction industry. Int. J. Project Manage. **24**, 24–37 (2006)
2. Wi, H., Jung, M.: Modeling and analysis of project performance factors in an extended project-oriented virtual organization (EProVO). Expert Syst. Appl. **37**, 1143–1151 (2010)
3. Borgatti, S.: Identifying sets of key players in a social network. Comput. Math. Organ. Theory **12**, 21–34 (2006)
4. Huemann, M., Keegan, A., Turner, J.R.: Human resource management in the project-oriented company: a review. Int. J. Project Manage. **25**, 315–323 (2007)
5. Pettersen, M.: Selecting project managers: An integrated list of predictors. Project Manage. J. **22**, 21–26 (1991)

6. Alavi, M., Tiwana, A.: Knowledge integration in virtual teams: The potential role of KMS. J. Am. Soc. Inform. Sci. Technol. **53**, 1029–1037 (2002)
7. Akgün, A.E., Byrne, J., Keskin, H., Lynn, G.S., Imamoglu, S.Z.: Knowledge networks in new product development projects: a transactive memory perspective. Inf. Manage. **42**, 1105–1120 (2005)
8. Büchel, B., Raub, S.: Building knowledge-creating value networks. Eur. Manage. J. **20**, 587–596 (2002)
9. Wi, H., Oh, S., Mun, J., Jung, M.: A team formation model based on knowledge and collaboration. Expert Syst. Appl. **36**, 9121–9134 (2009)
10. Higson, C., Briginshaw, J.: Valuing Internet business. Bus. Strategy Rev. **11**, 10–20 (2000)
11. Brainard, W.C., Tobin, J.: Pitfalls in financial model building. Am. Econ. Rev. **58**, 99–122 (1968)
12. Luthy, D.H.: Intellectual capital and its measurement. In: Proceedings of the Asian Pacific Interdisciplinary Research in Accounting Conference (APIRA), Osaka, Japan, pp. 16-17. Citeseer, (1998)
13. Wilkins, J., Van Wegen, B., De Hoog, R.: Understanding and valuing knowledge assets: overview and method. Expert Syst. Appl. **13**, 55–72 (1997)
14. Sveiby, K.E.: The New Organizational Wealth: Managing & Measuring Knowledge-Based Assets. Berrett-Koehler Publishers, San Francisco (1997)
15. Edvinsson, L., Malone, M.S.: Intellectual Capital: Realizing Your Company's True Value by Finding its Hidden Brainpower. Harper Business, New York (1997)
16. Dekker, R., De Hoog, R.: The monetary value of knowledge assets: a micro approach. Expert Syst. Appl. **18**, 111–124 (2000)
17. Liebowitz, J., Wright, K.: Does measuring knowledge make "cents"? Expert Syst. Appl. **17**, 99–103 (1999)
18. Wu, L.-C., Ong, C.-S., Hsu, Y.-W.: Knowledge-based organization evaluation. Decis. Support Syst. **45**, 541–549 (2008)
19. Wi, H., Mun, J., Oh, S., Jung, M.: Modeling and analysis of project team formation factors in a project-oriented virtual organization (ProVO). Expert Syst. Appl. **36**, 5775–5783 (2009)
20. Pereira, S.C., Soares, A.L.: Improving the quality of collaboration requirements for information management through social networks analysis. Int. J. Inf. Manage. **27**, 86–103 (2007)
21. Marsden, P.V.: Egocentric and sociocentric measures of network centrality. Soc. Netw. **24**, 407–422 (2002)
22. Wasserman, S., Faust, K.: Social Network Analysis: Methods and Applications. Cambridge university press Cambridge (1994)
23. Chen, L., Gable, G., Hu, H.: Communication and organizational social networks: a simulation model. Comput. Math. Organ. Theor. **18**, 1–20 (2012)
24. Zou, G., Yilmaz, L.: Dynamics of knowledge creation in global participatory science communities: open innovation communities from a network perspective. Comput. Math. Organ. Theor. **17**, 35–58 (2011)
25. Beaudry, C., Schiffauerova, A.: Impacts of collaboration and network indicators on patent quality: the case of Canadian nanotechnology innovation. Eur. Manage. J. **29**, 362–376 (2011)
26. Ynalvez, M.A., Shrum, W.M.: Professional networks, scientific collaboration, and publication productivity in resource-constrained research institutions in a developing country. Res. Policy **40**, 204–216 (2011)
27. Juran, J., Godfrey, A.B.: Quality Handbook. McGraw-Hill, New York (1999). Republished
28. Ahire, S.L.: Management science-total quality management interfaces: an integrative framework. Interfaces **27**, 91–105 (1997)
29. Tennant, G.: Six Sigma: SPC and TQM in Manufacturing and Services. Gower Publishing Ltd, Aldershot (2001)

30. Yang, S.J.H., Chen, I.Y.L.: A social network-based system for supporting interactive collaboration in knowledge sharing over peer-to-peer network. Int. J. Hum. Comput. Stud. **66**, 36–50 (2008)
31. Kruglinski, D., Wingo, S., Shepherd, G.: Inside Visual C++ 6.0. Microsoft press (2005)
32. Fuzzy Logic Design Tools; Xfuzzy 3, IMSE-CNM (2012) http://www2.imse-cnm.csis.es/xfuzzy/
33. Zhang, Y., Qiao, L.: Virtual Instrument Software Development Environment: Labwindows/Cvi 6.0 Programming Guide. Mechanical industry press, Beijing (2002)

Process Drama Based Information Management for Assessment and Classification in Learning

Heidi-Tuulia Eklund and Patrik Eklund

Abstract In this chapter we present a formal description of information management for assessment and classification in learning. The description is supported by a structure related to drama process for learning. Our logic follows the idea of invoking uncertainties using underlying categories, and the language of processes in 'drama process' is taken to be BPMN (Business Process Modelling Notation).

Keywords Classroom · Drama · Formal logic · Process modelling

1 Introduction

Assessment and assessment scales are widely used, and in a broad range of areas within the public and private domain. In this chapter we focus on assessment of individual capacities, and on conditions and behaviour as appearing in education. The underlying motivations include needs for quality assurance of learning and teaching, in turn including specific assessment scales to support quality assurance. Perception and performance is typically observed and assessed within modules of particular syllabi of courses.

Assessment scales, assessment and evaluation must not be confused, and indeed one of the objectives of this chapter is the formally describe and distinguish between these concepts. There is also a semantic distinction between rating scales, scorecards and assessment scales, so that e.g. assessment and grading must not be

H.-T. Eklund (✉)
Department of Art and Culture Studies, University of Jyväskylä, City of Salo, Finland
e-mail: heidi-tuulia.eklund@salo.fi

P. Eklund
Department of Computing Science, Umeå University, Umeå, Sweden
e-mail: peklund@cs.umu.se

W. Pedrycz and S.-M. Chen (eds.), *Social Networks: A Framework*
of Computational Intelligence, Studies in Computational Intelligence 526,
DOI: 10.1007/978-3-319-02993-1_15, © Springer International Publishing Switzerland 2014

confused. We also avoid the use of information characterizations like being 'qualitative' or 'quantitative', as the borderline between them is sometimes blurred, and consensual definitions for them do not exist.

Our justification for working with assessment and assessment scales in education is three-fold. Firstly, we investigate why and how assessments are carried out, and what the content in the scales ought to be in order to optimally support given quality assurance tasks. Secondly, and focusing on the content, we provide a formal structure of assessment scales involving proper typing of items and data. This potentially increases data quality and indeed aims at avoiding ambiguities in data gathering and information observation processes. Thirdly, quality assurance and related decision-making is process-oriented, where process tokens are substitutions of variables in scales, and data objects are formulated as terms arising from the underlying signature representing the formal language of assessment scales.

Assessment of data in scales frequently appears as scoring, where data is numeric and assessment functions are various forms of the weighted sum. Linear models are easier to comprehend, and are therefore very popular. However, in presence of nominal data, it often looks doubtful to convert nominal to numeric, and doing that quite intuitively with linearity appearing ad hoc. Scorecards are logically very poor. In our framework, assessment functions are drawn from syntactic operators in the underlying signature, and typing allows also for ordinal and indeed for nominal data as well. For learning and parameter estimation purposes, the concept of gradient is then generalized, even without presence of traditional metric spaces to estimate the numeric value of errors, and generalized learning algorithms can still be developed.

Parameter estimation and learning algorithms often involve iterations of gradient descents in one form or another. Gradient descent in analysis, and when working with real-valued and differentiable functions, involves classical derivatives. However, for other types of valuations appearing in functions used in assessments, more symbolically treated gradients are needed. Reinforcement learning is a typical situation where differentiability cannot be used as such and as based on function valuation, but rather an approximations of differentiations in closed loops. The solution in reinforcement learning is e.g. sometimes to 'binarize' the gradient. A similar strategy is adopted in this chapter for assessment scales and their items, where item types are not only numerical, but also ordinal and even nominal.

Derivation rules are not necessarily arrived at using limits in connections with differentiations, but derivation rules are mostly defined. The justification of these definitions must involve notions of 'nominal' or 'ordinal' gradients, and are confirmed by their adaptation in decision-making, and as based on assessment of learning and education.

Gradients can be arranged not only in relation to terms over a signature, but also as related to sentences. We will in this chapter prefer to use first-order like logic, even if equational style of λ-calculus style logics also would be appropriate. We are very formal and sometimes even 'non-traditional' in our logic presentation, in particular concerning the distinctions between operators and function symbols, terms, predicate symbols and predicates, and sentences involving logic operations over predicates using terms.

If we would have ω as a binary operator for doing a joint estimation of a student's performance capacity to solve problems and make decisions, then the term

$$t_{performance} = \omega\left(x_{Solving\ Problems}, x_{Making\ Decisions}\right)$$

would be the syntactic 'report' for that estimation, and could be further used within a predicate, additionally involving valuations or 'reports' as provided by a corresponding term for perceptions,

$$P\left(t_{performance}, t_{perception}\right).$$

The symbolic partial derivatives

$$\frac{\delta P}{\delta x}, \frac{\delta P}{\delta \omega}, \frac{\delta \omega}{\delta x}$$

should all be enabled by suitable definitions, also involving multiplication or conjunction of nominal values so as to enable derivation rules like

$$\frac{\delta P}{\delta x} = \frac{\delta P}{\delta \omega} \otimes \frac{\delta \omega}{\delta x}.$$

Error functions, in the absence of traditional metrics, similarly must involve topology like metrizability [17], so that symbolic error functions E, involving integrals based on disjunctions, can be formulated, and gradients be calculated with respect to such error functions. Techniques to allow neural learning in propositional logic using relations between addition on the real line and disjunction in a truth interval was originally developed in [8, 20].

Note the distinction between the symbolic and syntactic derivatives and the semantic ordinary derivatives. The typing of ω is according to $\omega : s_1 \times s_2 \to s$, where s_1 is the type for the problem-solving item, and s_2 is the type for the decision-making item. The output type is s. The algebras of all these types could e.g. be the set $\mathbb{R}$ of real numbers, so that $\mathfrak{A}(s) = \mathfrak{A}(s) = \mathfrak{A}(s) = \mathbb{R}$. The algebra of ω can then, if desired, be assigned to a partially differentiable mapping: $\mathfrak{A}(\omega)\ \mathbb{R} \times \mathbb{R} \to \mathbb{R}$

Note also how $\frac{\delta}{\delta x}$ syntactically appears in a similar role as λx in a lambda expression, so $\frac{\delta}{\delta x}$ itself should not be seen as an operator. This is more clearly seen if we would add a syntactic addition $\oplus : s \times s \to s$, with $\mathfrak{A}(\oplus)$ being the ordinary addition for natural numbers. Then it makes no sense to define $\mathfrak{A}\left(\frac{\delta}{\delta x}\right) = \frac{\partial}{\partial \mathfrak{A}(x)}$, but a definition like

$$\mathfrak{A}\left(\frac{\delta\omega(x, y) \oplus \omega(x, y)}{\delta x}\right) = \frac{\partial\mathfrak{A}(\omega)(\mathfrak{A}(x), \mathfrak{A}(y))}{\partial\mathfrak{A}(x)} + \frac{\partial\mathfrak{A}(\omega)(\mathfrak{A}(x), \mathfrak{A}(y))}{\partial\mathfrak{A}(x)}$$

could be justified.

Concerning sentences, our standpoint is that the partial derivative of a sentence like

$$P \leftarrow P_1 \& \ldots \& P_m$$

in the first-order case will also make sense, even if implication will not be treated as a logical operator, but rather as appearing as built into a sentence functor. For propositional logic we will see that the situation is different, as uncertainty resides in propositional variables only. In this special case, learning techniques like those applied in [20] can be applied.

This reveals our standpoint concerning the formal distinction between predicate and sentence. The truth degree of a predicate is a value subject to symbolic differentiation, whereas the validity of a sentence is outside the scope of further operational manipulation. Sentences appear in proof rules, proof sequences and argumentation, and their 'dynamics' is based only on the terms and predicates appearing in them. Terms over signature are in this chapter formally produced by term monads [14–16, 26].

The role and impact of parameter learning in scales and assessment functions is related to the intertwined development of curricula on national, regional, municipal and school levels. Optimal assignment of training levels adapted to student skills and performance requires parameter estimations. Converting nominal and ordinal values to numerics and ad hoc linear models will over simplify and hide the underlying operators and their logic.

In this chapter we present a formal model of assessment scales intended for use in education, and show how generalized parameter learning procedures can be imposed. Our application focus is on teaching music in secondary school.

Several approaches and methods throughout history touches upon the subject of logic. However, syntax and semantics often become confused in these contexts, mainly since the treatment of logic is not mathematically very strict until early twentieth century in mathematics, and pedagogy oriented viewpoints never deal properly with formal logic. Mill's [36] distinction between connotation and annotation has received surprisingly little attention after the kick-off of modern mathematical logic starting from late 19th century with Frege's arithmetics [23] and his notions on 'Sinn und Bedeutung' [24], and from there on ending up with Hilbert's and Bernays' monumental work on mathematical logic [30, 31].

In pedagogy, Dewey viewed education and learning as a social and interactive process, and was also concerned with logic [4], but his treatment was not mathematical enough to provide new ideas for logicians being active in the 1940s and 1950s. Dewey saw the shortcomings of logic at that time, and that formal mathematical logic as a language with mostly symbolic structures wasn't rich enough to model information scopes and content of propositions in education and learning processes. However, he was apparently unable to provide that required enrichment related to 'word' and 'sentence', in order to provide desired modelling e.g. of diagrams.

Piaget, promoting knowledge being acquired through interaction, was also weak on the distinction between syntax and semantics. In [39], Piaget wrote about 'thought psychology' and the psychological nature of logical operations, saying that *How far a psychological explanation of intelligence is possible depends on the way in which logical operations are interpreted: are they the reflection of an already formed reality or the expression of a genuine activity?* It may seem that Piaget on 'already formed reality' explicitly refers to logical semantics, and by "expression of a genuine activity" to syntax, but this is not so. In fact, Piaget continues *the logician then proceeds as does the geometer with the space that he constructs deductively, while the psychologist can be likened to the physicist, who measures space in the real world.* Piaget almost wants to exclude logic from psychology, and not to include it.

Other pedagogists, like Maria Montessori in her method focusing on initiative and natural ability, Reggio Emilia with a focus on early education, Rudolph Steiner on learning by imitation, and Lev Vygotsky with psychology involved in thought and language, all provide partly overlapping models of 'information and process' as appearing in cognition and learning, but without more formal considerations of information and process as represented in a logic language.

2 Social Architecture and Learning Processes

Social architecture is, on the one hand, logically appearing in social communication and in considerations of a diversity of individual logics or even in 'social logic'. Formally this is enabled by logic being in communication, having transformations or morphisms, and as being between various individual or individualized logics [19]. On the other hand, social architecture is a structure of social information, decision and choice [13]. Concerning decision-making, consensus plays an important role and consensus models related to our approach was initiated in [5, 6, 18], and further developed in [22, 40].

When speaking of a *set* of actors, we are more formally speaking about a rather complicated conglomerate of individuals and stakeholders, organized also hierarchically. Allowing dynamics between sets of actors appears in public areas like health, education and culture, but also in public authorities and private organization. In this chapter we focus on actor dynamics in education, and in particular as related to the logic and information management of assessment and classification in learning, based on process drama techniques and methodologies [3, 28, 38, 41, 42]. Historically, Shakespeare's theory of drama [32] is also interesting in these respects as some say Shakespeare was indeed concerned with dramatic theory, while others say he had no theoretical ambitions at all. Clearly, Shakespeare is understood as a poet, but he was obviously also much concerned with theatrical considerations. His literary part is more of "information, knowledge and logic", where the theatrical part is more about "process and logic".

Assessment and standardization in education should be inspired, at least partly, by ground breaking work in health and social care, where the structure, and even part of the content, of WHO's ICF (International Classification of Function) is expected to provide some suggestions also for *Classification of Learning*. Assessment of learning can be compared with assessment of health involving both success and risk factors, and thus based on both salutogenic as well as iatrogenic viewpoints of wellness. Learning is obviously more about cognitive skills and knowledge, attitude and growth in emotional areas, and psychomotor skills.

The distinction between an individual and the social environment, where the individual is active, has been well recognized in the education literature, but at the same poorly modeled from information and logic point of view. What is 'social' logically? What is 'individual' socially? What is 'individual in social'? Do we describe humans and human action by numerical means or logical means, or both? The numerical means is viewing 'social' as something close to 'population', and we then observe it from the outside, and we mostly quantify in doing so. We may, however also see 'social' as individuality using methods in life, in a social environment, being transcendent about truth and what is right, and being subsistent about ones existence. Viewing 'individual' in this way, 'social' is then even more tricky. We may e.g. ask ourselves the following question: If three persons communicate, is there really something "monadologically" [34] like "three persons as one", or is it really "three times two person", and in all combinations? Believing that social is something moving from that "combinations of two" to something "truly three", then we also speak more of social networks as emergent.

In this chapter 'individual' is then mostly 'teacher', and 'student' and 'social' is mostly 'classroom' and 'school'.

Numerics cannot do all this modeling alone. We need also logics, and we need ontology, classification, assessment, and so on and so forth. This means we also view Computational Intelligence (CI) more broadly, i.e., clearly covering (fuzzy) logic, neural/linear, Bayesian, and much more, for the 'numerical types', but also allowing for 'symbolic computational intelligence' involving e.g. more nominally oriented types.

There is also the phenomenology around 'social and the web (of linkage)'. Web ontology is logically rather poor. Web ontology is often connected with Description Logic (DL), and DL is said to be kind of a 'sublogic' to first-order logic. In [15] we clearly show how DL in fact is a lambda-calculus, and nothing like first-order, as the existential quantifier in DL comes from a relation, which in turn comes from the powerset monad, where the powerset functor can be modeled within the so called superceding signature, so that 'clauses' in DL eventually end up as lambda terms. DL is also suggested to be the logic for SNOMED, the terminology by IHTSDO (International Health Terminology Standards Development Organization). These development are still very premature, and we believe that the underlying logic must become much more elaborate to cover assessment and classification in health, and this indeed mirrors that belief in particular also over to 'learning and social'.

Analyzing the structure of entire social entities requires all that ontology and formal logics. If the underlying models are too simple and too set-theoretic only, we claim that sophisticated numerical algorithms in CI cannot recover very much, unless we indeed focus more on conceptually appealing and practically sound information technology, for which this chapter is an initial suggestion of an underlying logic and process model of 'social', using drama process as a learning method, and as a 'social technology'.

In doing so we expect to provide and promote a better formal understanding of what 'social networks', 'social media', "me and the web", "we and the web", and so on, really mean, and how it should be properly modeled so as to enable fruitful and powerful analytics and decision-support development, both numerically as well as logically.

On learning processes we may lean e.g. on the *Experiential Learning Theory* (ELT) model as developed by Kolb [33], involving the so called Kolb's learning styles including the items

- Concrete Experience (CE) and *feeling* it,
- Reflective Observation (RO), i.e., *watching* it,
- Abstract Conceptualization (AC) and *thinking* about it,
- Active Experimentation (AE) and actually *doing* it.

appearing in kind of an outer loop, and representing measurable information basically appearing in and modeled by terms over the signature. Transformation between items

- Diverging, i.e., *feel and watch*, between CE and RO
- Assimilating, i.e., *watch and think*, between RO and AC
- Converging, i.e., *think and do*, between AC and AE
- Accomodating, i.e., *do and feel*, between AE and CE

are then part of learning and teaching, and involves related reinforcements. Kolb's model does not embrace or distinguish between items and transformation being 'individual' or 'social', so the group dimension and related interoperability is basically missing in Kolb's model, as is any consideration of syntax and semantics of information. However, despite these deficits, Kolb's model may serve as providing names for some items and transformations which we can aim to enhance in our more formal framework.

3 Assessment Scales

In Kolb's learning styles we may see t_{CE}, t_{RO}, t_{AC} and t_{AE} as terms carrying values which can be used as such and in predicates, but also in assessment scales as more basic terms in higher level scales and assessments. The respective transformation can be seen as categorical (not necessarily natural) transformations $\eta_{CE,RO}^{\sigma,\tau}$, $\eta_{RO,AC}^{\sigma,\tau}, \eta_{AC,AE}^{\sigma,\tau}, \eta_{AE,CE}^{\sigma,\tau} : \mathsf{T}_\Sigma \to \mathsf{T}_\Sigma$, where the sets of terms $\mathsf{T}_\Sigma X_S$ are described more formally in Sect. 4. Here a particular student would represent the index σ, and τ would be represented by a teacher.

We are now in a position to create terms e.g. looking like

$$u_{\sigma_1,\sigma_2}^{\tau} = \omega_{Converging}\left(\left(\eta_{AE,CE_{X_S}}^{\tau}\right)_s \left(t_{performance}^{\sigma_1}\right), \left(\eta_{AE,CE_{X_S}}^{\tau}\right)_s \left(t_{perception}^{\sigma_2}\right)\right)$$

where

$$\left(\eta_{AE,CE_{X_S}}^{\tau}\right)_s : \mathsf{T}_{\Sigma,s}X_S \to \mathsf{T}_{\Sigma,s}X_S$$

is a morphisms that takes a term $t_{performance}^{\sigma_1}$, representing some valuation or 'report' about how student σ_1 performs with respect to AC, and transforms it to $\left(\eta_{AE,CE_{X_S}}^{\tau}\right)_s \left(t_{performance}^{\sigma_1}\right)$, representing the valuation of the outcome performance at AE. The assessment $u_{\sigma_1,\sigma_2}^{\tau}$ is then (part of) a teacher's observation of a students presence and receptivity involving factors and scales, where various 'factors' are included in scales for 'presence' and 'receptivity' [12].

Here we may now extend these notations involving individuals to explicitly cover groups and social context. Indeed, $u_{\sigma_1,\sigma_2}^{\tau}$ can be seen more generally like $u_{\{\sigma_1,\sigma_2\}}^{\{\tau\}}$, i.e., one teacher working with a group of two students, which immediately opens up the question of possible specifications for $u_{\{\sigma_2\}}^{\{\tau,\sigma_1\}}$, $u_{\{\sigma_1\}}^{\{\tau,\sigma_2\}}$, and $u_{\{\sigma_1,...,\sigma_n\}}^{\{\tau_1,...,\tau_m\}}$, which in turn would lead to more general considerations about the meaning of $u_{Classroom}^{Teacher}$. The information modelling of such terms is, however, outside the scope of this chapter.

This now also clearly shows how scales can be arranged as terms over a signature, where sorts and operators also capture and represent various types of uncertainties, and operators representing uncertain computation using those factors, rather that computing with uncertain factors. A factor is uncertain if it in turn arises from an uncertain computation. This signature, together with its term monad, provides the underlying fuzzy terms [15] for the *Logic of Assessment*.

The relation between educational and health assessment is also interesting. Educational assessment differs from health assessment basically only in content, but the information structures and decision procedures use similar logic languages and underlying algebraic structures. As mention before, assessment and evaluation are different, and 'doing' an assessment is different from 'using' an assessment.

Obviously, it is not clear how to arrive at a core set of assessment scales for teaching and learning related to performance and capabilities. Scales for skill with suitable activity are important, as is assessing social and individual performance. Inclusion and attitude are factors, and cerebration is usually seen as important in its own right. Circumstance and health are 'environmental factors', and family situations sometimes play a significant role.

Figure 1 illustrates a possible configuration of a minimal set of assessment scales, or rather, the set of subsets of assessment scales, where clearly some specific factors e.g. arising from and within Kolb's model can be built in. For assessing individual activity we typically include cognitive capacity, mental presence, ability to listen and be concentrated, and so on. Many, even if not all, of

Fig. 1 Subareas of
assessments [12]

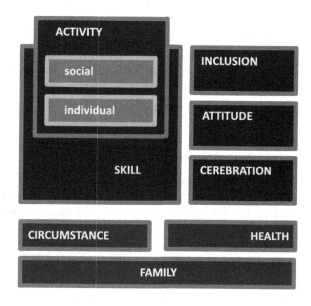

these specific factors and items can be seen as related to Kolb's model. Judgment is also interesting, and judgment is, on the one hand, part of skills, and, on the other hand, part of cerebration. Cerebration is related with intellection and mentation, and more formal specifications depend on the particular application at hand. In many specific applications, the design of factors and items, and the selection of operators in assessment, is indeed a matter of choice given the target application. Assessment in health and social care is similar.

Gradation and involvement of uncertainties e.g. using ordinal scales can be illuminated by using ICF as an example. In ICF, the set of logical truth degrees are the following:

xxx.0 NO problem	(none, absent, negligible,...)	0–4 %
xxx.1 MILD problem	(slight, low,...)	5–24 %
xxx.2 MODERATE problem	(medium, fair,...)	25–49 %
xxx.3 SEVERE problem	(high, extreme,...)	50–95 %
xxx.4 COMPLETE problem	(total, f...)	96–100 %
xxx.8 not specified		
xxx.9 not applicable		

Obviously ICF does not cover all specific items needed for assessment in education, and ICF, in its core content, also goes beyond what is needed in assessment in education. ICF's chapters on body functions and structures are partly related to basic capacities needed in learning, but are outside the core scope of learning assessment. Further, ICF is intended to register and recover impairments and decline in function rather than excellence and skill, and in doing so, the focus is more on adults, the working population, and older persons. However, some ICF chapters may serve as examples for further enrichment of the ontology of factors as discussed in the context of Kolb's model, like those appearing under the ICF component *participation and activation*:

- Learning and applying knowledge
- Communication
- Domestic life
- Interpersonal interactions and relationships.

The general items, typically appearing under activity, inclusion and attitude (Fig. 1), but also as related to skill, like those under chapter *applying knowledge* (ICF codes d160-d179), may serve as names for item groups used in specific assessments, but are too general to serve as items as such. Items d175 for *solving problems*, and d177 for *making decisions*, are then typical in these respects, and corresponding variables could appear in individual assessments like

$$u_{\sigma_1}^{\tau} = \omega_{Converging}\left(\left(\eta_{AE,CE_{X_S}}^{\tau}\right)_s \left(x_{ICF:d175}^{\sigma_1}\right), \left(\eta_{AE,CE_{X_S}}^{\tau}\right)_s \left(x_{ICF:d177}^{\sigma_1}\right)\right)$$

with $u_{\sigma_1}^{\tau}$ then being influenced more by the ICF information model than by Kolb's model.

4 The Formal Logic Language for Representing Assessment Scale Based Decision-Making

In this section we provide the formal logic modelling of information management for assessment and classification in learning.

4.1 Categorical Prerequisites

This subsection does not aim at being self-contained with respect to categorical notions, but rather to introduce the notion which is used in subsequent categorical constructions. For more detail on categorical notions, see e.g. [1]. We will, however, provide detail for some categories and constructions which are specific to our framework, and not so common in the categorical reference literature.

The category of sets, Set consists of sets as objects and functions as morphisms, together with the ordinary composition of functions and the identity function. For including uncertainty into an underlying category we use the Goguen category Set($\mathfrak{L}$) [27], $\mathfrak{L}$ being a lattice with some required properties, like being completely distributive or residuated. Objects are pairs (X, α), where X is an object of Set and

$\alpha : X \to Q$ is a function. The morphisms $(X, \alpha) \xrightarrow{f} (Y, \beta)$ are Set-morphisms $X \xrightarrow{f} Y$ satisfying $\alpha \leq \beta \circ f$. The composition of morphisms is defined as composition of Set-morphisms.

A (covariant) *functor* $F : C \to D$ between categories is a mapping that assigns each C-object A to a D-object $F(A)$ and each C-morphism $A \xrightarrow{f} B$ to a D-morphism $F(A) \xrightarrow{F(f)} F(B)$, such that $F(f \circ g) = F(f) \circ F(g)$ and $F(\mathrm{id}_A) = \mathrm{id}_{F(A)}$. Composition of functors is denoted $G \circ F : C \to E$ and the identity functor is written $\mathrm{id}_C : C \to C$. The (covariant) powerset functor $P : \mathrm{Set} \to \mathrm{Set}$ is the typical example of a functor, and is defined by PA being the powerset of A, i.e., the set of subsets of A, and $Pf(X)$, for $X \subseteq A$, being the image of X under f, i.e., $Pf(X) = \{f(x) | x \in X\}$. A contravariant functor $F : C \to D$ maps to each C-morphism $A \xrightarrow{f} B$ a D-morphism $F(B) \longrightarrow F(f)F(A)$, and for the contravariant powerset functor $\overline{P} : \mathrm{Set} \to \mathrm{Set}$ we have $\overline{P}A = PA$ and $\overline{P}f(Y) = \{x \in X | \exists y \in Y : f(x) = y\}$.

A *natural transformation* $\tau : F \to G$ between functors assigns to each C-object A a D-morphism $\tau_A : FA \to GA$ such that $Gf \circ \tau_A = \tau_B \circ Ff$, for any $f : A \to B$. It is easy to see that $\eta : \mathrm{id}_{\mathrm{Set}} \to P$ given by $\eta_X(x) = \{x\}$, and $\mu : P \circ P \to P$ given by $\mu_X(\mathcal{B}) = \bigcup \mathcal{B} (= \bigcup_{B \in \mathcal{B}} B)$ are natural transformations.

A *monad* (or triple, or algebraic theory) over a category C is written as $\mathbf{F} = (F, \eta, \mu)$, where $F : C \to C$ is a (covariant) functor, and $\eta : \mathrm{id} \to F$ and $\mu : F \circ F \to F$ are natural transformations for which $\mu \circ F\mu = \mu \circ \mu F$ and $\mu \circ F\eta = \mu \circ \eta F = \mathrm{id}_F$ hold.

The many-valued covariant powerset functor L for a completely distributive lattice $\mathfrak{L} = (\mathfrak{L}, \vee, \wedge)$ is obtained by $LX = L^X$, i.e. the set of functions (or $\mathfrak{L}$-sets) $\alpha : X \to L$, and following [27], for a morphism $f : X \to Y$ in Set, by defining $Lf(\alpha)(y) = \vee_{f(x)=y} \alpha(x)$. Further, if we define $\eta_X : X \to LX$ by

$$\eta_X(x)(x') = \begin{pmatrix} \top & \text{if } x = x' \\ \bot & \text{otherwise} \end{pmatrix}$$

and $\mu : L \circ L \to L$ by

$$\mu_X(\mathcal{M})(x) = \underset{\alpha \in LX}{\vee} A(x) \wedge \mathcal{M}(\alpha)$$

then $\mathbf{L} = (L, \eta, \mu)$ is a monad.

In the one-sorted (and crisp) case for signatures we typically work in Set, but in the many-valued (and crisp) case we need the 'sorted category of sets' for the many-sorted term functor. If S is a set of sorts (or types), then, for a category C, we write C_S for the product category $\prod_S C$. The objects of C_S are tuples $(X_s)_{s \in S}$ such that $X_s \in Ob(C)$ for all $s \in S$. We will also use X_s as a shorthand notation for these tuples. The morphisms between objects $(X_s)_{s \in S}$ and $(Y_s)_{s \in S}$ are tuples $(f_s)_{s \in S}$ such that $f_s \in Hom_C(X_s, Y_s)$ for all $s \in S$, and similarly, f_s is a shorthand notation.

4.2 Signatures and Terms

A many-sorted signature $\Sigma = (S, \Omega)$ over Set consists of a set S of sorts (or types), and a set Ω of operators, which should be seen as a tuple $(\Omega_s)_{s \in S}$, where Ω_s is an object in Set. Operators in Ω_s are then syntactically written as $\omega : s_1 \times \ldots \times s_n \to s$, and are also said to be n-ary operators. The 0-ary operators $\omega :\to s$ are called constant operators, or constants.

The set of terms over the signature Σ is produced by the term functor $T_\Sigma : \text{Set}_S \to \text{Set}_S$. Specifically, $T_{\Sigma,s} : \text{Set}_S \to \text{Set}$ produces all terms of type s. The term functor can be extended to a monad, which means e.g. that substitutions can be composed. The formal construction of the many-sorted term monad appears in [15].

In order to illuminate these constructions, let $\text{NAT} = (\{\text{nat}\}, \{0 :\to \text{nat}, \text{succ} : \text{nat} \to \text{nat}\})$, be the signature for natural numbers. Then the set $T_{\text{NAT}}X_{\text{nat}}$ of all terms over NAT is

$$X_{\text{nat}} \cup \{0, \text{succ}(0), \text{succ}(\text{succ}(0)), \ldots\} \cup \{\text{succ}(x), \text{succ}(\text{succ}(x)), \ldots | x \in X_{\text{nat}}\}.$$

We can now also formally comprehend the fundamental distinction between 'computing with fuzzy' and 'fuzzy computing'. The former means composing the many-valued covariant powerset functor L with the term functor T, to provide the composition $L \circ T$, and the latter means having the term functor $T : \text{Set}(\mathfrak{Q}) \to \mathfrak{Set}(\mathfrak{Q})$ running over the Goguen category $\text{Set}(\mathfrak{Q})$. Uncertainties in terms, with uncertainty values in a lattice or a quantale $\mathfrak{Q}$, arise from uncertainty annotations in operations. See [15, 16] for more detail.

The distinction between 'computing with fuzzy' and 'fuzzy computing' can be illustrated as follows. 'Computing with fuzzy' typically means arranging target information, e.g. as represented by values of terms, in forms of fuzzy sets $\alpha : \{t_1, \ldots, t_n\} \to \mathfrak{Q}$, where $\alpha(t_i)$ is like an annotation of a uncertainty value to a crisp term t_i. 'Fuzzy computing' means annotating operators and even variables with uncertainties so that the overall uncertainty of a crisp term like

$$\omega \left(x_{Solving\ Problems}, x_{Making\ Decisions} \right)$$

over Set is a fuzzy term like

$$(\omega, \alpha) \left(\left(x_{Solving\ Problems}, \alpha_1 \right), \left(x_{Making\ Decisions}, \alpha_1 \right) \right).$$

Here α can be seen as the uncertainty attached with the observer using the operator ω to provide an assessment. The uncertainty of the assessments arises now "from the inside", rather than being ad hoc like in the case of 'computing with fuzzy'. This approach has been used also for modelling uncertainty within social choice [13] and for assessment scales in old age psychiatry [9, 10].

4.3 Sentences

In this subsection we will define the concept of 'sentence' in a first-order setting. We initiate the constructions in the crisp case, i.e., using Set as our underlying category. Let $\Sigma_0 = (S_0, \Omega_0)$ be a signature over Set, and T_{Σ_0} be the term monad over Set_{S_0}. For the variables in X_{S_0}, the set of terms $\mathsf{T}_{\Sigma_0} X_{S_0}$, as an object of Set_{S_0}, then correspond to the 'ground terms' in the sense of [35]. In the programming language *Prolog* it is common to implement the logic language basically as 'untyped', i.e., S_0 is a one-pointed set. The single sort in this one-pointed set could be denoted s_0. In the end, what is important is how we attach an algebra to that single sort, i.e., to specify $\mathfrak{A}(s_0)$. Usually the algebra is the so called *Herbrand universe*. Classical logic programming then speaks about 'predicates' but keep these outside the scope of operators in a signature, so that predicates are really untyped, and truth values are attached not until in semantics. At the same time, however, truth and falsity is treated both syntactically and semantically, so leaving these constants as untyped creates a strange situation, from formal point of view. Classical logic programming therefore sometimes intertwines syntax and semantics.

In order to introduce predicates as operators in a separate signature, and then composing that resulting 'predicate' functor with the term functor, we assume that Σ contains a sort bool, which does not appear in connection with any operator in Ω, i.e., we set $S = S_0 \cup \{\mathrm{bool}\}$, $\mathrm{bool} \notin S_0$, and $\Omega = \Omega_0$. This means that $\mathsf{T}_{\Sigma,\mathrm{bool}} X_S = X_{\mathrm{bool}}$, and for any substitution $\sigma_{X_S} : X_S \to \mathsf{T}_\Sigma X_S$, we have $(\sigma_{X_S})_{\mathrm{bool}}(x) = x$ for all $x \in X_{\mathrm{bool}}$. The composition of the 'predicate' functor with T_Σ is intuitively expected to be the desired 'predicates as terms' functor.

We can now also separate propositional logic from predicate logic, and also decide whether or not to include negation. The key effect in doing this arrangement is that implication becomes 'sentential' where as conjunction (and negation, if included) produces terms from terms.

To proceed towards this goal, let $\Sigma_{PL} = (S_{PL}, \Omega_{PL})$ be the underlying *propositional logic* signature, where $S_{PL} = S$, and $\Omega = \{\mathsf{F}, \mathsf{T} :\to \mathrm{bool}, \& : \mathrm{bool} \times \mathrm{bool} \to \mathrm{bool}, \neg : \mathrm{bool} \to \mathrm{bool}\} \cup \{\mathsf{P}_i : s_{i_1} \times \cdots \times s_{i_n} \to \mathrm{bool} \,|\, i \in I, s_{i_j} \in S\}$. Similarly as bool leading to no additional terms, except for additional variables being terms, when using Σ, the sorts in S_{PL}, other than bool, will produce no additional terms other than variables.

In order to cover the case including 'Horn clauses' and also separating the form of the 'head' and 'body' in those clauses, we introduce the notation $\Sigma_{PL \setminus \neg}$ for the signature where the operator $\neg$ is removed, and $\Sigma_{PL \setminus \neg, \&}$ for the signature where both $\neg$ and $\&$ are removed.

It is convenient also to use set functors disabling the use of variables of particular sort. For S being the set of sorts with bool as a distinguished sort in S we write $\phi^{S \setminus \mathrm{bool}} : \mathrm{Set}_S \to \mathrm{Set}_S$ for the functors defined as $\phi^{S \setminus \mathrm{bool}}(X_s)_{s \in S} = (X'_s)_{s \in S}$, where $X'_s = X_s$ except for $X'_s = \emptyset$. Similarly we define the functor $\phi^{\mathrm{bool}} : \mathrm{Set}_S \to \mathrm{Set}_S$ as $\phi^{\mathrm{bool}}(X_s)_{s \in S} = (X'_s)_{s \in S}$, where $X'_s = \emptyset$ except for $X'_{\mathrm{bool}} = X_{\mathrm{bool}}$.

The set of 'ground terms' over Σ is now

$$\left(\mathsf{T}_\Sigma \; \circ \; \phi^{S \backslash \mathrm{bool}} \right) (X_\mathrm{s})_{\mathrm{s} \in S},$$

and the set of propositional logic formulas is

$$\left(\mathsf{T}_{\Sigma_{PL}} \; \circ \; \phi^{\mathrm{bool}} \right) (X_\mathrm{s})_{\mathrm{s} \in S}.$$

Note how

$$\left(\mathsf{T}_{\Sigma_{PL \backslash \neg, \&}} \; \circ \; \phi^{\mathrm{bool}} \right) (X_\mathrm{s})_{\mathrm{s} \in S}$$

is what we usually see as the two-valued set of boolean truth values.

Sentences in propositional logic is now obviously given by the functor

$$\mathsf{Sen}_{PL} = \mathsf{T}_{\Sigma_{PL}} \; \circ \; \phi^{\mathrm{bool}},$$

and sentences in 'Horn clause logic' can now be given by the functor

$$\mathsf{Sen}_{HCL} = \left(\mathsf{T}_{\Sigma_{PL \backslash \neg, \&}} \times \left(\mathsf{T}_{\Sigma_{PL \backslash \neg}} \; \circ \; \mathsf{T}_\Sigma \right) \right) \; \circ \; \phi^{\mathrm{bool}}$$

so that $(h, b) \in \mathsf{Sen}_{HCL}(X_\mathrm{s})_{\mathrm{s} \in S}$ means that h is an 'atom' and b is a conjunction of 'atoms'. Further, (h, T) is a 'fact', (F, b) is a 'goal clause', and (F, T) is a 'failure'.

A *rule base*, or a *guideline* for decision-support, is then a set of sentences produced by the selected sentence functor.

5 Symbolic Reinforcement Learning

As mentioned before, $\frac{\delta}{\delta x}$ itself is not an operator but '$\frac{\delta}{\delta}$' can be seen as a transformation

$$\Delta^\mathrm{s}_{\mathrm{s}'} = \frac{\delta_\mathrm{s}}{\delta_{\mathrm{s}'}} : \mathsf{T}_{\Sigma, \mathrm{s}} X_S \times X_{\mathrm{s}'} \to \mathsf{T}_{\Sigma, \mathrm{s}} X_S$$

where $\frac{\delta_\mathrm{s}}{\delta_{\mathrm{s}'}}(t, x)$ is written traditionally as $\frac{\delta_\mathrm{s} t}{\delta_{\mathrm{s}'} x}$. Note that we would intuitively expect t to be of some form $\omega(t_1, \ldots, x, \ldots, t_n)$, where $\omega : \mathrm{s}_1 \times \ldots \times \mathrm{s}' \times \ldots \times \mathrm{s}_n \to \mathrm{s}$, or else t is of form $\omega(t_1, \ldots, t_n)$, so that x appears, and is properly typed, within some subterm t^*, say, of type s'', within the list $t_1, \ldots, t_n$.

The possibility to attach derivation rules can be illuminated by adding a 'multiplication' operator $\otimes : nat \times nat \to nat$ into some extension of the NAT signature. We could then add a derivation rule like

$$\frac{\delta_{\mathrm{nat}} x \otimes x}{\delta_{\mathrm{nat}} x} = \mathrm{succ}\big(\mathrm{succ}(0)\big) \otimes x.$$

Fig. 2 The architecture of
GARIC

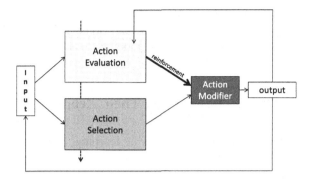

Then the expression

$$\frac{\delta_s t}{\delta_{s'} x} = \frac{\delta_s t}{\delta_{s''} t^{ast}} \otimes \frac{\delta_{s''} t^{ast}}{\delta_{s'} x}$$

would also make sense. Further, updating parameters would be seen in its traditional form

$$x^{new} = x^{old} \oplus \Delta^s_{s'} E(x)$$

where the error function is build up using the operators in the underlying signature.

This provides a general view of 'symbolic gradient descent', which clearly collapses to traditional gradient descent when the algebras of respective operators are the traditional ones. However, non-traditional algebras can be selected, if justified, and more generalized learning rules can be adopted [8, 20].

Communication and 'socially emerging' now plays an important role in this context, and information transfer can even be anticipated in the expressions $\frac{\delta_s t}{\delta_{s''} t^*}$ and $\frac{\delta_{s''} t^*}{\delta_{s'} x}$. The types s, s' and s'', could be annotated, respectively, to 'authority', 'teacher' and 'student', where communication between students, students and teachers, between teachers, and between school and education authority, can be seen as token flow in process models (Sect. 6), where data objects, on the one hand, is manifested as data and information residing in terms over a suitable signature, and, on the other hand, as rules, maintained, updated and used e.g. as guidelines by schools and authorities. Reinforcement involving parameter estimation in such a process model is typically as seen in the model of GARIC [2] (Fig. 2).

The communication patterns are obviously very complicated, and it is not clear what we mean by 'communicating individuals'. Our take is that individuals maintain, and dynamically so, their own underlying logics, either formally or informally, and may alter and change it at will and "on the fly" prior to or during argumentation in a social context. This means that 'individuals communicating' is 'individual logics communication'. This sounds almost like completely impossible to model, but an fascinating efforts of doing that has been made in [29]. As shown in Sect. 4, uncertainty resides in various components within terms, with a number of functions providing those required assignments, then also distinguishing e.g.

between lateral movement of functions or modifying function shapes [21], which can be done analytically, but can also be adapted symbolically by adding various rules for 'lateral moves' or 'modifications of shapes'.

6 Process Drama and Their Related BPMN Processes

Process drama is a method of teaching, and also a 'process' of and 'tool' in teaching. Process drama has emerged as an approach for exploring the multifaceted nature of learning.

In this chapter our fundamental example is teaching history of classical music in secondary school, using process drama, and turning Mozart's biography and life story into a basic set of facts, underlying a basic script for a 'life process' or 'milestones in life' [7]. This enables 'telling the story' about Mozart, and this is the phase where facts are allowed to become accompanied by fictive ingredients, not changing or modifying the facts, but enhancing and enriching the fact. Student activation is one of the most important ingredients in the drama process, and further, observation and assessment of student presence and receptivity. The observation is based on formal representation of logic language based information structures, and the assessment is similarly related to inference mechanisms in that logic.

The connection between 'drama' and 'social' is also very interesting and apparent, and as social choice theory embraces lots of logic, it may be expected that so does drama and drama process theory. Modern aspects of eLearning and social media leads also quickly to a notion of 'distributed drama', something that social media seems not yet to be able to model properly so as to internalize it and provide it as a tool for their users. The learning process description [11], encoded in BPMN, is shown in Fig. 3, with the subprocess described in Fig. 4 is based structure of Freytag's dramatic structure [25]. Nuthall's [37] process descriptions in the *base model* and *"Design Experiment"* have some resemblance with Fig. 4, but Nuthall's process descriptions do not follow any process language, and, in [37], there are no data or information models supporting are being annotated to the process descriptions.

Note also how and where the overall process correlates with the GARIC architecture for reinforcement learning, where the Action Evaluation happens and is operated in the curriculum task forces, and Action Selection results from the course evaluations. Action Modifications take place in the enriched course planning and specifically during the teacher's (τ) storytelling, where the interaction with the students (σ) take place.

Our intuition here is also that parameter tuning and gradient descent techniques apply more to terms, $\frac{\delta t}{\delta x}$, within the education provider, and more to sentences, $\frac{\delta P}{\delta x}$, as appearing e.g. in guidelines and managed by the authority.

Note also that a term t, being a syntactic term in the sense of being a "term over terms, inductively", may be very shallow and even a constant in its most primitive form. In this case a term is comparable to a 'value'. However, a complicated term

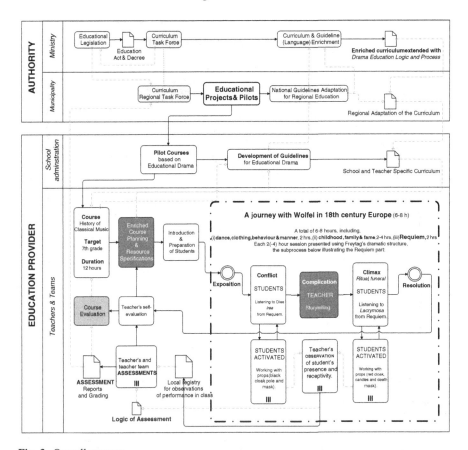

Fig. 3 Overall process

should not be seen as a value, and syntactically it isn't, but actually embraces all the subterms, and subterms of subterms, etc. in it. In this case, such a complicated term is more like a 'report' or summary of assessments. There may be summaries for assessments in music in a school, with such a report denoted t^{music}_{school}. A municipality may then provide statements based on these reports, as a basis for reasoning and decisions within the municipality, and as to be forwarded to national authorities. Such statements are then predicates and could be denoted $P^{music}_{municipality}$, and so on and so forth.

BPMN process design aims at being close to human workflow and human resource flow. BPMN enables not just process information but in particular the communication of process information between participants in the process. BPMN diagrams build syntactically upon four basic categories of elements, namely Flow Objects, Connecting Objects, Artifacts and Swimlanes. Flow Objects, represented by Events, Activities and Gateways, define the behaviour of processes. Start and End are typical Event elements. Task and Sub-Process are the most common

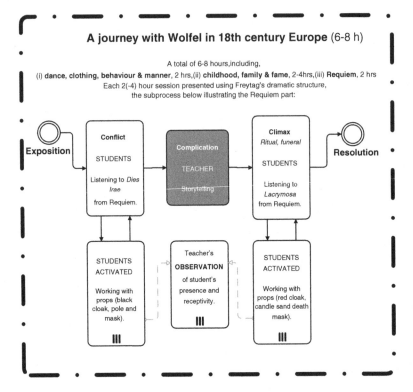

Fig. 4 Mozart's Requiem

Activities. There are three Connecting Objects, namely Sequence Flow, Message Flow and Association. Gateways, as Event elements, handle branching, forking, merging, and joining of paths. A Data Object is an Artifact, and Data Objects are seen to represent data, even if BPMN does not at all specify these representation formats or rules for such representations. In this chapter we view Data Objects as terms, and BPMN tokens as variable substitutions. BPMN provides no semantics but gives some hints where semantic issues eventually will become visible. A Sequence Flow is expected to *show the order that activities will be performed in a Process*, but nothing is said about overlaps. A Message Flow is expected to *show the flow of messages between two participants that are prepared to send and receive them*. Synchrony or asynchrony is intentionally not specified.

7 Conclusion and Future Work

We have suggested a model for assessment and classification of learning both in a individual context as well as in social framework. The underlying logic model being logically formal and formally categorical means that constructions must be comprehended using examples, and some illuminating constructions and examples have been produced in this chapter. However, readers are advised to come up with their own example in their favourite application domains, in order to fully understand to power and depth of these constructions.

Future work involves developments for 'fuzzy sentences' and from there on also involving inference rules and all other logic ingredients making this framework covering all aspects of a logic. Moreover, in doing so, we are able to provide a category of logics, where communication and dialogue between individuals and groups in a social context can be viewed and model using morphisms between between logics as adopted by respective individuals and groups. Terms like $u_{Classroom}^{Teacher}$ can indeed be seen in a narrow sense from classroom point of view only, but also within the whole reinforcement cycle including updating of curricula.

References

1. Adámek, J., Herrlich, H., Strecker, G.: Abstract and Concrete Categories. Wiley-Interscience, New York (1990)
2. Berenji, H.R., Khedkar, P.: Learning and tuning fuzzy logic controllers through reinforcements. IEEE Trans. Neural Netw. **3**(5), 724–740 (1992)
3. Bolton, G. M.: Towards a Theory of Drama in Education. Longman, London (1979)
4. Dewey, J.: Logic: The Theory of Inquiry. Henry Holt & Co, New York (1938)
5. Carlsson, C., Ehrenberg, D., Eklund, P., Fedrizzi, M., Gustafsson, P., Merkuryeva, G., Riissanen, T., Ventre, A.G.S.: Consensus in distributed soft environments. Eur. J. Oper. Res. **61**, 165–185 (1992)
6. Ehrenberg, D., Eklund, P., Fedrizzi, M., Ventre, A. G. S.: enhancing consensus in distributed environments. In: Caianiello, E. R. (eds.) Third Italian Workshop on Parallel Architectures and Neural Networks, pp. 403–406. World Scientific Publishing Co., Singapore (1990)
7. Eklund, H.-T.: Wolfelin matkassa 1700-luvun Euroopassa (A journey with Wolfel in 18th century Europe), unpublished manuscript, in Finnish (2008)
8. Eklund, P.: Network size versus preprocessing. In: Yager, R., Zadeh, L. (eds.) Fuzzy Sets, Neural Networks and Soft Computing, pp. 250–264. Van Nostrand Reinhold, New York (1994)
9. Eklund, P.: Signatures for assessment, diagnosis and decision-making in ageing. In: Hüllermeier, E., Kruse, R., Hoffmann, F. (eds.) IPMU 2010, Part II, CCIS **81**, pp. 271–279. Springer, Berlin (2010)
10. Eklund, P.: The logic and ontology of assessment of conditions in older people. In: Guo, P., Pedrycz, W. (eds.) Human-Centric Decision-Making Models for Social Sciences, Springer, Berlin, Heidelberg (2013)
11. Eklund, H.-T., Eklund, P.: Logic and process drama. In: Presented at the 7th Nordic Congress Drama Boreal, Reykjavík, Iceland, 6–10 August, 2012
12. Eklund, H.-T., Eklund, P.: Logical modelling of process drama for quality assurance of teaching and learning, (2013, Submitted)
13. Eklund, P., Fedrizzi, M., Helgesson, R.: Monadic social choice, multicriteria and multiagent decision making with applications to economic and social sciences. In: Hošková-Mayerová, Š., Kacprzyk, J., Maturo, A., Ventre, A. (eds.) Studies in Fuzziness and Soft Computing, Springer, Berlin, Heidelberg (2013)
14. Eklund, P., Gähler, W.: Fuzzy filter functors and convergence, applications of category theory to fuzzy subsets. In: Rodabaugh, S. E., Klement, E. P., Höhle, E. P. (eds.) Theory and Decision Library B, pp. 109–136. Kluwer, Dordrecht (1992)

15. Eklund, P., Galán, M., Helgesson, R., Kortelainen, J.: Fuzzy terms, Fuzzy Sets and Systems, Elsevier (2013, in press)
16. Eklund, P., Galán, M. A., Ojeda-Aciego, M., Valverde, A.: Set functors and generalised terms. In: Proceedings IPMU 2000, 8th Information Processing and Management of Uncertainty in Knowledge-Based Systems Conference III, pp. 1595–1599 (2000)
17. Eklund, P., Georgsson, F.: Unravelling the thrill of metric image spaces, discrete geometry for computer imagery. In: Bertrand,G., Couprie, M., Perroton, L. (eds.) Lecture Notes in Computer Science 1568, pp. 275–285. Springer, Berlin, Heidelberg (1999)
18. Eklund, P., Gustafsson, P., Lindholm, P., Tony Riissanen.: An Architecture for Topological Concensus Reaching. In: Proceedingsof the AIRO91, Riva del Garda,pp. 110–113 September 18–20 (1991)
19. Eklund, P., Helgesson, R.: Monadic extensions of institutions. Fuzzy Sets Syst. **161**, 2354–2368 (2010)
20. Eklund, P., Klawonn, F.: Neural fuzzy logic programming. IEEE Trans. Neural Netw. **3**(5), 815–818 (1992)
21. Eklund, P., Klawonn, F., Nauck, D.: Distributing errors in neural control. In: Proceedings of the IIZUKA-92, Iizuka, , pp. 1139–1142. Japan (1992)
22. Eklund, P., Rusinowska, A., de Swart, H.C.: Consensus reaching in committees. Eur. J. Oper. Res. **178**, 185–193 (2007)
23. Frege, G.: Begriffsschrift, eine der Arithmetischen Nachgebildete Formelsprache des Reinen Denkens, Halle, 1879. In: van Heijenoort, J. (ed.) English translation in: From Frege to Gödel, a Source Book in Mathematical Logic, pp. 1–82. Harvard University Press, Cambridge (1967)
24. Frege, G.: Über Sinn und Bedeutung. Zeitschrift für Philosophie und philosophische Kritik **100**, 25–50 (1892)
25. Freytag, G.: Technique of the Drama: An Exposition of Dramatic Composition and Art, 3rd edn, Scott, Foresman and company, Chicago, (1900)
26. Gähler, W.: Monads and convergence. In: Generalized functions, convergences structures, and their applications, pp. 29–46. Plenum Press, New York (1988)
27. Goguen, J.: L-fuzzy sets. J. Math. Anal. Appl. **18**, 145–174 (1967)
28. Heathcote, D., Bolton, G.: Drama for Learning. Heinemann, Portsmouth (1995)
29. Helgesson, R.: Generalized general logics, Ph.D. thesis, Umeå University, Umeå, (2013)
30. Hilbert, D., Bernays, P.: Grundlagen der Mathematik I, Die Grundlehren der mathematischen Wissenschaften **40**, Springer (1934)
31. Hilbert, D., Bernays, P.:Grundlagen der Mathematik II, Die Grundlehren der mathematischen Wissenschaften **50**, Springer (1939)
32. Kiernan, P.: Shakespeare's Theory of Drama. Cambridge University Press, Cambridge (1996)
33. Kolb, D. A.: Experiential Learning: Experience as the Source of Learning and Development. Prentice-Hall, New Jersey (1984)
34. Leibniz, G. W.: Lehrsätze über die Monadologie imgleichen von Gott und seiner Existenz, seinen Eigenschafften, und von der Seele des Menschen (1714). Heinrich Köhler (1720)
35. Lloyd, J. W.: Foundations of logic programming. Springer, Berlin (1984)
36. Mill, J. S.: A System of Logic, Ratiocinative and Inductive (1843)
37. Nuthall, G.: Relating classroom teaching to student learning: A critical analysis of why research has failed to bridge the theory-practice gap. Harvard Education. Rev. **74**, 273–306 (2004)
38. Toole, J. O'.: The Process of Drama: Negotiating Art and Meaning. Routledge, New York (1992)
39. Piaget, J.: The Psychology of Intelligence. Routledge & Paul, London (1950)
40. Rusinowska, A., Berghammer, R., Eklund, P., van der Rijt, J.-W., Roubens, M., de Swart, H.: Social software for coalition formation. In: de Swart, H., Orlowska, E., Schmidt, G., Roubens, M. (eds.) Theory and Applications of Relational Structures as Knowledge Instruments II, LNAI 4342, pp. 1–30, Springer (2006)
41. Taylor, P.: The Drama Classroom. Routledge, London (2000)
42. Wagner, B. J.: Dorothy Heathcote - Drama as a Learning Medium. National Education Association, Washington, DC (1976)

Social Network Computation for Life Cycle Analysis by Intelligence Perspective

Wang-Kun Chen and Ping Wang

Abstract In this study, a method is proposed to calculate the product life cycle by fuzzy intelligent scheme. The product life cycle is estimated by three factors: carbon footprint, environmental management system, and environmental performance evaluation. The social network and mechanism of product life cycle assessment can be considered from the three perspectives. (1) Benefits, including organizational interests, individual interests; (2) Cost: including time, financial resources, and negative influence; (3) Difficulty in execution, including the ability of the employees, the support from a company's senior management. Each question can be answered on a five-point scale or directly from subjective judgment. The operational definition of various assessment criteria is also considered. There are nine different strategies proposed within the three factors. The final verification of these strategies was made by comparing their comprehensive evaluation factors. In practice, the approach is relatively straightforward. It also takes into account the product life cycle related to social networks and the impact of subjective and objective factors.

Keywords Social network · Life cycle analysis · Intelligent computation · Fuzzy theory

W.-K. Chen (✉)
Department of Environment and Property Management, Jinwen University
of Science and Technology, New Taipei, Taiwan
e-mail: wangkun@just.edu.tw

P. Wang
Department of Civil Engineering, Chien Hsin University of Science
and Technology, Taoyuan, Taiwan

W. Pedrycz and S.-M. Chen (eds.), *Social Networks: A Framework*
of Computational Intelligence, Studies in Computational Intelligence 526,
DOI: 10.1007/978-3-319-02993-1_16, © Springer International Publishing Switzerland 2014

1 Life Cycle: An Important Social Network

1.1 Social Network

No Human beings can live without community. Interaction of a man with society results in an expansion of interpersonal networks. This can be called social network. A social structure is constructed by a set of individuals or organizations, and connected by tangible or intangible links. Social networks may affect our daily lives. We are often unconsciously affected by the network that we do not realize.

Social network is a tool to realize the feature of this field. It provides rules to examine the composition of a complete societal structure. Researchers have attempted to use a variety of scientific methods to demonstrate the correlation between these networks. Because social networks can be classified as large-scale, medium scale and small scale, it is necessary to develop methodology in terms of the different scales. The large scale social networks may be in global level, compared with medium scale as state-to-state and the small scale as in the community and the family. It is necessary to build a framework to explain all the above scales [6].

A social network, which is an interdisciplinary field of social science, is highly related to the subjects like statistics, psychology, sociology, and geography etc. Mathematical models help us construct theoretical relationship between human behavior and social network observed. The knowledge of social network has been well developed from systematic research by social scientists. Therefore, the terms must be used with rigor. Typical terms in social network, such as the social groups like tribe or family or the social category like gender and education, should be considered more carefully.

In general, social network is invoked to express interpersonal relationship, either by qualitative or quantitative way. A remarkable progress has been made in its theoretical development. It is useful not only to provide more insight into the variation of a social system, but also to facilitate decision-making for social policy.

1.2 Social Network Analysis

The development of information science contributes to the new field of social network analysis. Contemporary sociology has taken into account the analysis of social network where resorting to mathematics is thought as paradigm. The research on network science is helpful to analyze social network in terms of complex forms of networks.

In addition to being a very useful tool in management science, social network analysis introduces many techniques into modern sociology. It also shows its influence in the fields of economics, geography, politics, social psychology, and communication studies etc.

A basic assumption for calculating social network is that there is a social structure in the society. Social interaction with the society, which can be tracked, determines the type of social structure. The society can be divided into several social units. Among the units, there exists a bond connecting each other by various societal contacts. Each social unit follows certain rules of social behavior to contact with each other. The analysis of social network requires a clear understanding of these behaviors.

Analysis of social network mainly comes from a macro point of view, so it may be criticized for not considering individual perception. Another criticism is the dynamics of social event that has to be considered is neglected in the current calculating process. Therefore, the model developed in this study has to take into account the dynamic and static changes. In addition to sociological theory, computational intelligence perspective may be assistant to find a solution to this dilemma. A detailed derivation of the calculation will be introduced in the following pages.

Network analysis concept is introduced in this research to cope with complex process of analysis. It helps us clearly differentiate the types of relation, such as singular one or complex one, in social network [7, 8, 9].

1.3 Life Cycle in the Social Network

Another social phenomenon concerned is what kind of social network will present when a product (or substance) enters a social network or someone's life. What will be the morphological status of this material in the social network? Or where will it be and what will it become finally? How will the social networks determine the flow of these materials? In addition, will it stay at a stationary status, or moving in a dynamic way in the social network? The difficulty is increased in analysis when these problems are considered. Consequently, another research area referred as Life-cycle analysis (LCA) arises.

Due to global climate change, the life cycle of product gains more attention by the public. A product with short life cycle is to waste the resources earth. As a result products should be designed to have longer life cycle from the viewpoint of environmental protection. The distribution of products in a market is influenced by many factors. For example, consumer psychology, which is an invisible network, is a factor has to be considered in the analysis of products LCA.

The most popular topics in the field of environmental issues, including "green building", "environmental impact assessment", "greenhouse gases", "carbon footprint"… are all associated with life cycle assessment. It refers to the analysis of environmental impact caused by a product in different stages of production, use, disposal, recycling, and re-use. It involves three main fronts as product, cost and the environment. Therefore, it has many applications in different management domain. It involves various management aspects, such as finance and administration. All levels are required to do considerable social network integration (Fig. 1).

Fig. 1 The application
network of life cycle analysis

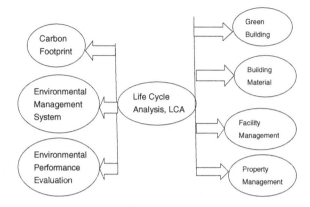

2 Life Cycle Analysis for a Sustainable Development Society

LCA is also identified as life-cycle assessment, life cycle analysis, eco-balance analysis, or cradle-to-grave analysis. The meaning of "from-cradle-to-grave" is specially referred to the process of distribution, use, repair and maintenance, and disposal or recycling. From the sustainable development point of view, the life cycle of material has to be taken care in order to avoid unnecessary disturbance of environment. Thus it also helps the decision maker to avoid a narrow outlook on environmental concerns.

2.1 The Concept of LCA

Life Cycle Assessment refers to the analytical assessment of environmental impact of a product from different stages of production, use, waste and recycling. The so-called environmental impact includes energy use, resource consumption, and pollution emissions. Therefore, LCA is a complex assessment to analyze the relationship between product and ecological environment. It involves different theories, assessment techniques and standards. These also constitute the social network of LCA.

The major criteria in life cycle assessment are made based on the following

- International Standards Organization, ISO 14000 environmental management standards.
- Carbon footprint.
- Environmental performance evaluation, EPE.
- Environmental impact assessment indicators.
- Green building and its assessment indicators.

- Environmental quality monitoring and technology, mainly in the air, water, and waste.
- The theory and technology of greenhouse gases and waste reduction.
- Other necessary theory and practice.

2.2 The Concept of Sustainable Development Society

The theoretical approach of LCA for a sustainable society is necessary. In this section, how to calculate the LCA from the perspective of social network analysis will be introduced.

LCA is a method to evaluate environmental impacts related to all the stages of a product's existence. It helps us to build a sustainable development society because it compiles a list of relevant energy and material inputs. It also evaluates the possible impacts related to identified inputs and releases; and interprets the consequences to help make a more informed judgment.

Social network analysis has been found in various applications [1, 2, 3]. Network analysis indeed has also been applied to many different fields of academic disciplines, and practical applications such as preventing money laundering and terrorism. Although there has been much progress achieved by previous researchers [4] in extending the use to systematic social network analysis [5] network analysis has not been used in the LCA study for sustainable society.

Social phenomena is conceived primarily by assumption and investigated through the properties of relations between and within units, instead of the properties of these units themselves. This is the concept of social interaction that has to be understood in the approach of social network study. The life cycle analysis is a new approach to perform integrated sustainability assessment and impact analysis through the product's life in the society.

2.3 LCA as a Social Network Analysis

LCA also contains a lot of social networks. From the social science aspect, these social networks include the network of economics, sociology, biology, anthropology, geography, communication, information, social psychology, social culture, and organizational behavior study etc. The four steps related to social networks are described in the following.

The first step is to set a goal and scope in order to make a relevant evaluation process, and to assure that the results of assessment are effective and applicable.

Secondly, a life-cycle inventory analysis is invoked, which includes two categories of data collection and inventory calculation. The process is mainly to construct the quantitative system of environmental inputs and outputs.

The third step is the impact evaluation of LCA. It is essentially an application of LCA results to assess the degree of environmental impact in the product's life cycle.

The fourth step is the interpretation of life cycle. It is mainly a stage to combine the inventory analysis and impact assessment results according to the purpose and scope defined. Additionally, the conclusions will be provided to decision makers. Therefore it can be used as a reference for selection of low-polluting materials, improvement of production process, or enhancement of product design and production decisions.

3 Traditional Method to Calculate LCA by Statistical Analysis

In this section, how to use statistical methods to calculate LCA will be discussed. A few commonly used indicators, including carbon footprint, environmental management systems, and environmental performance evaluation are introduced. These tools can help us perform the LCA social network analysis.

3.1 Life Cycle Assessment

Life cycle assessment can be divided into four parts: Goal and scope definition (GSD), Life cycle inventory analysis (LCIn), Life cycle impact assessment, LCIm) and Life cycle interpretation (LCIt). The purpose of GSD is to provide a clear definition of the goal and scope of assessment task. The LCIn includes data collection and calculation procedure. The purpose of this stage is to quantify the overall production system via the analysis of various inputs and outputs. The purpose of LCIm stage is to quantitatively evaluate the extent of environmental impact during the product's life cycle. The final stage, LCIt, aims to integrate the results of the inventory analysis and impact assessment, and uses it as the basis of internal reference or directs application to improve the production. The above steps are shown in Fig. 2.

LCA is a method, which aims to conduct a detailed analysis of environmental impact of a product by the four procedures of definition, investigation, evaluation and interpretation. The assessment stage through the whole life cycle includes: energy consumption, water utilization, potential effects of global warming, ozone depletion potential, acid rain potential, eutrophication potential, and photochemical pollution potential etc. There have been different methods of investigation, computation and presentation used in the assessment. Therefore, the calculation of life-cycle is also involved in the social network analysis.

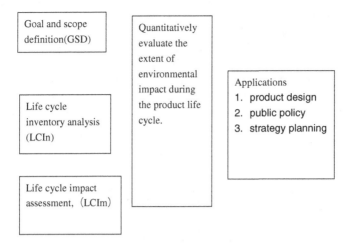

Fig. 2 Framework of the social network, direction, and application of LCA (by ISO 14040)

3.2 Carbon Footprint Analysis

The so-called carbon footprint is direct and indirect carbon emissions due to human activities. Carbon footprint may also be defined as direct and indirect emission of carbon dioxide of an activity or a product from raw materials to the final stage of the product throughout the life cycle. Carbon footprint can be represented by the daily release amount of greenhouse gases used by each person, family or company. The carbon dioxide is a kind of greenhouse gas; it can be used to measure the impact of human activities on the environment. The current carbon footprint application can be divided into four levels of personal carbon footprint, product carbon footprint, corporate carbon footprint, and country or city carbon footprint. It is the different scales of social networks we are concerned about.

Personal carbon footprint is the process estimating the carbon emissions caused by daily life of everyone in food, clothing, housing, and transportation. Two approaches commonly used by international communities is the top-down and bottom-up estimation method. The top-down approach is to calculate the carbon footprint of average profile among families or income groups in a country. It is based on the household income and expenditure survey, and supplemented by the analysis of environmental inputs. While the bottom-up method is to use a carbon footprint calculator to estimate the actual consumption in accordance with the individual's daily life and traffic patterns.

The products carbon footprint is the greenhouse gas emissions on a single product manufacture due to the fuel use and process. It concerns the entire "cradle-to-grave" process, from use to disposal stages. Corporate carbon footprint in addition includes the non-productive activities. For example, the carbon emissions of the relevant investment are also required to expose the scope of the corporate carbon footprint. The carbon footprint of the country or the city is mainly focused on the emissions generated by the entire consumption of the country's overall materials and energy.

The purpose of carbon footprint calculation is to understand how carbon is generated because the generation of carbon is inevitable by human activities and production. The ultimate goal is to make equilibrium between carbon generation and natural balance. The generation of carbon footprint is owing to the flow of material in social networks. Due to the complexity of the social network, we need to take into account the different calculation methods to estimate. It is necessary to introduce intelligent system in the study.

3.3 Environmental Management System

The purpose of environmental management is to establish a system to enable enterprises to identify possible environmental problems and make improvements throughout the product life cycle of production, sales, product use and waste, as well as to reduce the impact on the environment. In the process of reducing environmental impact, it is often accompanied by the improvement of production, the reduction of energy depletion and the waste of raw materials, therefore the environmental management system at the same time not only enhances the production efficiency, but also increases the revenue of enterprise. ISO 14000 is a series of standards developed for corporate environmental management. It includes environmental management systems, ISO 14001, ISO 14004, and environmental management tools, such as product life cycle assessment, corporate environmental report, and green mark, etc.

The ISO 14000 series of international environmental management standard contains much practical experience and research results in the field of environmental management in recent years. It includes a series of international standards such as the environmental management system (EMS), environmental audit (EA), life cycle assessment (LCA) and environmental standards (ES).

ISO plans to promulgate more than 100 issued of ISO 14000 series of standards, now 24 of them are in a process of drafting, and five of them has been officially released. The five standards are divided into two categories: one is the basic requirements of environmental management, and the other is the auditing standards of environmental certification. The uses of this standard are completely different from other environmental quality standards or emission standards. First, it is a voluntary standard; second, it is a management standard providing a set of standardized environmental management system and management methods for all types of organizations.

ISO14000 offers the following benefits:

- to obtain "green pass" of international trade
- to enhance competitiveness and expand market share
- to build up excellent corporate image

- to improve product performance, manufacturing "green products"
- to reform process equipment and reduce energy consumption
- to prevent pollution, and to protect environment
- to avoid economic losses caused by environmental issues
- to improve employees' environmental literacy
- to improve the internal management level
- to reduce environmental risks and achieve the sustainable business.

3.4 Environmental Performance Evaluation

Another assessment indicator of social network is environmental performance evaluation. It is a procedure that systematically and continuously makes measurement and evaluation on environmental performance of an organization. A formal announcement of environmental performance evaluation from ISO14031 was issued in October 1999, targeting the organization's management system, operating system, and even the surrounding environmental conditions. The environmental management performance of the organization is the main focus of environmental performance evaluation. Considering cost-effectiveness, it is possible to focus on important environmental parameters to establish continuous monitoring system. The evaluation results are distributed to the stakeholders through the social network.

The spirit of environmental management system requires continuous improvement. And the most important issue of environmental performance evaluation is to select the appropriate environmental performance indicators. The choice of indicators is relevant to the follow-up performance assessment of its effectiveness. The presented EPE indicators or information selected within the organization, whether qualitative or quantitative, should be simple and easy to understand. The indicators also reflect the properties of the organization and its scope. According to the size of the object and purpose of the scope for assessment, the environment indicators can be divided into environmental condition indicators (ECIs) and environmental performance indicators (EPIs). The EPI can be divided into management performance indicators (MPIs) and operation performance indicators (OPIs). Therefore, the evaluator can assess the organization's operating system and management system, or the outside environment for each organization (Fig. 3).

Selection of environmental performance indicators includes the state of environmental condition indicators (ECIs), management performance indicators (MPIs), and operating performance indicators (OPIs), which will be elaborated on as follow.

Environmental condition indicators (ECIs): This is the most fundamental issue to be considered in the three indicators of environmental performance evaluation. It provides necessary information to help organizations choose appropriate environmental management system and indicators of operating system. The environmental situation around the organizations includes local area as well as global

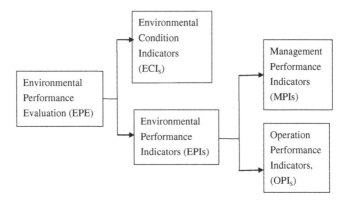

Fig. 3 Structure of the environmental performance evaluation

environmental conditions. The scope includes air, water, land, plants, animals, and even human health.

Management performance indicators (MPIs): This indicator is a direct reflection of the efforts of the management performance in the organization. This is also the most direct source of performance indicators that can be found in the environmental management system. The organization's environmental policy, objectives and targets can be used, and MPIs can also be used to evaluate whether the environmental performance indicators have achieved the goal or not.

Operating performance indicators (OPIs): The operating system of an organization includes the design and operation of the plant, equipment, raw materials and energy supply and demand system. Therefore, the indicators of operating system were categorized as (1) material: process recycling and reuse of materials, (2) energy resources (including fuel), (3) products: Main products or by-products, (4) pollutants: air, water, waste, toxics, noise and so on.

These three indicators can also be used as the performance indicators in the implementation of environmental management system (ISO 14001).

4 A Framework of LCA by Fuzzy Theory

4.1 Social Network Analysis of LCA, CF, EMS, and EPE

The analysis of product life cycle through social networks is affected by carbon footprint, environmental performance evaluation, and environmental management systems. The operational definition, including carbon footprint, environmental performance evaluation, and environmental management system of a product life cycle in the social mechanism, was established. Among them there are eight mechanisms available for evaluation. They are derived from the analysis discussed in previous section.

Life cycle analysis requires the whole process within the organizations, such as product design, production, use, end-of-life and recycling, to be controlled without affecting the environment. In the specification of TC207, which is a technical document for product of environmental impact, the life cycle assessment of the products is specifically requested. It is not only to evaluate the LCA at each stage, but also to help a company improve their environmental quality (Table 1).

4.2 Evaluation Model of LCA, CF, EMS, and EPE

Based on the above discussion, the social network and mechanism of product life cycle assessment can be considered from the three perspectives. (1) benefits, including organizational interests, individual interests (2) cost: including time, financial resources, and negative influence (3) difficulty in execution, including the ability of the employees, the support from a company's senior management. Each question in this table can be answered on a five-point scale or directly from subjective judgment. The operational definition of various assessment criteria are shown in Table 2.

A horizontal tree diagram can be established based on the above assessment mechanism. The fuzzy operation system can be introduced into this framework for further evaluation of the optimal results (Fig. 4).

4.3 Implementation of Fuzzy Calculation and Procedures

The applications of fuzzy computing theory to the case of LCA are described in following sections.

(1) Define the linguistic variables and its triangular fuzzy number

First, the individual evaluation factors are calculated by assessment of fuzzy measurement. Quintile distinction is used to measure, that is, "low", "low", "medium", "high", and "high". As for the good and bad performance of this factor, another scale is used to measure, that is, "absolutely not agree", "do not agree", "not sure", "agree", and "absolutely agree". Because personal knowledge is different, the larger number represents the degree which is closer to "high" and "strongly agree". Otherwise, it is closer to the extents of "low" and "strongly disagree".

(2) Fuzzy evaluation of the weighting factor

The first step of evaluation is to carry out the assessment of relative significance of each factor at each level of the tree structure. The various experts subjectively

Table 1 Operational definition and social network mechanisms of product life cycle

Social network	Mechanism	Operational definition	Reason for evaluation
Carbon footprint	The lowest carbon footprint	Use the material with the lowest carbon footprint	The organization could use these two strategies to enhance life cycle of product according to its interest
	The reasonable carbon footprint	Use the material with reasonable carbon footprint	
Environmental management System	National	environmental standard	The environmental level set up by the government
	The organization could use	environmental management system with three different levels according to its interest. Each strategy will cause various life cycle of the product	
		Environmental accounting	An accounting tool to make internal business decisions, mainly for environmental management activities
Corporation		environmental policy	The environment policy established by the corporation
Environmental performance Evaluation	Environmental condition indicators	The environmental condition of the corporation, mainly referring to the environmental factors	The organization could use environmental performance evaluation method with three different levels. Each of them will cause various life cycle of the product
	Management performance indicators	The management performance of the corporation, mainly to the management level	
	Operating performance indicators	The operating performance of the corporation, mainly to the employees	

Table 2 Assessment index and its operational definition

Objective	Mechanism	Rating indicator	Operational definition
Maximum benefits	Organizational interests	Increasing resources	Will it increase the resources of the organization when import of this program?
		Increasing efficiency	Will it increase the efficiency of the organization's work when import of this program?
		Promoting organizational reputation	Will it increase the reputation of the organization when import of this program?
		Promoting organizations identity	Will it increase the organization's identity of the staff when import of this program?
	Individual interests	Personal wages	Will the import of this program increase the personal salary of an employee?
		Personal learning	Will it be beneficial to employees' personal learning when import of this program?
		Personal growth	Will it be beneficial to employees' personal growth when import of this program?
Lowest cost	Time	Leading time	What is the length of leading time needed to import this program?
		Execution time	What is the length of execution time needed to import this program?
		Waiting time	What is the length of waiting time needed to import this program?
	Financial resources	Overall expenses	What are the overall expenses of investment needed to import this program?
		Monitoring costs	What are the monitoring costs of investment needed to import this program?
		Training costs	What is the training costs of investment needed to import this program?
		Communication costs	What is the communication costs of investment needed to import this program?
	Negative influence	Working pressure	Will it increase the pressure of employees when import of this program,?
		Slow decision	Will it slow the decision-making of company when import of this program?
		Organizational conflict	Will it result in the company's organizational conflict when import of this program?

(continued)

Table 2 (continued)

Objective	Mechanism	Rating indicator	Operational definition
Easy to implement	Regulatory requirements	International standard	Is it an international standard for the imported program?
		Domestic regulations	Is it a domestic laws or regulations for the imported program?
		Company policy	Is it a company's internal regulations or policies for the imported program?
	Technical maturity	Market available	Is it a proven technology already on the market for the imported program?
		Staff competency	Does the employee have the capability for the implementation of the imported program?
		Company capability	Does the company have the capability for the planning and execution of the imported program?
	Degrees of difficulty	Unexpected difficulties	Will there be any unexpected difficulty for the execution of this program?
		Company consensus	Will there be adequate organizational consensus for the implementation of this program?
		Social consensus	Will there be adequate social consensus for the implementation of this program?

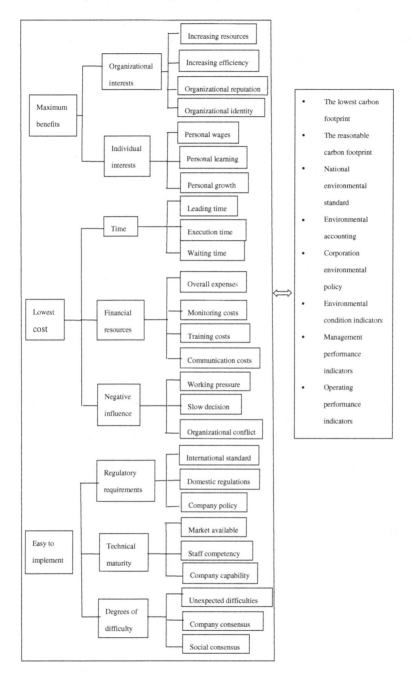

Fig. 4 A horizontal tree diagram for the LCA social network

determine the linguistic variables for each factor as $F_{ij} = (LF_{ij}, MF_{ij}, UF_{ij})$. The comprehensive weighting factor F can be obtained by adopting the "Fuzzy addition "and" Fuzzy multiplication" as the following.

$$
\begin{aligned}
\overline{F}_j &= \frac{1}{n}\left(\sum_{i=1}^{n} LF_{ij}, \sum_{i=1}^{n} MF_{ij}, \sum_{i=1}^{n} UF_{ij}\right) \\
&= \left(\frac{1}{n}\sum_{i=1}^{n} LF_{ij}, \frac{1}{n}\sum_{i=1}^{n} MF_{ij}, \frac{1}{n}\sum_{i=1}^{n} UF_{ij}\right) \\
&= \left(\overline{LF}_{ij}, \overline{MF}_{ij}, \overline{UF}_{ij}\right)
\end{aligned}
\tag{1}
$$

Secondly, obtain the relative importance of the factors between all levels through the inverse-fuzzy and standardization.

$$
DF_j = \left|\left(\overline{UF}_j - \overline{LF}_j\right) + \left(\overline{MF}_j - \overline{LF}_j\right)\right|/3 + \overline{LF}_j
\tag{2}
$$

After obtaining the non-fuzzy value, apply the normalization procedures to calculate by the following formula.

$$
DF_j = DF_j / \sum_{j=1}^{m} DF_j
\tag{3}
$$

(3) Measure of the performance of evaluation factors

A method considering the maximum benefit, lowest cost, and most likely performance is adopted to measure the impact of all factors. The linguistic variables is used in the study to measure the extent, that is "strongly not agree", "not agree", "uncertain", "agree" to "strongly agree". In actual operation, the linguistic variables of the invited experts have to change to the triangular fuzzy numbers.

(4) Comprehensive evaluation of individual factors

After processing the factors above, a group of fuzzy triangular scale, Rij, is obtained to represent the performance of a factor. Because the influence of each factors for assessing the control and coordination mechanisms is different, the equation $R_{ij} = (LR_{ij}, MR_{ij}, UR_{ij})$ has to multiplied by the weight factor DF_{ij} so as to obtain the value of contribution for each factor on the overall evaluation, that is,

$$
\begin{aligned}
E_{ij} &= DF_i \otimes R_{ij} \\
&= \left(LE_{ij}, ME_{ij}, UE_{ij}\right)
\end{aligned}
\tag{4}
$$

where $\otimes$ denotes multiplying operation. Representing the opinion of experts, E_{ij} denotes the contribution of the jth factor to the performance of the overall evaluation mechanism. When two or more experts involved in assessing, the average

value method can be used to calculate the results of expert judgment, the formula
is as follows:

$$Em_j = \frac{1}{n}\left(E_{1j} \oplus E_{2j} \oplus \cdots E_{ij} \oplus \cdots E_{nj}\right) \forall j \tag{5}$$

where $\oplus$ denotes fuzzy addition operation. E_{mj} are the results of comprehensive
evaluation by experts on the jth factor.

4.4 Fuzzy Comprehensive Evaluation

If there are m factors, then the overall performance assessment of the value is the
cumulative contribution of all m factors, which can be formulated as

$$Sm = \sum_{j=1}^{m} Em_j \tag{6}$$

where, Sm is the expert's evaluation results of the overall performance results. If
the performance value is higher, then it has better suitability of the mechanism.

5 Case Study of LCA by Computational Intelligence Perspective

5.1 Case Study of LCA Social Network Calculation

It is feasible to take the advantage of fuzzy intelligent system in life cycle
assessment, including the assessment of CF, LCA and EPE. A comprehensive
evaluation results obtained through fuzzy intelligent calculation methods is possible when a variety of options are available. A case study for a building is
conducted as an example because the materials and equipments for construction as
well as the facilities installed for living function have their respective life cycles.
From the view point of LCA, the entire building development is merely a continuous consumption process of the earth's resources. The construction works are
all energy consumption, such as foundation excavation, building construction,
waste disposal, use of building materials, remodeling works, annual regular
maintenance work after completion of the construction, as well as dismantling
when reaching its lifetime limit, etc.

From the life-cycle perspective, the energy consumption of buildings can be
roughly divided into the following main stages: 1. building materials production
and transportation, 2. constructing, 3. daily use, 4. renovation, 5. demolition and
waste treatment, 6. building materials recycling. The life cycle of a building can

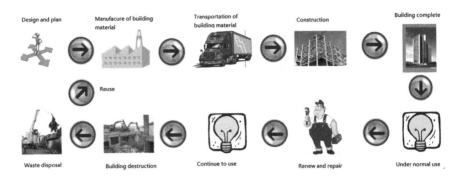

Fig. 5 The flow chart of building life cycle

also be divided into the stages of planning and construction, use, demolition, and waste. The main consideration in LCA is the assessment of carbon dioxide emission. It is possible to gain a more favorable outcome of the building LCA by the fuzzy intelligent algorithm. The flow chart of building life cycle is shown in Fig. 5.

5.2 Fuzzy Intelligent LCA Model of a Building

In this study an attempt is made to conduct LCA while achieving three goals, that is, the largest benefits, the lowest cost, and the easiest implementation. There are three levels under the goal of evaluation index. Shown in Table 3 are the results of overall applicability of the third level factors (Fig. 6).

The proposed programs are shown in Table 4. In Table 5, A1 represents The lowest carbon footprint; A2 represents The reasonable carbon footprint; A3 represents National environmental standard; A4 Environmental accounting; A5 represents Corporation environmental policy; A6 represents Environmental condition indicators; A7 represents Management performance indicators; and A8 represents Operating performance indicators.

The value of the performance of various options can be found after the de-fuzzification process. Table 4 is the results of overall applicability of the second level factors. The sorting of the applicability for various programs are listed in Table 4. As revealed in Table 4 from a cost perspective, option A5 is a better result. From the maximum benefit perspective, scenario A3 is better. From the implementation perspective, the program A8 is the easiest to perform.

Table 5 shows the results of overall applicability of the second level factors. In Table 5, A1 represents organizational interests; A2 represents individual interests; A3 represents time; A4 represents financial resources; A5 represents negative influence; A6 represents Regulatory requirements; A7 represents Technical maturity; and A8 represents degrees of difficulty.

Table 3 The results of overall applicability of the third level factors

Objective	The valuation indicators of the second level	Evaluation index of the third level	(LRj, MRj, URj)			Weighting value	(LEj, MEj, UEj)		
Maximum benefits	Organizational interests	Increasing resources	40	50	60	0.0554	2.216	2.77	3.324
Lowest cost		Increasing efficiency	40	50	60	0.0665	2.66	3.325	3.99
		Promoting organizational reputation	60	80	80	0.0859	5.154	6.872	6.872
		Promoting organizations identity	60	80	80	0.0693	4.158	5.544	5.544
	Individual interests	Personal wages	20	20	40	0.0309	0.618	0.618	1.236
		Personal learning	80	100	100	0.0634	5.072	6.34	6.34
		Personal growth	60	80	80	0.0683	4.098	5.464	5.464
Lowest cost	Time	Leading time	60	80	80	0.0345	2.07	2.76	2.76
		Execution time	80	100	100	0.0373	2.984	3.73	3.73
		Waiting time	60	80	80	0.0191	1.146	1.528	1.528
	Financial resources	Overall expenses	80	100	100	0.069	5.52	6.9	6.9
		Monitoring costs	60	80	80	0.0411	2.466	3.288	3.288
		Training costs	60	80	80	0.0345	2.07	2.76	2.76
		Communication costs	60	80	80	0.0197	1.182	1.576	1.576
	Negative influence	Working pressure	40	50	60	0.0217	0.868	1.085	1.302
		Slow decision	60	80	80	0.0396	2.376	3.168	3.168
		Organizational conflict	40	50	60	0.033	1.32	1.65	1.98

(continued)

Table 3 (continued)

Objective	The valuation indicators of the second level	Evaluation index of the third level	(LRj, MRj, URj)			Weighting value	(LEj, MEj, UEj)		
Implementation of the most vulnerable	Regulatory requirements	International standard	60	80	80	0.0335	2.01	2.68	2.68
		Domestic regulations	40	50	60	0.0317	1.268	1.585	1.902
Maximum benefits		Company policy	40	50	60	0.0229	0.916	1.145	1.374
	Technical maturity	Market available	80	100	100	0.0328	2.624	3.28	3.28
		Staff competency	60	80	80	0.0164	0.984	1.312	1.312
		Company capability	60	80	80	0.0221	1.326	1.768	1.768
	Degrees of diffic	Unexpected difficulties	60	80	80	0.0131	0.786	1.048	1.048
		Company consensus	60	80	80	0.0176	1.056	1.408	1.408
		Social consensus	60	80	80	0.019	1.14	1.52	1.52
						1	58.088	75.124	78.054

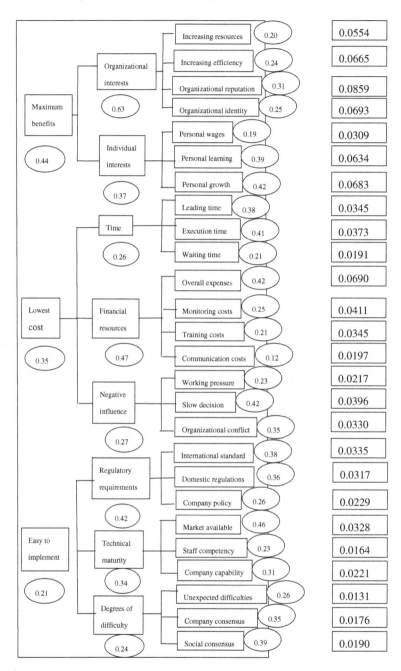

Fig. 6 Calculating results of the social network weighting factor for a building material in its life cycle

Table 4 The results of overall applicability of the second level factors

Category	Program	Maximum benefits			Lowest cost			Easy implementation			Overall	Sequence
CF	A1	40	51	56	60	66	77	23	31	42	53.64	8
		49			67.6			32				
	A2	52	63	75	54	59	68	50	63	71	61.59	5
		63.3			60.3			61.3				
EMS	A3	68	74	80	33	42	55	70	79	85	61.34	6
		74			43.3			78				
	A4	50	55	64	52	56	64	65	72	74	59.71	7
		56.3			57.3			70.3				
	A5	54	63	72	72	79	83	73	80	85	73.03	2
		63			78			79.3				
EPE	A6	53	62	69	61	68	75	78	76	82	67.9	4
		61.3			68			78.6				
	A7	61	65	72	68	74	76	65	72	79	70.19	3
		66			72.6			72				
	A8	68	73	79	71	78	82	73	80	85	76.2	1
		73.3			77			79.3				

Table 5 The results of overall applicability of the second level factors

Category	Proposed program	A1	A2	A3	A4	A5	A6	A7	A8
CF	Lowest carbon footprint	52	47	69	67	50	35	34	31
	Reasonable carbon footprint	65	61	57	62	60	57	61	63
EMS	National environmental standard(NES)	73	78	41	44	47	74	77	79
	Environmental accounting(EA)	52	57	53	54	59	67	71	72
	Corporation environmental policy(CEP)	60	64	62	67	69	74	76	78
EPE	Environmental condition indicators(ECIs)	60	62	68	64	69	75	76	79
	Management performance indicators(MPIs)	65	67	72	71	73	69	71	74
	Operating performance indicators(OPIs)	71	74	76	76	79	74	78	83

6 Conclusions

In this study, a method is proposed to calculate the product life cycle by fuzzy intelligent scheme. The product life cycle is estimated by three factors: carbon footprint, environmental management system, and environmental performance evaluation. Each factor is composed of various indicators of social network. Fuzzy operation logic is developed in this study for choosing the optimum indicator. The influence on life cycle assessment in different cases has been evaluated from three aspects, that is, maximum benefit, lowest cost, and easiest execution.

There are nine different strategies proposed within the three factors. The final verification of these strategies was made by comparing their comprehensive evaluation factors. In practice, the approach is relatively straightforward. It also takes into account the product life cycle related to social networks and the impact of subjective and objective factors. This preliminary study has provided useful information for further in-depth analysis in the future.

References

1. Anheier, H.K., Gerhardsen, J., Romo, F.P.: Forms of capital and social structure of fields: examining Bourdieu's social topography. Am. J. Sociol. **100**, 859–903 (1995)
2. De Nooy, W.: Fields and networks: correspondence analysis and social network analysis in the framework of field theory. Poetics **31**, 305–327 (2003)
3. Senekal, B.A.: Die afrikaanse literêre sisteem: 'n eksperimentele benadering met behulp van sosiale-netwerk-analise (SNA), LitNet Akademies **9**(3) (2012)
4. Ronald, S.B.: Brokerage and Closure: An Introduction to Social Capital. Oxford University Press, Oxford (2007)
5. Pinheiro, C.A.R.: Social Network Analysis in Telecommunications. Wiley, New York. p. 4 (2011)
6. Burt, R.S.: Structural Holes: The Social Structure of Competition. Harvard University Press, Cambridge. ISBN 978-0-674-84371-4 (1995)
7. Freeman, L.: The Development of Social Network Analysis: A Study in the Sociology of Science. Empirical Press, Vancouver, ISBN 1-59457-714-5 (2004)
8. Freeman, Linton: The Development of Social Network Analysis. Empirical Press, Vancouver (2006)
9. Wasserman, S., Faust, K.: Social network analysis in the social and behavioral sciences. In: Social Network Analysis: Methods and Applications. Cambridge University Press, Cambridge. pp. 1–27 (1994)

Analyzing Tweet Cluster Using Standard Fuzzy C Means Clustering

Soumya Banerjee, Youakim Badr and Eiman Tamah Al-Shammari

Abstract Since the inception of Web 2.0. the effort of socializing, interacting and referencing has been substantially enhanced.this is completely aided through the various means of social network expansions like blogging, public chat rooms and social networking sites such as Facebook, Twitter etc. Behavior on these websites leaves an electronic trail of social activity which can be analyzed and valuable information can be discerned. The development of such analysis has become phenomenal to foster psychological analysis, behavioral modeling and even commercializing the business activities under those paradigms itself. Therefore, micro-blogging service *Tweeter* recently has gained much interest to social network community with the trend of its Follower/Following Relationship, Mentions, trends, retweet, Twitter Lists etc. and the result of such impact could be realized while investigating diversified tweet clusters under the same community and under the same relevant discussion of topic. This chapter initiates a novel idea to analyze the random tweet cluster and its relevant trend through computational intelligence e.g. through *Standard Fuzzy C Means clustering*. The idea solicits and introduces a better method of clustering with more number of actually found dynamic clusters. Results have been evaluated with broader implication of analysis and research in futuristic Tweeter network.

Keywords Tweeter · Cluster analysis · Standard fuzzy C means algorithm · Blogging · C mean clustering

S. Banerjee (✉)
Department of Computer Science, Birla Institute of Technology, Mesra, India
e-mail: dr.soumya@ieee.org

Y. Badr
National Institute of Applied Sciences (INSA-Lyon), Villeurbanne, France
e-mail: youakim.badr@insa-lyon.fr

E. T. Al-Shammari
College of Computing Science and Engineering, Kuwait University, Kuwait City, Kuwait
e-mail: eiman.tamah@gmail.com
URL: http://www.isc.ku.edu.kw/eiman/

W. Pedrycz and S.-M. Chen (eds.), *Social Networks: A Framework of Computational Intelligence*, Studies in Computational Intelligence 526, DOI: 10.1007/978-3-319-02993-1_17, © Springer International Publishing Switzerland 2014

1 Introduction

They say,—"*Man is not an island*" because every man has the inherent need to socialize for survival. People interact with their close friends; they may interact with other people with whom they share common interests which have greatly been aided by the advent of the various means to expand ones social circle like blogging, public chat rooms and social networking sites such as Facebook, Twitter etc. Behaviour on these websites leaves an electronic trail of social activity which can be analysed and valuable information can be discerned. Analysing the tremendous amounts of information flow on these social networks provides an opportunity to study the human psyche for sociologists and since people are spending so much time on these sites, it has obviously bright commercial prospects for advertising and e-commerce. Individual behaviour of the actors in the social network can be used to categorize their interest and it can be used to deliver relevant advertisements, and information that is relevant to them based on their behaviour. Quite recently browser plug-ins to discover information retrieved by friends, as well as cooperative web search tools has been developed.

Considering the relevance of trend analysis (includes **Follower/Following Relationship, Mentions, Trends, Retweet, Twitter Lists**) on micro-blogging service Twitter (it is an online social networking service and micro-blogging service that enables its users to send and read text-based posts of up to 140 characters, known as "tweets") and its users and to investigate meaningful results focussing mainly to explore users relevance to some particular search query, several contemporary clustering algorithms have been developed. Subsequently, it has been observed that all the associated tweet parameters like tweet frequency per week, following count, follower count, retweet expectancy with in-degree and out degree of tweet directions, become more uncertain and random, hence a novel method of deterministic clustering application has been initiated in this chapter envisaging *Standard Fuzzy C mean* as generic clustering approach. This method is comparatively fresher in the social network analysis domain with *better number of clusters, and degree of fuzziness*.

The outline of the chapter is as follows:

Section 2 describes the detailed parameters services and theoretical studies belong to tweeter domain. Section 3 introduces the recent breakthrough of tweeter analysis and inference using different computational intelligence approaches. Section 4 elaborated about the trend of tweet data and acquisition process followed by relevance of *Standard Fuzzy C means clustering* approaches in Sect. 4.1. Section 4.3 gives an idea to develop proposed model of Fuzzy C means on tweeter data set and discusses the experimental results. A comparison with conventional clustering with tweet data set using Fuzzy c means also has been presented in Sect. 4.1. Section 5 presents the gist of the chapter and also points the future emerging trend of social network analysis deploying soft computing methods. Standard recent references have been provided for the readers.

1.1 The Social Network

They say, "*Man is not an island*" because every man has the inherent need to socialize for survival. People interact with their close friends; they may interact with other people with whom they share common interests which have greatly been aided by the advent of the various means to expand ones social circle like blogging, public chat rooms and social networking sites such as Facebook, Twitter etc. Behaviour on these websites leaves an electronic trail of social activity which can be analysed and valuable information can be discerned. Analysing the tremendous amounts of information flow on these social networks provides an opportunity to study the human psyche for sociologists and since people are spending so much time on these sites, it has obviously bright commercial prospects for advertising and e-commerce. Individual behaviour of the actors in the social network can be used to categorize their interest and it can be used to deliver relevant advertisements, and information that is relevant to them based on their behaviour. Quite recently browser plug-ins to discover information retrieved by friends, as well as cooperative web search tools [1] have been developed.

More formally, social network analysis can be defined as [2],

> The disciplined enquiry into the patterning of relationships amongst social actors, as well as the patterning of relationships among actors at different levels of analysis (such as persons and groups).

1.2 Twitter

It is an online social networking service and micro blogging service that enables its users to send and read text-based posts of up to 140 characters, known as "tweets". It was created in March 2006 by Jack Dorsey and launched that July. The service rapidly gained worldwide popularity, with over 140 million active users as of 2012, generating over 340 millions tweets daily and handling over 1.6 billion search queries per day. It has been described as "the SMS of the Internet". Unregistered users can read the tweets, while registered users can post tweets through the website interface, SMS, or a range of apps for mobile devices. Tweets are publicly visible by *default*; however, senders can restrict message delivery to just their followers.

Follower/Following Relationship As a social network, twitter revolves around the principle of followers. When a user chooses to follow another twitter user, that user's tweets appear in reverse chronological order on the former user's twitter page/timeline. If one follows 20 people, he will see a mix of tweets scrolling down the page: breakfast-cereal updates, interesting new links, music recommendations, even musings on the future of education.

Mentions When a user wants to directly address someone who he follows or he is interested in, he can mention that particular user. This is done by prefixing the unique screen name corresponding to the particular user to be mentioned with the '@' symbol. Interactions between users on twitter take place through these mentions. Users are sent additional notifications in addition to appearance of that particular tweet on his timeline, so that they can respond them.

Trends Users can also contribute to an ongoing popular conversation between the tweets by the Trends feature of twitter. A user can contribute to a topic of discussion by prefixing a '#' symbol before the particular topic. Trending topics are mentioned at a more frequent rate by the users as compared to other topics.

Retweets Retweets are the means of information flow on twitter. When a user follows another user, his tweets appear on his timeline. If the user finds the tweet of someone he follows to be interesting enough, he can disseminate that tweet to his followers as well by retweeting that information, by which the original tweet will appear on his followers timelines as well. This is usually done by prefixing 'RT @ < screen_name of the original tweeter > or by just pressing the retweet button.

Twitter Lists A list is a curated group of twitter users. When a user clicks to view a list, he will see a stream of tweets from only the users included in that group. He need not follow a user to add them to a list. Similarly, following someone else's list does not mean that he follows all users in that list. Rather, the user follows the list itself.

1.3 Literature Review

Much research has already been carried out on twitter analysis. Here we review some of the important research works which give us a better understanding on the subject.

"Who Says What to Whom on Twitter" by Shaomei Wu et al. [3]. This chapter gives us an insight on the several long standing questions in media communications research, in the context of the microblogging service twitter, regarding the production, ow, and consumption of information. Lists, a feahas become increasinglyture of Twitter, are exploited to distinguish between elite users (celebrities, bloggers, and representatives of media outlets and other formal organizations) and ordinary users. Based on this classification, a striking concentration of attention on Twitter was found, in that roughly 50 % of URLs consumed are generated by just 20 K elite users, where the media produces the most information, but celebrities are the most followed. Significant homophily within categories was also found: celebrities listen to celebrities while bloggers listen to bloggers etc.

A longstanding objective of media communications research is encapsulated by what is known as Lasswells maxim: who says what to whom in what channel with what effect [4]. Lasswells maxim has proven difficult to answer in part because it is generally difficult to observe information owse in large populations, and in part because different channels have very different attributes and effects. As a result, theories of communications have tended to focus either on "mass" communication or on "interpersonal" communication [5].

Recent changes in technology, however, have increasingly undermined the validity of the mass versus interpersonal dichotomy itself. On the one hand, over the past few decades mass communication has experienced a proliferation of new channels, including cable television, satellite radio etc. Meanwhile, in the opposite direction interpersonal communication has become increasingly applied through personal blogs, social networking sites etc. Together, these two trends have greatly obscured the historical distinction between mass and interpersonal communications, leading some scholars to refer instead to "mass personal" communications [6]. A striking illustration of this erosion, for example, the top ten most-followed users on twitter are not corporations or media organizations, but individual people, mostly celebrities.

This chapter makes three main contributions:

- Introducing a method for classifying users using twitter Lists into "elite" and ordinary users.
- Investigating the ow of information among these categories.
- Finding that different categories of users emphasize different types of content, and that different content types exhibit dramatically different characteristic life spans, ranging from less than a day to months.

Twitter analysis was carried out with the help of Twitter Follower Graph, Twitter Firehose and Twitter lists. A Twitter Follower Graph is a directed network characterized by highly skewed distributions both of in-degree (#followers) and out-degree (#friends). The follower graph is also characterized by extremely low reciprocity—the most followed individuals do not follow many others. Twitter Firehose refers to the complete stream of all tweets. Twitter lists are methods of classifying sets of users into topical or other categories, and thereby to better organize and/or filter incoming tweets. To create a Twitter List, a user provides a name (required) and description (optional) for the list, and decides whether the new list is public (anyone can view and subscribe to this list) or private (only the list creator can view or subscribe to this list). Once a list is created, the users can add/edit/delete list members.

1.3.1 Who Listens to Whom

Clearly, ordinary users on Twitter are receiving their information from many thousands of distinct sources, most of which are not traditional media organizations-even though media outlets are by far the most active users on twitter, only about 15 % of tweets received by ordinary users are received directly from the media. Equally interesting, however, is that in spite of this fragmentation, it remains the case that 20 K elite users, comprising less than 0.05 % of the user population, attract almost 50 % of all attention within twitter. Thus, while attention that was formerly restricted to mass media channels are now shared amongst other "elites".

1.3.2 Who Listens to What

First, for ordinary users, the majority of appearances of URLs after the initial introduction derive not from retweeting, but rather from reintroduction, where this result is especially pronounced for long-lived URLs. Second, however, for URLs introduced by elite users, the result is somewhat the opposite—that is, they are more likely to be retweeted than reintroduced, even for URLs that persist for weeks. Although unsurprising that elite users generate more retweets than ordinary users, the size of the difference is nevertheless striking, and suggests that in spite of the dominant result above that content lifespan is determined to a large extent by the type of content, the source of its origin also impacts its persistence, at least on average—a result that is consistent with previous findings [6].

"What is Twitter, a Social Network or a News Media?" by H. Kwak et al. [7]. The entire twitter site was crawled and 41.7 M user profiles, 1.47 B social relations, 4,262 trending topics and 106 million tweets was obtained. In its follower-following topology analysis a non-power-law follower distribution, a short effective diameter, and low reciprocity were found, which all mark a deviation from know characteristics of human social networks [8]. To identify 'influentials' on twitter users were ranked by number of follower, by page rank and by retweets.

Tweets of top trending topics were analysed. A closer look at retweets revealed that any retweeted tweet is to reach an average of 1,000 users no matter what the number of followers is of the original tweet. Once retweeted, a tweet gets retweeted almost instantly on next hops, signifying fast diffusion of information after the 1st retweet.

Various analyses that were done include:-

Basic analysis Analysis of graph between number of followings (/followers) v/s complimentary cumulative distribution function.

Followers v/s tweets Co-relation between the number of followers and that of written tweets was gauged [9].

Reciprocity It was found that Twitter shows a low level of reciprocity; 77.9 % of user pairs with any link between them are connected one way, and only 22.1 % have reciprocal relationship between them.

Degree of Separation Stanley Milgram's famous 'six degree of freedom' experiment [10] reports that any two people could be connected on average within six hops from each other. The main difference between the other networks and Twitter is the directed nature of twitter relationship. Thus a path from a user to another may follow different hops or not exist in reverse direction. The average path length was found to be 4.12 which is quite short for the network of twitter site and is the opposite of the expectation on a directed graph. It was noted that information is to flow over less than five or fewer hops between 93.5 % of user pairs, if it is to, taking fewer hops than on other known social networks.

Homophily The tendency that a contact between similar people occurs at a higher rate than among similar people.

Ranking Twitter users Users are ranked by the number of followers and by page rank and these two rankings were found to be similar. Ranking by retweets differs from the previous two indicating a gap in the influence inferred from the number of followers and that from the popularity of ones tweets.

Impact of a Retweet Retweet is an effective means to relay the information beyond adjacent neighbours. On twitter people acquire information not always directly from those they follow but often via retweets. Individual users have the power to dictate which information is important and should spread by the form of retweet, which collectively determines the importance of the original tweet. A closer look at retweets reveals that any retweeted tweets is to reach an average of 1,000 users no matter what the number of followers is of the original tweet. Once retweeted, a tweet gets retweeted almost instantly on the 2nd, 3rd, and 4th hops away from the source, signifying fast diffusion of information after the 1st retweet.

Related Works Aside from the above, a number of recent chapters have examined information diffusion on twitter. In a similar vein to Kwak et al. [7], Cha et al. [11] compared three measures of influence number of followers, number of retweets, and number of mentions and also found that the most followed users did not necessarily score highest on the other measures. Weng et al. [12] compared number of followers and page rank with a modified page-rank measure which accounted for topic, again finding that ranking depended on the influence measure. Bakshy et al. [6] studied the distribution of retweet cascades on twitter, finding that although users with large follower counts and past success in triggering cascades were on average more likely to trigger large cascades in the future, these features are in general poor predictors of future cascade size.

1.4 Our Work

In this project we have performed data analysis on Twitter users and have used the results obtained to find out users relevant to some particular search query and find out their characteristics. We performed cluster analysis on the user data representative of activity parameters of individual users to identify various groups of users present on twitter. The size of the data set downloaded includes 250 K users, adjacency lists of approximately 1 K users and tweets over a period of 1 month for 50 K users for 30 days between 21st March and 21st April. We used Java over a Windows 7 platform and extensively used twitter4j Java library for downloading the data. The data was downloaded into an Oracle Express Edition 10 g database. Over The work included writing coding several modules to download preliminary data about users, their adjacency lists and tweets, as well as other modules to keep the database fairly up to date in terms of whether the users downloaded have been suspended or have deactivated their accounts or not and so on.

1.5 Outline of Contribution

The work done during the course of this project can be broadly categorized into three parts:

- Downloading the data set
- Cluster Analysis on the data set
- Application developed based on results obtained from data analysis.

2 Theoretical Studies

2.1 Graphs

A graph G is a collection of two sets, a set of edges E and a set of vertices V. The vertices are a mathematical abstraction of some real world entity which is connected by an edge $\{v_i, v_j\}$. The pair of vertices connected by an edge may be ordered or unordered.

If the edges are unordered then the graph is un-directed, and if the edges are ordered, there is a sense of direction in the edges and the graph is said to be directed.

Two vertices which are connected by an edge are said to be neighbours of each other or adjacent to each other. Graphs are often represented in terms of adjacency matrices or adjacency lists.

An **adjacency matrix** of a graph with N nodes is an N × N matrix in which the entry at the ith row and jth column is 1 if node i is connected to node j and 0 otherwise.

An **adjacency list** of a node contains a list of all the nodes to which it is connected.

In a directed graph, the number of edges terminating in a node is referred to as the **in-degree** of the graph and the number of edges originating from a node is called as the **out-degree**. In case of undirected graphs the number of edges incident on a node is simply referred to as the degree of the node, without any notion of in or out.

It should be noted that a node is connected to another node in a directed graph only if there is an edge from the former to the latter node Fig. 1.

Fig. 1 An illustrative
directed graph

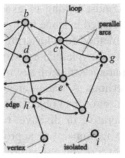

actor – vertex, node
relation – line, edge, ar
tie
arc = directed line, $(a, a$
a is the *initial* vertex,
d is the *terminal* vertex.
edge = undirected line, (
c and *d* are *end* vertices.

2.2 The Twitter Social Network as a Graph

In most of the literature related to social network analysis, the social network is
viewed as a graph. The users are viewed as the nodes of the graph and the
relationship of friendship/mutual interest is modeled as an edge between the two
nodes. Analysis on these kinds of graphs is usually done in terms of in-degree and
out-degree of various nodes, centrality measures of various kinds and so on.

Twitter is different from other social networks such as Facebook in the sense
that twitter is a directed graph. If a person A follows person B it means that there is
a directed edge from B to A (in accordance with the flow of information), whereas
in other networking sites interactions are based on friend requests being accepted
and a bidirectional flow of information is established.

2.3 Cluster Analysis

The process of grouping a set of physical or abstract objects into classes of similar
objects is called clustering. A cluster is a collection of data objects that are similar
to one another within the same cluster and are dissimilar to the objects in other
clusters. It is often more desirable to proceed in the reverse direction: First partition the set of data into groups based on data similarity (e.g., using clustering),
and then assign labels to the relatively small number of groups. Additional
advantages of such a clustering-based process are that it is adaptable to changes
and helps single out useful features that distinguish different groups.

Many clustering algorithms exist in the literature. It is difficult to provide a
crisp categorization of clustering methods because these categories may overlap,
so that a method may have features from several categories. In general, the major
clustering methods can be classified into the following categories.

Partitioning methods Given a database of "n" objects or data tuples, a partitioning method constructs k partitions of the data, where each partition represents
a cluster and k ≤ n. That is, it classifies the data into k groups, which together
satisfy the following requirements: (1) each group must contain at least one object,

and (2) each object must belong to exactly one group. Notice that the second requirement can be relaxed in some fuzzy partitioning techniques. We have used in this project a fuzzy method for clustering keeping in mind that user behaviour on social networks cant be explained by assuming that a user present in one cluster wont exhibit any characteristics of any other cluster.

Hierarchical methods A hierarchical method creates a hierarchical decomposition of the given set of data objects. A hierarchical method can be classified as being either agglomerative or divisive, based on how the hierarchical decomposition is formed. The agglomerative approach, also called the bottom-up approach, starts with each object forming a separate group. It successively merges the objects or groups that are close to one another, until all of the groups are merged into one (the topmost level of the hierarchy), or until a termination condition holds. The divisive approach, also called the top-down approach, starts with all of the objects in the same cluster. In each successive iteration, a cluster is split up into smaller clusters, until eventually each object is in one cluster, or until a termination condition holds.

Density-based methods Most partitioning methods cluster objects based on the distance between objects. Such methods can find only spherical-shaped clusters and encounter difficulty at discovering clusters of arbitrary shapes. Other clustering methods have been developed based on the notion of density. Their general idea is to continue growing the given cluster as long as the density (number of objects or data points) in the "neighbourhood" exceeds some threshold; that is, for each data point within a given cluster, the neighbourhood of a given radius has to contain at least a minimum number of points. Such a method can be used to filter out noise and discover clusters of arbitrary shape.

Clustering Optimality Even though clustering prevents the explicit need to know the various labels of classes present in a data set and the properties of these underlying classes, however, there may be a need to know the number of clusters to segregate the data set into beforehand. Arbitrarily choosing the number of clusters as a fixed parameter to any clustering algorithm can lead to problems of under-clustering or over-clustering as depicted in Fig. 2.

Dunn's Validity Index [14] This technique [14] is based on the idea of identifying the cluster sets that are compact and well separated. For any partition of clusters, where c_i represent the i-cluster of such partition, the Dunns validation index, D, could be calculated with the following formula:

$$D = \min_{1 \leq i \leq n} \left\{ \min_{1 \leq j \leq n, i \neq j} \left\{ \frac{d(c_i, c_j)}{\max d'(c_k)} \right\} \right\} \tag{1}$$

where $d(c_i, c_j)$—distance between clusters c_i, and c_j (intercluster distance); $d'(c_k)$—intracluster distance of cluster c_k, n—number of clusters. The minimum is calculating for number of clusters defined by the mentioned partition. The main goal of the measure is to maximize the intercluster distances and minimize the intracluster distances. Therefore, the number of cluster that maximizes D is taken as the optimal number of the clusters.

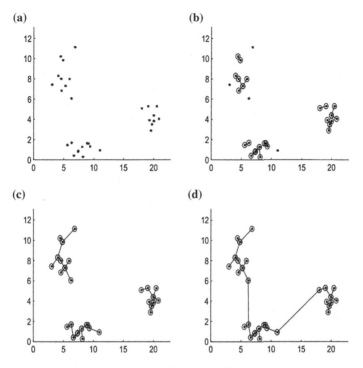

Fig. 2 Over, under and optimal clustering illustration [13]

Davies-Bouldin Validity Index [15] This index [15] is a function of the ratio of the sum of within-cluster scatter to between-cluster separation where n—number of clusters, S_n—average distance of all objects from the cluster to their cluster centre, $S(Q_i, Q_j)$—distance between clusters centres. Hence the ratio is small if the clusters are compact and far from each other. Consequently, Davies-Bouldin index has a low value for a good clustering.

$$DB = \frac{1}{n} \sum_{i=1}^{n} \max_{i \neq j} \left\{ \frac{S_n(Q_i) + S_n(Q_j)}{S(Q_i, Q_j)} \right\} \tag{2}$$

Silhouette Validation Method [16] The Silhouette validation technique [16] calculates the silhouette width for each sample, average silhouette width for each cluster and overall average silhouette width for a total data set. Using this approach each cluster could be represented by so-called silhouette, which is based on the comparison of its tightness and separation. The average silhouette width could be applied for evaluation of clustering validity and also could be used to decide how good the number of selected clusters.

To construct the silhouettes $S(i)$ the following formula is used:

$$S(i) = \frac{(b(i) - a(i))}{\max \{a(i), b(i)\}} \tag{3}$$

where $a(i)$—average dissimilarity of i-object to all other objects in the same cluster; $b(i)$ minimum of average dissimilarity of i-object to all objects in other cluster (in the closest cluster). It is followed from the formula that $-1 \leq S(i) \leq 1$. If silhouette value is close to one, it means that sample is well-clustered and it was assigned to a very appropriate cluster. If silhouette value is about zero, it means that that sample could be assign to another closest cluster as well, and the sample lies equally far away from both clusters. If silhouette value is close to one, it means that sample is misclassified and is merely somewhere in between the clusters. The overall average silhouette width for the entire plot is simply the average of the $S(i)$ for all objects in the whole dataset.

The largest overall average silhouette indicates the best clustering (number of cluster). Therefore, the number of cluster with maximum overall average silhouette width is taken as the optimal number of the clusters.

3 Data Model

3.1 Preliminary Details

To analyze the social network on twitter we downloaded data using the twitter4j JAVA library and the downloaded data was stored in an Oracle Express Edition 10 g database. Twitter API provides the following major types of data about a user as long as he is not protected, that is he does not wish to share his personal information and his tweets etc. To download information from the twitter servers, an application has to be registered with twitter. Users are made to register with the application after which the application is provided with an Access Token, Token Secret, Consumer Token and Consumer Secret. Combining these four keys, an application can make 350 requests per hour to the twitter servers. We made 25 users to register with our application and used their Access Token and Secret in a round robin fashion to download data continuously from the twitter users. The method of authenticating of an app by a user so that the app can download data on behalf of the user is OAUTH Authentication mechanism. Refer to Appendix I for more details.

Information about the user In twitter every user has a unique screen name and a unique numerical ID. In addition to this information about the people he follows and the people he is followed by, his tweets can also be downloaded. Other types of data can also be downloaded about the user, which are given in Appendix II.

- Name
- Screen Name
- Follower count
- Following count
- Follower list

Column Name	Data Type	Nullable	Default	Primary Key
USER_ID	NUMBER	No	-	1
SNO	NUMBER	Yes	-	-
SCREEN_NAME	VARCHAR2(100)	Yes	-	-
TWT_FRQ	NUMBER	Yes	-	-
STATURE	NUMBER	Yes	-	-
FLAG	NUMBER	Yes	-	-
FOLLOWING	NUMBER	No	-	-
FOLLOWERS	NUMBER	No	-	-
CLASS	NUMBER	Yes	-	-
RT_EXPECTED	NUMBER	Yes	-	-
				1 - 10

Fig. 3 The MUI structure

- Following list
- All tweets from a particular user
- Location

Information about a tweet:

- Tweet content
- Tweet id
- Tweeter id
- Location
- Time
- Mentions
- In reply to
- Trends
- Hashtags
- Number of retweets

3.2 Database Design

All the information that can be downloaded is not all that popular or considered to be a metric of a persons social influence on the site. We intuitively chose to discard several features of the data about a user and the tweets so that only relevant patterns were mined and also due to computational and storage concerns and came up with the following database schema for efficient storage and programming convenience Fig. 3.

The Main User Index (MUI) The MUI contains is a central index of the entire user data collected. It has the following structure:-

Column Name	Data Type	Nullable	Default	Primary Key
USER_ID	NUMBER	No	-	1
SNO	NUMBER	Yes	-	-
REL	VARCHAR2(2)	Yes	٬	-
MBA	NUMBER	Yes	-	-
MBB	NUMBER	Yes	-	-
RTA	NUMBER	Yes	-	-
RTB	NUMBER	Yes	-	-
				1 - 7

Fig. 4 Structure of the adjacency table for each user

User_id the unique user id of a user. It is the primary key of the table. Even though a person might change his unique screen name or handle as they call it on twitter, his user id always remains the same.

Screen_name is the unique screen name of a user. He can change his screen name to something new subject to one condition that it is not in use by any other user.

Twt_Frq Short for tweeting frequency, it is a measure of how activity the user is on twitter. It represents the number of tweets that a user makes per week. Mathematically, it is the number of tweets the user has made since he joined twitter divided by the number of weeks that the user has spent on twitter.

Following The number of users that the user is following.

Followers The number of users that the user is followed by.

RT_Expected It is the expected number of retweets that a tweet made by a user will get when he tweets. The term shall be explained in more detail in the coming chapters.

Class The class label to which the user belongs to. It is determined after the cluster analysis.

Flag The flag is an indicator to the programming modules about the extent of information that has been downloaded about the user.

$Flag = 1$ indicates that the user has just been entered into the MUI and his tweets and adjacency tables have not been downloaded.

$Flag = 2$ indicates that the adjacency list has been downloaded but the tweet table has not been downloaded.

$Flag = 3$ indicates that the tweets of the particular user for a period of the past one month have been downloaded as well and no further information about the user needs to be downloaded.

Adjacency Table ADJ < user_id > The ADJ < user_id > table corresponding to the adjacency list and other details of the user bearing the user id < user_id > is kept in this table. It has the structure as shown in Fig. 4:

Column Name	Data Type	Nullable	Default	Primary Key
TID	NUMBER	No	-	1
SNO	NUMBER	Yes	-	-
TXT	VARCHAR2(1000)	Yes	-	-
D_O_T	VARCHAR2(20)	Yes	-	-
RT	NUMBER	Yes	-	-
ORIGINAL	VARCHAR2(50)	Yes	-	-
ORG_ID	NUMBER	Yes	-	-
SOURCE	VARCHAR2(200)	Yes	-	-
				1 - 8

Fig. 5 Structure of the tweet table

Rel It represents the kind of relationship that the user A having the user table has with those users present in the table. Let B be some user present in the adjacency table of A

Rel = *1* indicates that A follows

Rel = *2* indicates that B follows A

Rel = *3* A follows B and B follows A

RTA: Number of tweets of A that B has retweeted.

RTB: Number of tweets of B that A has retweeted.

MBA: Number of tweets of A in which B is mentioned.

MBB: Number of tweets of B in which A is mentioned.

RTA, RTB, MBA, MBB establish the strength of relationship between A and B.

Tweet Table (TWT < user_id >) TWT_<user_id> contains the tweets of the user having the user id < user_id > indicated in the name of the table. The structure of this table is as indicated in the Fig. 5 shown:

TID: Represents the unique tweet id of the tweet. It is in chronological order. A tweet which is tweeted at an earlier time will have a lower tweet is as compared to a tweet at a later time Fig. 5.

TXT: The text content of the tweet.

D_O_T: The date on which the tweet was made.

RT: Contains the number of Retweets of that particular tweet.

ORIGINAL: In case the tweet itself has been retweeted, this field contains the user_id of the original tweet owner. In case the user to which the tweet table belongs is the original owner of the tweet, the field is set to −1.

ORIG_ID: It contains the original tweet id, in case the tweet is a retweeted tweet.

3.3 Approach to Download a Cohesive Data Set

The approach to download the data to fill up the tables described in the previous section is quite interesting. We have used the underlying connectedness feature of the social network. There are basically two programming modules:

Module I: Filling up of the MUI and the creation and filling up of adjacency tables of the corresponding users.

Module II: Creation of the Tweet table of each user and filling it up.

Module I is concerned with the filling up of the MUI and Adjacency tables of the individual users as previously mentioned. This is done by taking a seed user 'A' as input and inserting it into the MUI as the first user. The list of people that A follows and is followed by is inserted into the adjacency table of A keeping the data consistent with the definitions of fields presented in the previous section. These users are inserted into the MUI as well. The process that was initially applied to A (that is, downloading the adjacency table and inserting into MUI) is now performed on the new entries of MUI as well, thus forming a recursive procedure to download all the required information of all the users present on twitter. The method can potentially download all those users present on twitter who are followed by even a single user. Theoretically, it has been predicted in that the degrees of separation or the number of hops between any two vertices (users) on the twitter social graph is somewhere between six and eight on an average [10]. Thus if this recursive procedure is completed for six degrees of separation, we can download the almost the whole of Twitter users in a connected, cohesive fashion in which relationships between users are well defined—in terms of directedness.

So far we have been able to exceed the second degree neighbors corresponding to the initial seed user and are moving on to complete the third degree neighbors. In the process we have already downloaded 250 K users which is a miniscule proportion of the total number of users present on twitter which, by the way, is growing as you read this document.

The module designed by us is intended to download users continuously as till the entire twitter user data is not downloaded. Doing that, however, is practically impossible. Hence, the program also has measures to restart after being shut down without letting the progress made in previous runs of the program in filling up the database go to waste. Also, in case the users that are already present in the MUI get suspended or de-activate their account between the time they are first inserted into the MUI and the time the adjacent table is created, the program has been designed to carry on working to download data for other users gracefully, without throwing any unexpected results or halting on finding this discrepancy. A formal presentation of the algorithm used is given below:

Algorithm 1 Algorithm used to fill Main User Index and Adjacency Tables

---

INPUT

init_user: initial seed user

count: count contains the value from where the creation/ filling up of the adjacency table has to begin, initial value = 1

VARIABLES

Hits: number of hits remaining corresponding to an access token-secret pair

Threshold: minimum number of remaining hits after which the access token-secret pair is changed

PROCEDURE

   procedure

   Initialize Twitter instance using access token and run parallel thread shown separately

      if seed user exists and is not protected **then**

         Proceed

      else

         Run program again with a different seed user

      end if

      while count < number of users in the MUI **do**

         if user suspended or protected or not found **then**

            Skip user

            count = count +1

         else

            Proceed

         end if

         Create adjacent table for the current user and fill with details of followers and friends

         Enter the users obtained in the main database also

         count= count + 1

      end while

   end procedure

   procedure

      while true **do**

         if hits¡ threshold **then**

            Change the access token-secret pair

         end if

      end while

   end procedure

Module II is concerned with the downloading of tweets of individual users. The approach to this is quite simple as compared to creating adjacency tables of individual users and filling up the MUI. Twitter API offers a paging feature by which up to 20 tweets can be downloaded with a single request. By increasing the page number of the tweets to be downloaded, more and more tweets can be downloaded. This also provides resume support to downloading the data. By storing the number of pages on a secondary storage device, download of tweets can be resumed from any given page number. For our analysis purpose we have downloaded tweets for the last one month of the user or a minimum of 50 tweets per user—whichever is larger.

3.4 Challenges Faced

As already described in the previous section, downloading data from the twitter servers has its share of challenges. A brief summary of these challenges has been presented below:

Limited Hits Per Hour Corresponding to every user registered with the app that we have registered with twitter, we get an Access Token and Access Secret. Corresponding to each such pair of credentials that enables us to download data, there is a limit of 350 hits per hour. For downloading the massive dataset that is required to perform any kind of meaningful analysis to be performed, such a limit is obviously detrimental. To solve this problem, we made 25 users to register with our app and stored it in our database in a table. We used a synchronized JAVA thread which continuously monitored the current token-secret pair being used for remaining hits and recycled old unused token-secret pairs for subsequent usage while the old pair gets rejuvenated.

Users Getting Suspended by Twitter or Users De-activating Their Accounts Users who do not comply with the policies of usage as prescribed by twitter are suspended by twitter and at times users also de-activate their accounts. When such users are present in the database they represent useless data and the effort put into download their data also goes to waste. Whenever users are downloaded and they are processed subsequently, they have to be checked for these criteria before the processing can proceed.

Twitter Servers Misbehaving Twitter servers are frequently over burdened with requests from normal users and other applications like our very own application and they may respond unpredictably at times or they may not respond at all. This might cause the program to behave unpredictably or the database getting corrupted due to scrupulous/inconsistent data.

4 Cluster Analysis, Results, and Discussions

4.1 Cluster Analysis

The importance of clustering has been discussed at length in Sect. 2.3. The behavior of people on twitter cannot always be of the same kind. It is intuitively wrong to label people according to the distinct classes/clusters and expect them to behave rigidly according to their cluster properties. This calls for a fuzzy analysis of the scenario in which the boundaries between the clusters are fuzzified for analysis and suitably defuzzified when required. We use Bezdek's Fuzzy c–Means clustering algorithm [17] for our project.

Each object i is described by a feature vector X containing r attributes.

$$x_i = [x_{i1}, x_{i2}, \ldots, x_{ir}]$$

Here our object is the user. Each user is characterized by his tweet frequency, following count, follower count and retweet expectancy. Thus value of $r = 4$.

Suppose there are N users, then all the feature vectors are organized into an $N \times r$ two dimensional vector called the feature vector.

Suppose the data is to be clustered into c classes.

Let $X = x_1, x_2, \ldots, x_n$ be a set of given data. A fuzzy pseudo partition or fuzzy c-partition of X is a family of fuzzy subsets of X, denoted by $P = A_1, A_2, \ldots, A_c$, which satisfies A_i

$$\sum_{i=1}^{c} A_i(x_k) = 1 \qquad (4)$$

For all $k \in N_n$ and

$$0 < \sum_{k=1}^{n} A_i(x_k) < n \qquad (5)$$

For all $I \in N_c$, where c is a positive integer.

Given a set of data $X = x_1, x_2, \ldots, x_n$ where x_k, in general is a vector

$$x_k = [x_{k1}, x_{k2}, \ldots, x_{kp}] \in R^p \qquad (6)$$

For all $k \in N_n$, the problem of fuzzy clustering is to find a fuzzy pseudo partition and the associated cluster centres by which the surface of the data is represented as best as possible. This requires some criterion expressing the general idea that associations (in the sense described by the criterion) be strong within clusters and weak between clusters.

To solve the problem of fuzzy clustering, we need to formulate this criterion in terms of a performance index. Usually the performance index is based upon cluster centres. Given a pseudo partition $P = A_1, A_2, \ldots, A_c$, the c cluster centres $v_1, v_2, \ldots, v_n$ associated with the partition are calculated by the formula

$$v_i = \frac{\sum_{k=1}^{n} [A_i(x_k)]^m x_k}{\sum_{k=1}^{n} [A_i(x_k)]^m} \qquad (7)$$

For all $I \in N_c$, where $m > 1$ is a real number that governs the influence of membership grades.

4.1.1 FUZZY c-Means Algorithm

The algorithm is based on the assumption that the desired number of clusters c is given and, in addition, a particular distance, a real number $m \in (1, \infty)$, and a small positive number, serving as a stopping criterion, are chosen.

Step 1.   Let t = 0. Select an initial fuzzy pseudo partition $P^{(0)}$

Step 2.   Calculate the c cluster centres $v_i^{(t)}, \ldots, v_c^{(t)}$ for $P^{(t)}$ and the chosen value of m

Step 3.   Update $P^{(t)}$ by the following procedure: For each $(x_k) \in X$, if $(x_k) \in X$, if $|x_k - v_i^{(t)}| > 0$ for all $I \in N_c$, then define

$$A_i^{(t+1)}(x_k) = \left[\sum_{j=1}^{c} \left(\frac{||x_k - v_i^{(t)}||^2}{v||x_k - v_j^{(t)}||^2} \right)^{\frac{1}{m-1}} \right]^{-1} \tag{8}$$

Step 4.   Compare $P^{(t)}$ and $P^{(t+1)}$. If $|P^{(t+1)} - P^{(t)}| \leq \varepsilon$, then stop; otherwise increase t by one and return to Step 2

In **Step 4**, $|P^{(t+1)} - P^{(t)}|$ denotes a distance between P(t + 1) and P(t) in the space $R^{n \times c}$

Distance measure that was used by us is given as under

$$|p^{t+1} - pt| = \max_{i \in N_c, k \in N_c} |A_i^{(t+1)} x_k - A_i^{(t)} x_k| \tag{9}$$

Since the optimal number of clusters was not known, we ran the clustering algorithm for different values of c. The optimality of the cluster values was made on the basis of a modified Dunns index. The Dunns index that we used in the analysis is described below:

Characteristics of the Dataset Used The number of users that was used for clustering was 4730. The number is a negligible proportion of the users present on Twitter, but the variation in results obtained when higher numbers of users were considered was not found to be significantly appreciable. The users are clustered on the basis of their following attributes:

Tweet Frequency per week (TWT_FRQ)

Following count

Follower count

RT_expectancy

Retweet_Expectancy We have defined Retweet Expectancy statistically as :

RT_expectancy = Sum of retweets of all tweets originally owned by the user/ Total number of tweets made

Modified Dunn's Index (D*) We observed that the dataset under consideration was extremely dense and the use of the normal Dunns index illustrated previously in this document in Sect. 2.3 had a tendency to produce unrealistically high performance indices for low values of the number of clusters, c. This can be attributed to the high density of data in the 4-D attribute space considered for analysis for our study.

We define D* as,

$$D^* = \frac{minimum\ iter - cluster\ distance}{(maximum\ intracluster\ distance)^{1.5}} \tag{10}$$

The division by an additional factor of (max intra-cluster distance).5 favors a higher number of clusters as those values for which maximum intra-cluster distance is low which would be the case if the number of clusters are more and size of individual clusters small, will have higher values of D*.

Standardization of Data Before Clustering For standardization we used the z-score of the individual data attributes of an object. Standardization is done so that the relative importance of the individual attributes of a feature vector is not diminished and each attribute contributes equally in the cluster analysis, irrespective of the range of values that they assume.

$$z_i = \frac{x_i - x_m}{s}$$

where z_i is the z-score of the ith object's attribute

x_i is the actual value of the attribute

s is the absolute mean of the attribute

x_m is the mean of the attribute

The merits of using z-score as a means of standardizing the data attributes are well documented in [18].

The above algorithm was used to find out the fuzzy membership values of the users in various clusters and the results/knowledge gained was later used in the application software that we developed.

4.2 Standard Fuzzy system

An interval standard fuzzy system lets any one or all of the following kinds of uncertainties be quantified:

- Words that are used in antecedents and consequents of rules because words can mean different things to different people.
- Uncertain consequents because when rules are obtained from a group of experts, consequents will often be different for the same rule, i.e. the experts will not necessarily be in agreement.
- Membership function parameters because when those parameters are optimized using uncertain (noisy) training data, the parameters become uncertain.
- Noisy measurements—because very often it is such measurements that activate the FLS.

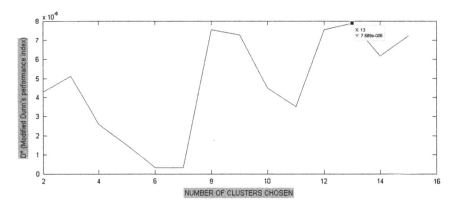

Fig. 6 Plot of D* versus number of clusters

4.3 Results

In order to find the optimal number of clusters, the algorithm was run for different values of c ranging from 2 to 15. **The optimal number of clusters was found out to be 13**. This result obtained by us is concurrent with results obtained in Fig. 6, however since ours is a fuzzy algorithm which is known to emulate real world scenario better, we expect to provide better results in the software developed by us.

Class wise analysis:

Class number	#Users	Mean twt_freq	Mean #following	Mean #followers	Mean retweet_exp
1	5	79.349	255758.0	481679.2	538.611
2	77	653.188	3439.324	52752.623	2.722
3	11	79.125	222.091	1262384.0	2243.990
4	2	4532.351	0.0	9418.5	0.158
5	1	13.335	757928.0	2461858.0	54.653
6	36	209.524	73749.0833	460809.944	16.583
7	3	24.365	560.0	1.2254E7	1006.365
8	855	273.134	1695.662	27713.814	2.0589
9	25	147.298	2855.4	684114	848.523
10	33	82.350	4357.515	3342592.939	202.057
11	1	6.477	139299.0	2.2401802E7	6265.263
12	3257	18.127	466.552	4885.822	1.292
13	927	102.46	2188.932	57915.190	7.534

Overall Features of the Data Set:

- Average Tweet Frequency: 68.782
- Average Follower Count: 64536.32
- Average Following Count: 1999.833
- Average RT_expectancy: 16.343

Remarks on the various classes:

Class 1:   Popular users with large and comparable following count, but low RT_expectancy distorted by some users with high RT_expectancy.

Class 2:   Cluster of common people, with comparable yet high follower and following count, high tweet frequency, but low RT_Expectancy.

Class 3:   Celebrity cluster characterized by low following count, high follower count and high RT_Expectancy.

Class 4:   Users characterized primarily by their zero following count, but high follower count. News Agencies etc. and some celebrities fall in this category.

Class 5:   High follower count, following and RT_Expectancy, but low tweet frequency. Possibly a celebrity cluster.

Class 6:   High follower, following, yet low RT_Expectancy.

Class 7:   Celebrity cluster, low following count, high follower count, below than average Tweet Frequency and high RT_expectancy.

Class 8:   Cluster of common people having almost no celebrities.

Class 9:   Celebrities who are highly active on Twitter.

Class 10:   Celebrities who have less popular than members of Class 9 but have comparable follower counts.

Class 11:   Includes just Lady Gaga and also contained other ridiculously popular celebrity users on Twitter such as Eminem, characterized by an astronomical follower count.

Class 12:   The largest chunk of users on Twitter fall into this category.These are people with low retweet expectancy, but still are addicted to Twitter.

Class 13:   An extension of Cluster 12 but the users have slightly better statistics in terms of sheer follower count.

5 Future Scope of Study

5.1 Trend Analysis

Trends as a useful way for people to find out what topics are being talked about around the world, right now. Trends can be viewed globally i.e. worldwide or locally i.e. specific countries and metropolitan locations. Twitter uses an algorithm that determines which topics are "trending" in the location which is selected. This algorithm identifies topics that are immediately popular, rather than topics that have been popular for a while or on a daily basis, to help people discover the hottest emerging topics of discussion on twitter. Some topics include a # (pound sign) before the word or phrase; this is called a hashtag and is included specifically in tweets to mark them as relating to a topic so that people can follow the conversation in search. Any user can participate in the trending topics by posting a tweet that contains the exact word or phrase as it appears in the trending topics list.

Trends can also be promoted. Promoted Trends began as an extension of Twitters Promoted Tweets platform, and are now a full-fledged product in their own right. With Promoted Trends, users see time-, context-, and event-sensitive trends promoted by Twitters' advertising partners. These paid Promoted Trends appear at the top of the Trending Topics list on Twitter and are clearly marked as "Promoted".

An analysis of the trending topics and their duration in the trending topics list can be useful for a social network analyst. By analysing the tweets of top trending topics, one can comment on their temporal behaviour and participation of various users. The period or duration that a topic appears in the top trending topics list can be taken as measure of the popularity of the corresponding trending topic. A trending topic does not last forever nor dies to never come back. A trending topic is considered inactive if there is no tweet on the topic for 24 h.

Kwak et al. [3] focused their attention on the user participation and active period of trends. Their analysis revealed that a large number of users participated in trending topics. It was found that long-lasting topics with an increasing number of tweets do not always bring in new users into the discussion. Also there existed core members generating many tweets over a long time period for particular trending topics. Crane and Sornette present a model that categorizes the response function in a social system [19]. Their model takes into consideration whether the factor behind an event is endogenous or exogenous and whether a user can spread the news about the event to others or not (critical or subcritical). Kwak et al. applied this classification methodology on the number of tweets and their times, and classify trending topic periods into the following four categories: exogenous subcritical, exogenous critical, endogenous subcritical and endogenous critical. Manual inspection of the topics that fall into the exogenous critical class revealed that they are mostly timely breaking news (headline news). The topics in the endogenous critical class are of more lasting nature: professional sports teams, cities, and brands. (persistent news). Those exogenous subcritical topics have hashtags, such as #thoughtsintheclub and #thingsihate, catching a limited subset of users attention and eventually dying out (ephemeral).

Given a dataset of users and their tweets, one can perform a simple trend analysis by extracting all the frequent hashtags. These hashtags can be thought of as trending topics among the users who participated within the dataset. Duration of the hashtags i.e. the time period between the first tweet containing that hashtag and the last (latest) tweet containing that hashtag, can be determined easily. Identifying users who tweet the most about a particular topic i.e. hashtag can help in determining for what topic the user is an 'influencer' of and whom all, that particular user 'influences'. Such analysis can help in finding influential users of a particular entity (here entity can refer to a group, organization, etc.).

5.2 Influential User Analysis

Users who are influencers of a particular entity (here entity may refer to a topic, a group, a person especially a celebrity or an organization) are termed as influential users. A user may be influential about more than one thing. Influential users are deemed as such because of the fact that they influence a lot of other people, i.e. anything they say (tweet) reaches a large audience most of the time. Such users have the ability to stir up a conversation (discussion) easily and also direct it. An analysis of the influential users based upon the influencing topics can be very useful, e.g. an organization may want to find users who are influential about their organization and provide perks to such users. Companies may also want to hire influential users who have the required skill set.

Klout [20] is one such site that identifies influential users. It also provides what all things the influential user influences about and who are the users that are influenced by that user. Each user who registers with Klout is given a score depending upon the users influence. Higher scores represent a wider and stronger sphere of influence of that user. Based on the score the user is classified into categories such as observers, conversationalists, celebrities etc. Klout also provides the user with various metrics such as true reach, network score etc. all of which are a measure of how popular the user is, how strong his network is etc.The model enables the auto-segmentation of the events and the characterization of tweets into two categories: (1) episodic tweets that respond specifically to the content in the segmentsof the events, and (2) steady tweets that respond generally about the events. By applying our method to two large sets of tweets in response to President Obama's speech on the Middle East in May 2011 and a Republican Primary debate in September 2011, we present what these tweets were about. We also reveal the nature and magnitude of the influences of the event on the tweets over its timeline. In a user study, we further show that users find the topics and the episodic tweets discovered by our method to be of higher quality and more interesting as compared to the state-of-the-art, with improvements in the range of 18–41 [21].

The analysis of influential users can be done by analysing the tweets of the user, his/her tweet frequency (how frequently the user tweets), numbers of followers, numbers of friends, retweet count etc. A popularity measure based on all these parameters can be created and calculated for all the users in question (users available in the dataset). The analysis should be spread out i.e. the activity of the user from the start (date of creation) must be taken into consideration. Also the analysis should incorporate changes that occur at a later point in time, i.e. it should be dynamic. Relationships are examined between the user and his/her friends and followers. Their influence also has an affect on the current user on which the analysis is being performed i.e. the activity of the surrounding environment (surrounding users friends and followers) also plays a part in the analysis.

5.3 Tweet Sources Analysis

Associated with each tweet (status) is its source. The source specifies from where the tweet was actually tweeted. Some of the sources of the tweets are: Web (World Wide Web), Android, Ovi by Nokia, etc. Any application that is used by a user (or on his/her behalf) to post updates, is also considered a source. An analysis of the tweets can reveal how much tweets are done via each source which will help in identifying popular sources of tweets. This analysis can prove useful from a market point of view as advertisers can use such data to divide their attention based on these sources, i.e. if more tweets are done via a source then it will be a good strategy to place advertisements on that source.

5.4 Personalization and Recommendation System

Social Network analysis is an approach to the study of human social interactions. In this era of emerging Web 3.0, personalization is the niche of the market. Information filtering is a technology in response to this challenge of information overload in general. Based on a profile of user interests and preferences, systems recommend items that may be of interest or value to the user. Clique formation plays a central role in recommendation systems, as it is able to recommend information that has not been rated before and accommodates the individual differences between users. Information filtering has been applied in various domains, such as email, news, and web search [22]. By analysing the flow of information on twitter and modelling a graph where in the users/entities are nodes and the link between them being edges, we can determine the closeness centrality, betweenness, centrality, Cliquishness, Homogeneity and clustering coefficients which shall eventually help us deduce group of people sharing similar interests and ease the objective of personalized recommendation task.

Analysing the graph thus formed; eventually the users/entities with similar interests can be clustered to form cliques. With this the interest of any user/entity can be studied and recommended to other nodes in the same clique.

A.1 6   Appendix I: OAuth

Twitter uses the open authentication standard OAuth for authentication. It is an authentication protocol that allows users to approve application to act on their behalf without sharing their password. Before OAuth, basic authentication was used which required that each request be accompanied with the user name and password combination of a user. This was a security issue in that the application had to store the user name and password for the users which is not safe from a users viewpoint. OAuth eliminates this need by introducing the concept of access tokens. Each application which wants to use OAuth feature must first register itself

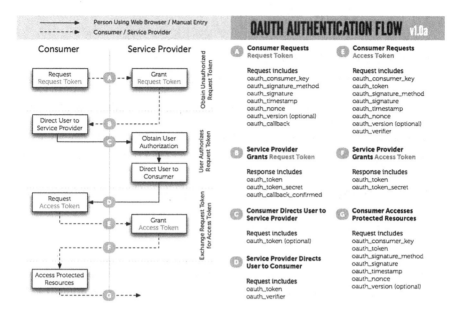

Fig. 7 OATH authentication flow

with twitter, whereupon it will be provided with a unique consumer token which is a token and secret pair. Users are authenticated in the following way:

1. The consumer (application) requests a request token from Twitter using the consumer token etc.
2. When the service provider (Twitter) grants the request token, the consumer directs the user to the service provider.
3. The user is authorized by the service provider (in case of twitter logging in followed by granting rights to the consumer/application) and directed back to the consumer.
4. The consumer then requests for the 'access token' for that user by exchanging the request token for an access token.
5. The service provider then supplies the consumer with the access token which consists of a token and secret pair.
6. The consumer can then access protected resources by including the access token with each request made on the behalf of the user.

This process is secure as the password if the user is not provided to the application; instead the consumer stores access tokens for the various authenticated users. Every request to the API the application makes is signed using the access token key/secret pair. Access tokens currently do not expire. The user has the power to revoke the access of the application, in which case the access token for that particular will no longer be valid. Also when Twitter admin suspends the application then the access token will become invalid Fig. 7.

This diagram illustrates further:

A.2 7   Appendix II: Twitter4j

Twitter4j is the java library for the Twitter API in this project. It is open sourced and created by Yusuke Yamamoto. It provides an easy integration of the Java applications with the Twitter service. Some of its features include—built-in OAuth support, zero dependency: no additional jars required, Android platform etc. Twitter4J is thread-safe and method calls can be made concurrently. It has methods which allow us to use all the three API of the Twitter API namely REST API, SEARCH API and STREAMING API. Twitter4j has methods to access the various resources given below:

1. Timeline Resources—getting the timeline (stream of tweets posted by the user and his/her friends and retweets) of a user, getting mentions etc.
2. Tweet Resources—showing status (latest tweet) of a user, updating the status of a user, retweeting a status etc.
3. User Resources—getting information about a particular user which includes user id, screen name, date of creation of the user, follower count, following count, statuses count, checking if the user is a verified user or protected user etc., searching users etc.
4. Friends and Followers Resources—getting user ids of the followers and friends of a particular user.
5. Friendship Resources creating/destroying friendship between two users, checking whether a friendship exists between a pair of users etc.
6. Account Resources getting account settings, verifying credentials of a user, getting rate limit status for a particular user etc.
7. Trends Resources getting current trends, daily trends etc.
8. Search API Resources the above resources accessed the REST API, the search API allows us to search for tweets based on various parameters such as presence of a keyword or a particular hashtag, by a particular user, mentioning a user, since a date, tweets until some date, around a location, retweets, containing URLs etc.

Twiiter4j provides few other resources not mentioned here.

Almost all the methods of Twitter4j are rate limited i.e. there is limit on how many methods calls for that method can be made during a time period. The rate limit is 350 requests/hour, and is different for the search API.

Almost all the methods require authentication i.e. the user making the request must be authenticated by Twitter or the application making requests on behalf of a user must do so for an authenticated user. There is no authentication required in the Search API.

The stream API methods provides for accessing live streaming tweets.

A.3 8   Appendix III: Tracing a Tweets Path

From a given point/node/user a tweets, which has been retweeted at least once, path can be traced. It can found out from where it reached (came) the given user and till where it travelled (went). Associated with each tweet (status) in the Twitter API is its unique tweet id, user id/screen name of the tweeting user, date of creation, retweeted count, and if it is a retweeted status (i.e. a retweet) then information about the original tweet id and the original twitter user who tweeted that status (i.e. introduced the content for the 1st time) can also be retrieved. So by storing these attributes of tweets for every tweet downloaded can help us in determining the path the tweet has travelled.

Only tweets with non-zero retweet count are taken under consideration in the tweet tracing, as a tweet with a retweet count of zero has only travelled to all the followers of the tweeting user and tracing such tweets path (having path length 1 only) is of no significance. The retweets of a tweet are the ones which are of importance as they can help us in analysing network activity surrounding a user. An analysis of the flow of all the tweets (excluding statuses with zero retweet count) of a particular user can be useful in identifying the various popular (frequent) routes.

In order to trace a tweets path, we need to download and store the tweets of the given user and also the information about the connected users and their tweets. The complexity of this method can be reduced by noting the fact that a retweet will only go to a follower and if that follower retweets the tweet (may already be a retweet) then it will reach his/her followers and so on. So there is only a need to store information (user and his/her tweets) about a followers followers and so on ignoring the friends of such followers. On the other hand, when a user is either a friend (following) or both a friend and follower of the given user, then the retweets (or simply tweets) that the given user receives are from such users only who, in turn, may have received the tweet (or a retweet) from similar users (i.e. who are either friends or both friends and followers) and so on.

Such reduction in the directed twitter network converts it or makes it look like a river like structure with many tributaries (the network can also be visualized as consisting of unidirectional pipelines in which tweets flow). The extent of 'flow' of a tweet in such tributaries or pipelines depends upon the retweeting of the tweet by users in the way. Retweeting in turn depends upon the content of the tweet i.e. better the content of a tweet better are the chances of it being retweeted. Popularity of the original user who tweeted the status and the popularity of the subsequent followers are also factors that affect the extent to which a tweet is retweeted.

References

1. Mislove, A., Gummadi, K.P., Druschel, P.: Exploiting social networks for Internet search. In: Proceedings of the 5th Workshop on Hot Topics in Networks (HotNets-V), Irvine, CA, November 2006
2. Breiger, R.L.: The analysis of social networks. In: Hardy, M., Nryman, A. (eds.) Handbook of Data Analysis, pp. 505–526. Sage Publications, London (2010)
3. Wu, S., Cornell University, USA, Jake, M., Hofman Yahoo! Research, NY, USA, Winter, A., Mason Yahoo! Research, NY, USA, Duncan, J., Watts Yahoo! Research, NY, USA, Who Says What to Whom on Twitter, 20th Annual World Wide Web Conference, ACM, Hyderabad, India (2011)
4. Lasswell, H.D.: The structure and function of communication in society. In: Bryson, L. (ed.) The Communication of Ideas, pp. 117–130. University of Illinois Press, Urbana (1948)
5. Walther, J.B., Carr, C.T., Choi, S.S.W., DeAndrea, D.C., Kim, J., Tong, S.T., Van Der Heide, B.: Interaction of interpersonal, peer, and media influence sources online. In: Papacharissi, Z. (ed.) A Networked Self: Identity, Community, and Culture on Social Network Sites, pp. 17–38. Routledge, New York (2010)
6. Bakshy, E., Hofman, J.M., Mason, W.A., Watts, D.J.: Identifying influencers on twitter. In: Fourth ACM International Conference on Web Search and Data Mining (WSDM), Hong Kong (2011)
7. Kwak, H., Lee, C., Park, H., Moon, S.: What is Twitter, a Social Network or a News Media? In: Proceedings of the 19th international conference on World Wide Web, pp. 591–600. ACM, New York (2010)
8. Newman, M.E.J., Park, J.: Why social networks are different from other types of networks. Phys. Rev. E **68**(3), 036122 (2003)
9. Hannak, A, et al.: Tweetin' in the rain: Exploring societal-scale effects on mood. In: Proceedings of the Sixth International AAAI Conference on Weblogs and Social Media (2013)
10. Milgram, S.: The small world problem. Psychol. Today **2**(1), 60–67 (1967)
11. Cha, M., Haddadi, H., Benevenuto, F., Gummad, K.P.: Measuring user influence on twitter: the million follower fallacy. In: 4th Int'l AAAI Conference on Weblogs and Social Media, Washington (2010)
12. Weng, J., Lim, E.P., Jiang, J., He, Q.: Twitterrank: finding topic-sensitive influential twitterers. In: Proceedings of the Third ACM International Conference on Web search and Data Mining, pp. 261–270. ACM, New York (2010)
13. Jung, Y., Park, H., Du, D.-Z., Drake, B.L.: A decision criterion for the optimal number of clusters in hierarchical clustering. J. Bioinf. **18**, S182–191 (2002)
14. Dunn, J.C.: Well separated clusters and optimal fuzzy partitions. J. Cybern. **4**, 95–104 (1974)
15. Davies, D.L., Bouldin, D.W.: A cluster separation measure. IEEE Trans. Pattern Anal. Machine Intell. **1**(4), 224–227 (1979)
16. Rousseeuw, P.J.: Silhouettes: a graphical aid to the interpretation and validation of cluster analysis. J. Comput. Appl. Math. **20**, 53–65 (1987)
17. Bezdek, J.C. et.al.: Bezdel's fuzzy c-means clustering algorithm. Comput. Geosci. **10**(2–3), 191–203 (1984)
18. Han, J., Kamber, M. : Data Mining Concepts and Techniques, 2nd edn. Elsevier, San Francisco (2006). ISBN 13: 978-1-55860-901-3
19. Crane, R., Sornette, D.: Robust dynamic classes revealed by measuring the response function of a social system. Proc. Nat. Acad. Sci. **105**(41), 15649–15653 (2008)
20. www.klout.com accessed 2nd July 2013
21. De Choudhury, M., Counts, S., Horvitz, E.: Major life changes and behavioral markers in social media: case of childbirth. In: Proceedings of the 16th ACM Conference on Computer Supported Cooperative Work (San Antonio, TX, USA, February 23–27, 2013). CSCW (2013)
22. Liu, J., Dolan, P., Pedersen, E.R.: Personalized news recommendation based on click behavior. In: Proceedings of the 15th International Conference on Intelligent User Interfaces ACM, IUI, pp. 31–40, (2010)

The Global Spread of Islamism: An Agent-Based Computer Model

Morgan L. Eichman, James A. Rolfsen, Mark J. Wierman,
John N. Mordeson and Terry D. Clark

Abstract We use an agent-based model to model a dynamic network that considers the rate at which Islamism will spread globally. We define Islamism as the organized political trend that seeks to solve modern political problems by referencing Muslim teachings. The trend is also associated with Radical Islam or Islamic Fundamentalism and is often revolutionary and violent in nature. The model assumes that Islamism spreads from state to state based on existing relations that replicate those defining global trade and communications. Islamism must diffuse through these existing networks. Since Islamism is inimical to western liberal values such as women's rights and social tolerance, the diffusion of Islamism is hindered by a strong commitment to western liberal values. We include all countries in the analysis, scored on the degree to which they are committed to Islamism and western liberal values.

Keywords Network analysis · Diffusion · Agent-based model · Islamism · Radical Islam · Western liberalism

1 The Spread of Islamism

Islamism is a contemporary movement that has emerged in reaction to globalization [1]. In contrast to Islam, which is a cultural system centered on faith and ethics, Islamism is an all encompassing social, political, and economic ideology based on religion. Viewing the modern world as a hindrance to the existence of an authentic Islamic community and believing that Islam itself is at risk, Islamism places itself in opposition to western liberalism as a manifestation of growing

M. L. Eichman · J. A. Rolfsen · M. J. Wierman (✉) · J. N. Mordeson · T. D. Clark
Creighton University, Omaha, NE, USA
e-mail: mwierman@creighton.edu

W. Pedrycz and S.-M. Chen (eds.), *Social Networks: A Framework*
of Computational Intelligence, Studies in Computational Intelligence 526,
DOI: 10.1007/978-3-319-02993-1_18, © Springer International Publishing Switzerland 2014

global interdependence. The movement directly confronts western liberal ideals of the separation of church and state, power-sharing, open elections [2, 3] and human rights, in particular those of women [4].

Not surprisingly, scholars have devoted considerable attention to the movement. Particular focus has been placed on the dynamics of the movement. Most of these studies have been qualitative, contextual descriptions of the growth of Islamism as a social movement either in reaction against modernism [5, 6] or as a search for a post-modernist identity [7]. As yet, however, no attempt has been made to consider the growth of Islamism in a holistic manner within both the context of the international system and in opposition to western liberalism.

This chapter is a purposeful attempt to fill that lacuna. We use an agent-based model to consider the assumptions under which the diffusion of Islamism across the international system is more likely. We represent the international system as a social network. Our model assumes that western liberalism and Islamism stand in opposition to one another. We further assume that a state is more susceptible to Islamism if it is strongly connected to other Islamist states. The number and strength of such connections in interaction with both the internal and global appeal of western liberalism determines the likelihood of the diffusion of Islamism across the global social network.

Our purpose is not to predict the actual likelihood of the global spread of Islamism. That would require that we have empirically verifiable data and that our model accounts for all interactions. Our purpose is a bit more modest. Moreover, it is better suited to the strengths of agent based models. We wish to identify the degree to which changes in parameters affecting the spread of Islamism are sensitive to modest changes. Such an analysis provides both scholars and policy analysts with insights into questions concerning the relative impact of such changes.

2 Diffusion Across Social Networks

Social networks map relationships between actors. A node represents an agent or player (in our chapter, a state), and edges represent connections between them. Such mappings permit us to consider the diffusion of various types of phenomena across social networks [8, 9]. Three strands of research have used social networks for these purposes.

The first is in the field of epidemiology. This research has made extensive use of scale-free degree distribution of nodes within a given network to study the spread of disease [10–13]. Degree distribution is the number of connections or edges a given node has to other nodes within the network. Common characteristics of scale-free networks include nodes that contain a higher degree than others. These nodes act as the hubs and are believed to be critical in the transmission of diseases in a given network. If a highly connected node is infected, the disease spreads at a faster rate than if nodes are infected randomly.

A second strand of research has focused on the spread of computer viruses [14]. This research has produced the idea of the Susceptible, Infected, Recovered (SIR) model. The SIR model demonstrates that nodes can contract a disease from infected neighbors and, once infected, they eventually recover or are removed from the system and can no longer infect others in the model [15]. The SIR model can be adapted to fit specific cases. For example, it is possible to allow for actors to make strategic choices about behavior rather than being randomly assigned with an attribute, such as being infected or not infected. Actors are able to consider the relative costs and benefits to behavior. They can also make decisions based on the proportion of neighbors choosing different behaviors [15]. The likelihood that a node will choose a certain behavior over another also might depend on the distribution of adopters at each point in time. Therefore, the volume of initial adopters in the model greatly impacts the dynamic behavior of the adopters as the model progresses in time [16].

The third strand of research is located in the discipline of economics [17]. Economic research examines different ways in which neighbors can affect each others' behavior [18]. The economic literature also examines the conditions under which equilibrium is reached in networks e.g., [5, 9, 19]. Equilibrium is accomplished when the model reaches a static state and the interactions between nodes prevent further diffusion with the network.

3 The Global Social Network

We contribute to the burgeoning research on diffusion in social networks by considering how Islamism spreads across the international system. Our approach makes use of a global social network defined by global trade and communications. We contend that Islamism must diffuse through these existing networks.

There are multiple domains in which countries interact with one another. We construct our global social network on the basis of inter-state exchanges on five dimensions: direct flights between countries, diplomatic and consular exchange [20], submarine cable connections [21], telephone minutes exchanged [22], and electricity exchanges [23–30]. Each of these is represented by a 196×196 matrix. If two countries, i and j, interact on the given dimension, then cell ij is coded "1". Otherwise, it is coded "0".

The adjacency matrix defining the global social network is the sum of these five matrixes. We use a dendrogram to organize the social network map into a hierarchical clustering. Three large clusters are apparent in the dendrogram, at Appendix 1. We have separated them with dashed vertical lines. The smallest of these clusters, on the left side of the dendrogram, cluster 1, contains seventeen states. It represents a highly connected set of states whose density is 0.9779. Cluster 1 includes the globe's dominant actors, among which are the United States, the United Kingdom, China, and Germany.

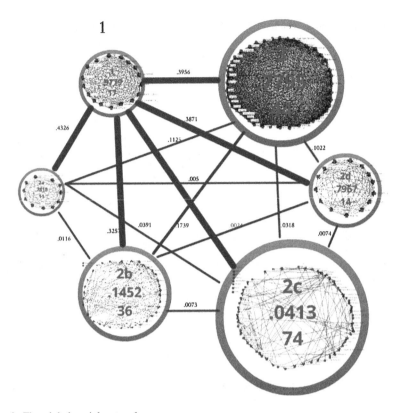

Fig. 1 The global social network

The next largest cluster, cluster 3, is on the right hand side of the dendrogram. Cluster 3 is relatively highly connected of 44 states, with an internal density of 0.7795. The cluster includes countries that are important regional actors, among which are Brazil, South Korea, Thailand, and Turkey.

The largest cluster, cluster 2, can be further divided into four subclusters (subclusters; 2a, 2b, 2c, and 2d). Subcluster 2a has an internal density of 0.7818. It consists of 11 states, to include Senegal, Gabon, Libya, and Kenya. Cluster 2b has an internal density of 0.1452. It consists of 36 states, among which include Belarus, Kosovo, Sudan, and Cambodia. Cluster 2c is the least interconnected subcluster with an internal density of 0.0413. It comprises 74 states of which 8 are isolates. Among the states in the subcluster are Niger, Chad, Fiji, and Samoa, as well as most of the international system's island states. Cluster 2d is the strongest interconnected subcluster and has an internal density of 0.7967. It comprises 14 states, including Honduras, Panama, Costa Rica, Peru, and Taiwan.

Figure 1 depicts this global social network. Countries are placed within their respective clusters and are depicted by nodes. The thick black lines, or edges, within clusters characterize the relationship between each node in the cluster.

Those between clusters represent the ties between states in each cluster. Their width represents the strength of those ties. All clusters are most strongly connected to cluster 1. The subclusters of cluster 2 are weakly related to each other and moderately related to cluster 3.

4 The Spread of Islamism within the Global Social Network: An Agent-Based Model

We construct an agent-based model to analyze the spread of Islamism within our global social network. The model conceptualizes the diffusion of Islamism as both inimical to and hindered by western liberal values. Thus, we assume that as one of them gains in influence within a state, the other necessarily wanes [31, 32]. While there is a middle ground within which both might reasonably coexist at relatively moderate levels of social influence within a state, coexistence is generally not possible. The same is true concerning the ties that a state has with others in the global system. As a state's neighbors move increasingly in one direction or the other, the state will be moved in that same direction. We do not consider the possibility that a state may sever its ties in our model, but we do account for the strength of a state's ties increasing or decreasing in intensity as a function of the degree of ideological affinity with its neighbors.

Our agent-based model begins with the global social network. All countries in the network are arrayed in accordance with their relations with one another. The strength of these relations (the edges), which we label M, is determined by the adjacency matrix defining the global social network.

Each country (node) has an initial score on each of two dimensions: western liberalism (L) and Islamism (I). Both scores are normalized on a scale from 1 to 10. We use the political rights scores reported in "Freedom in the World 2012: The Arab Uprisings and their Global Repercussions" [33] to score countries on western liberalism. Prior to being normalized, these scores range from 1 to 7. The higher the value, the more influence that western liberalism exerts. We use multiple sources to code Islamism's influence in a state on a scale from 0 to 4 [34–37]. A country with a Muslim population of 50 % or greater that has not adopted a state religion and has not declared itself either an Islamic state or a secular state is coded a 1. This would include countries such as Indonesia. While Indonesia has a Muslim population well over 50 %, it has neither adopted a state religion nor declared itself a secular state. Countries with a Muslim population of 50 % or greater that have officially declared themselves to be a secular state are coded 2. Countries that use Sharia or Quran as the basis of their legal systems are coded 3. Countries that have formally declared Islam the official state religion are coded 4. All other countries are coded 0. The scores assigned to each country are at Appendix 2.

The nodal characteristics (L and I) and the strength of ties between states (the edges, M) comprise the environment in our agent-based model. The rules that govern how states act within the environment are determined by two distinct sets of adjustable parameters. The first set of these parameters determines the *intra-state* trade-off between western liberalism and Islamism. There are three such parameters: western liberalism intolerance (LI), Islamism intolerance (II), and the speed factor (SF). The Intolerance value is the drive of one value to compete with or to suppress the existence of the other value. Western liberalism intolerance (LI), determines the degree to which it actively suppresses Islamism. Islamism intolerance (II), determines the degree to which Islamism actively suppresses western liberalism. It is generally thought that Islamism is far less tolerant; nonetheless, our model permits us to adjust both levels of tolerance independently of the other. The third intra-state parameter is the speed factor (SF), which regulates the rate of change caused by the competition between western liberalism and Islamism within states.

A second set of parameters regulates how states' commitments to western liberalism and Islamism are affected by those of other states with which they share ties in the international system. We call these diffusion variables. The diffusion variables include western liberalism assertiveness (LA), Islamism assertiveness (IA), the general diffusion rate (DR), western liberalism impetus (LImpetus) and Islamism impetus (IImpetus). The two assertiveness factors (LA and IA) determine the ability of a given value to project or proliferate beyond the borders of one state to another. These variables help determine the minimal pressure that needs to be applied to a country in order for change to occur. The general diffusion rate regulates how quickly the diffusion of values occurs throughout the system. If the General Diffusion Rate is 0, then the assertiveness variables have no effect. Finally, the impetus factors (LImpetus and IImpetus) are auxiliary variables intended to act as stabilizers in order to measure the effects that the other variables have on the model. In simplified terms, the impetus factors act as additional normalization mechanisms.

After all initial parameters are set, the model runs in time ticks. Each tick is repeated as often as the user wishes. A single time tick progresses as follows. First, the model calculates the diffusion of western liberalism and Islamism via the edges defining ties between states in the global social network. The model takes the current Islamism (I) and western liberalism (L) score of a particular state (STATE) and finds the "I" and "L" scores of all other states (state) it shares an M-value connection with. The following code accomplishes these tasks.

```
I,L = findI(STATE),findL(STATE)
TotI,TotL,TotM = 0,0,0
if STATE==state: continue
M = findM(STATE,state)
If M <= 0: continue
```

Next the model records the Islamism values (i) and western liberalism values (l) of states that have a tie (M-value connection) with the particular state. It uses these scores to calculate the average values of all states related to the particular state. It outputs the average weighted Islamism value (PeerI) and average weighted western liberalism value (PeerL) of a state's peers weighted by the M-value connection. The code is as follows.

```
i,l = findI(state),findL(state)
TotM += M
TotI += (M*i)
TotL += (M*l)
PeerI = TotI/TotM
PeerL = TotL/TotM
```

It is necessary to bind the limits of the model so the values assigned to the variables stay within a range of 0–10. Thus, we add the code,

```
PeerI = Bound(0,PeerI,10)
PeerL = Bound(0,PeerL,10)
```

The model then returns the difference of a state's peer values and the state's current values for both Islamism (IDif) and western liberalism (LDif). This is accomplished by the code,

```
IDif = PeerI - I
LDif = PeerL - L
return IDif, LDif
```

Next, the model calculates the real change that needs to occur to each existing value of a state during the diffusion process. First, the minimum amount of assertiveness (MA) is established by picking the smaller of the two assertiveness values, LA and IA. This value is applied when determining the amount of change Islamism undergoes (IChange) and the amount of change western liberalism undergoes (LChange). MA is only applied when determining the IChange or LChange if the IDif or LDif values are less than zero. IDif, LDif are redefined as the amount of peer pressure being diffused to connecting states. The code is as follows.

```
MA = min(LA,IA) # Minimum Assertiveness
  IDif,LDif = PeerPressure(state)
```

The Diffusion Rate (DR) is then applied to regulate the rate of diffusion. The new values are returned. The code is as follows.

```
if IDif >0: IChange = IDif*IA*DR
else:       IChange = IDif*MA*DR
if LDif >0: LChange = LDif*MA*DR
else:       LChange = LDif*MA*DR
return IChange,LChange
```

The next component of the equation calculates the change in values that results from competition between Islamism and western liberalism within a state. First the model establishes that the range of the I and L values are bound to an upper limit of 10.0 (maxint). Second, the amount of force Islamism and western liberalism has within a state when in competition with each other is calculated. This value is multiplied by the Speed Factor (SF) which regulates the rate at which Islamism and western liberalism compete. This returns the LChange and IChange values indicating the amount of competition within a given state. Competition causes the suppression of the weaker value, but not the proliferation of the stronger. The code is as follows.

```
maxint = MI
LC = L * LI
IC = I * II
if LC+IC <=0: LD = 0
else: LD = ((LC-IC)/(LC+IC))*(SF/2)
ID = -LD
LChange = LD * min(L,maxint-L)
IChange = ID * min(I,maxint-I)
LChange = min(LChange,LChange/2)
IChange = min(IChange,IChange/2)
return LChange,IChange
```

The final process of the model brings all of the values together. It assigns the newly generated diffusion changes and competition changes to the current I and L values in order to create the next iteration of the model. This is how the model progresses in time. The code is as follows.

```
i,l = findI(state),findL(state)
IChange,LChange = ValuesChange(state)
I = i + IChange
L = l + LChange
LChange2,IChange2 = Competition(L,I)
newI = I + IChange2
newL = L + LChange2
return newI,newL
```

5 Results of the Model

Given that western liberalism is more widely distributed across the international system, it seems reasonable to expect that western liberalism will dominate the globe in our model. (The average score on western liberalism for the 196 countries is 5.56 on a scale from 0 to 10 and the average Islamism score is 1.47 on a scale from 1 to 10.) Our expectation is heightened by the fact that countries scoring higher on western liberalism are strategically located in more central clusters. For example, the average score for western liberalism in cluster 1 is 7.04; and that for Islamism is 0.55. The reader will recall that cluster 1 contains the world's most connected states. Moreover, the average score on Islamism does not exceed that for western liberalism in any of the clusters.

Table 1 Parameter values

Parameters	Model 1	Model 2	Model 3
Islamism intolerance	0.4	0.5	0.5
Liberalism intolerance	0.2	0.2	0.4
Islamism assertiveness	0.4	0.4	0.4
Liberalism assertiveness	0.2	0.2	0.2
Islamism impetus	1.0	1.0	1.0
Liberalism impetus	1.0	1.0	1.0
Speed factor	0.3	0.3	0.3
General diffusion rate	0.2	0.2	0.2

Fig. 2 Graphical representation of model 1: equilibrium between Islamism and western liberalism

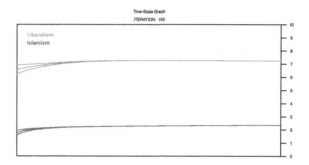

However, our expectations are not borne out by our model under several assumptions that have intuitive appeal. First, we assume that western liberalism is more tolerant of Islamism than Islamism is of western liberalism (Islamism Intolerance > Liberalism Intolerance). Second, we assume that Islamist states exert greater pressure on their neighbors than do western liberal states (Islamism Assertiveness > Liberalism Assertiveness). The values assigned for each of these parameters in our first model (Model 1) are presented in Table 1. While these initial values seem reasonable, we do not assert that they are empirically correct. Our intention is a bit more nuanced. We wish to determine the degree to which changes in these parameters result in changes in the diffusion of Islamism. That is, what factors are relatively more important in predicting the rate at which Islamism will diffuse globally?

Based on these initial parameters, Fig. 2 depicts the spread of Islamism globally over time as determined by our agent based model. The x-axis is the number of time ticks, or number of iterations of the model, from time $t = 0$ to $t = 150$, and the y-axis is the average Islamism or western liberalism score for all 196 countries from 0 to 10. The green line represents the change in the average level of liberalism over time, and the red line represents the change in the average level of Islamism over time. The parameter settings in Model 1 reach an equilibrium after 68 time ticks: western liberalism maintains global dominance. Moreover, it does so despite the greater intolerance and assertiveness scores of Islamism. This does not mean that all states trend toward western liberalism. In fact, as the model approaches the equilibrium, four states convert from predominantly western liberal

Fig. 3 Graphical
representation of model 2:
Islamism's dominance over
time given increased Islamist
intolerance

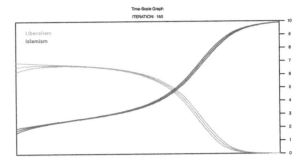

Fig. 4 Graphical
representation of model 3:
Western liberalism's
dominance over time given
increased intolerance on the
part of liberalism

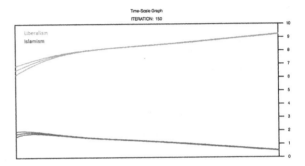

states to predominantly Islamist states. These states in the order in which they convert to Islamism are Niger, Senegal, Turkey and Mali.

Our results change when we modify the degree to which Islamism is more intolerant than western liberalism. (The parameter values are those for Model 2 in Table 1). The results are demonstrated in Fig. 3. When we raise the Islamism Intolerance parameter by 0.1, the results indicate a substantially different outcome. After 116 time ticks, all countries in the model convert to Islamism. The first to convert to Islamism are Niger, Senegal, Turkey, and Lebanon. The last countries to convert to Islamism are Guatemala, Nicaragua, St. Vincent and The Grenadines, and Honduras.

We can change the results once again by increasing the liberalism intolerance value by 0.2 (the parameter values are those listed for Model 3 in Table 1). Initially both Islamism and western liberalism start to rise, but after the ninth time tick Islamism gradually declines and western liberalism continues to rise. At the end of 150 time ticks, 29 states convert from Islamism to western liberalism including Egypt, Iraq, and Pakistan. Only two states, Qatar and Bahrain, remain dominated by Islamism Fig. 4.

6 Conclusions

Our agent-based model argues that Islamism is capable of overcoming western liberalism despite beginning with a significant disadvantage, to include fewer Islamist states that are located in less central positions within the international system. However, if Islamism is not significantly more intolerant than western liberalism (and all other parameters are held constant), then our agent based model argues that western liberalism can maintain global dominance.

Our models also reveal some interesting outcomes related to individual states. One would expect countries located in cluster 1 of the global social network map with a strong commitment to western liberal values to retain their ideals the longest. These include the United States, Germany, and the United Kingdom. However, it is the countries that are least connected within the global network that do so, including most of the world's island states. Thus, the marginally connected states of Nicaragua, St. Vincent and The Grenadines, and Honduras retain western liberal values longer than the better connected states of Belgium, Canada, and Denmark. Less surprising is our finding that countries that are highly connected to Islamist states are likely to become Islamist the fastest, despite any commitment they might have to western liberalism. These include Niger, Senegal, and Turkey.

Our models provide some initial insights into the dynamic of the global diffusion of Islamism. In future, we intend to exploit the model further to determine the conditions under which equilibrium between Islamism and western liberalism is likely. We also intend to make further refinements to the model. Among these are adjusting the strength of ties between states (M), to permit both finer measurement and changes over time, and adding an intra-state parameter accounting for demographic changes resulting from immigration and birth rates.

Appendix 1. Dendrogram Representing Ties Between States

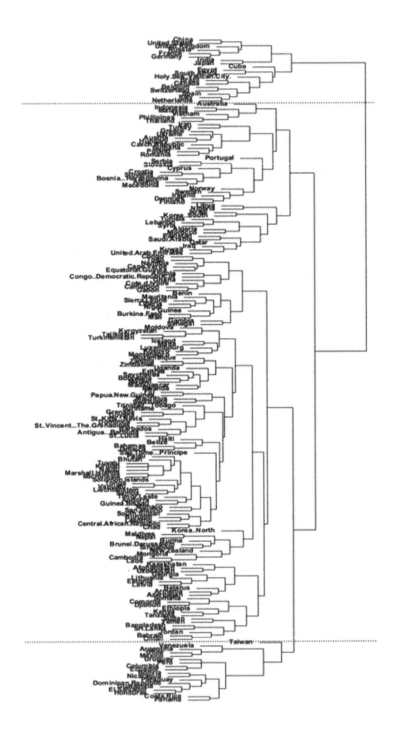

Appendix 2. Islamism Scores

Country	Islamism
Afghanistan	4
Albania	1
Algeria	4
Andorra	0
Angola	0
Antigua and Barbuda	0
Argentina	0
Armenia	0
Australia	0
Austria	0
Azerbaijan	2
Bahamas	0
Bahrain	4
Bangladesh	4
Barbados	0
Belarus	0
Belgium	0
Belize	0
Benin	0
Bhutan	0
Bolivia	0
Bosnia and Herzegovina	0
Botswana	0
Brazil	0
Brunei Darussalem	4
Bulgaria	0
Burkina Faso	2
Burma	0
Burundi	0
Cambodia	0
Cameroon	0
Canada	0
Cape Verde	0
Central African Republic	0
Chad	2
Chile	0
China	0
Columbia	0
Comoros	3
Congo	0
Congo, Democratic Republic of	0

(continued)

(continued)

Country	Islamism
Costa Rica	0
Cote d'Ivoire	0
Croatia	0
Cuba	0
Cyprus	0
Czech Republic	0
Denmark	0
Djibouti	2
Dominica	0
Dominican Republic	0
Ecuador	0
Egypt	4
El Salvador	0
Equatorial Guinea	0
Eritrea	0
Estonia	0
Ethiopia	0
Fiji	0
Finland	0
France	0
Gabon	0
Gambia	2
Georgia	0
Germany	0
Ghana	0
Greece	0
Grenada	0
Guatamala	0
Guinea	2
Guinea-Bissau	0
Guyana	0
Haiti	0
Holy See (Vatican City)	0
Honduras	0
Hungary	0
Iceland	0
India	0
Indonesia	1
Iran	4
Iraq	4
Ireland	0
Israel	0
Italy	0
Jamaica	0
Japan	0

(continued)

(continued)

Country	Islamism
Jordan	4
Kazakhstan	2
Kenya	0
Kiribati	0
Korea, North	0
Korea, South	0
Kosovo	2
Kuwait	4
Kyrgyzstan	2
Laos	0
Latvia	0
Lebanon	1
Lesotho	0
Liberia	0
Libya	4
Liechtenstein	0
Lithuania	0
Luxembourg	0
Macedonia	0
Madagascar	0
Malawi	0
Malaysia	4
Maldives	4
Mali	2
Malta	0
Marshall Islands	0
Mauritania	4
Mauritius	0
Mexico	0
Micronesia	0
Moldova	0
Monaco	0
Mongolia	0
Montenegro	0
Morocco	4
Mozambique	0
Namibia	0
Nauru	0
Nepal	0
Netherlands	0
New Zealand	0
Nicaragua	0
Niger	2
Nigeria	3
Norway	0

(continued)

(continued)

Country	Islamism
Oman	4
Pakistan	4
Palau	0
Panma	0
Papua New Guinea	0
Paraguay	0
Peru	0
Phillipines	0
Poland	0
Portugal	0
Qatar	4
Romania	0
Russia	0
Rwanda	0
St. Kitts and Nevis	0
St. Lucia	0
St. Vincent and The Grenadines	0
Samoa	0
San Marino	0
Sao Tome and Principe	0
Saudi Arabia	4
Senegal	2
Serbia	0
Seychelles	0
Sierra Leone	1
Singapore	0
Slovakia	0
Slovenia	0
Solomon Islands	0
Somalia	4
South Africa	0
South Sudan	0
Spain	0
Sri Lanka	0
Sudan	3
Suriname	0
Swaziland	0
Sweden	0
Switzerland	0
Syria	1
Taiwan	0
Tajikistan	2
Tanzania	0
Thailand	0
Timor-Leste	0

(continued)

(continued)

Country	Islamism
Togo	0
Tonga	0
Trinidad and Tobago	0
Tunisia	4
Turkey	2
Turkmenistan	2
Tuvalu	0
Uganda	0
Ukraine	0
United Arab Emirates	4
United Kingdom	0
United States	0
Uruguay	0
Uzbekistan	2
Vanuatu	0
Venezuela	0
Vietnam	0
Yemen	4
Zambia	0
Zimbabwe	0

References

1. Esposito, J.L.: Claiming the center: political Islam in transition. Harv. Int. Rev. **19**(2), 8–16 (1997)
2. Gould, C.C.: Golbalizing Democracy and Human Rights. Cambridge University Press, Cambridge (2004)
3. Tibi, B.: Political Islam, world politics, and Europe: democratic peace and Euro-Islam vs. Global Jihad. Routledge, New York (2008)
4. Marshall, P.: Radical Islam's rules: the worldwide spread of extreme Shari'a law. Rowman and Littlefield Publishers, INC, Maryland (2005)
5. Morris, A.D., McClurg Mueller, C.: Frontiers in Social Movement Theory. Yale University Press, London (1992)
6. Bayat, A.: Islamism and social movement theory. In: Third World Quarterly, vol. 26, pp. 891–908. Taylor and Francis Ltd., UK (2005)
7. Burgat, F., Dowell, W., Fernea, R.: The Islamic movement in North Africa. In: Middle East Studies Association Bulletin, vol. 28, p 238. Indiana University Press, London (1994)
8. Jackson, M.O., Yariv, L.: Diffusion on social networks. Économie Publique/Public economics **16**, 69–82 (2005)
9. Jackson, M.O.: Social and Economic Networks. Princeton University Press, New Jersey (2008)
10. Pastor-Satorras, R., Vespignani, A.: Epidemic spreading in scale-free networks. Phys. Rev. Lett. **86**, 3200–3203 (2001)

11. Pastor-Satorras, R., Vespignani, A.: Epidemic dynamics and endemic states in complex networks. Phy Rev E Stat Nonlinear Soft Matter Phy **63**, 1–8 (2001)
12. May, R.M., Lloyd, A.L.: Infections on scale-free networks. In: Physical Review E: Statistical, Nonlinear, and Soft Matter Physics, vol 64, pp. 1–6. University of Oxford, Oxford (2001)
13. Dezso, Z., Barabasi, A.-L.: Halting viruses in scale-free networks. In: Physical Review E. vol. 65, pp. 1–4. University of Notre Dame, Indiana (2002)
14. Newman, M. E. J., Forrest, S., Balthrop, J.: Email networks and the spread of computer viruses. Phys Rev E. **66**, 1–4 (2002)
15. Jackson, M.O., Yariv, L.: Diffusion on social networks. Économie Publique/Public economics **4**, 6 (2005)
16. Jackson, M.O., Yariv, L.: Diffusion on Social Networks. In: Économie Publique/Public economics **4**, 7 (2005)
17. Young, H.P.: The Diffusion of Innovations on Social Networks. John Hopkins University, Baltimore (2000)
18. Jackson, M.O., Yariv, L.: Diffusion on social networks. Économie Publique/Public economics **4**, 8 (2005)
19. Jackson, M.O., Yariv, L.: Diffusion on social networks. Économie Publique/Public economics **4**, 5 (2005)
20. Embassy Pages. Embassy Pages. www.embassypages.com (2012)
21. TeleGeography. Submarine Cable Map. http://www.submarinecablemap.com/ (2012)
22. TeleGeography. Global Traffic Map 2010. http://www.telegeography.com/assets/website/images/maps/global-traffic-map-2010/global-traffic-map-2010-l.jpg (2012)
23. Central Intelligence Agency Electricity—Consumption.*CIA*. https://www.cia.gov/library/publications/the-world-factbook/rankorder/2042rank.html
24. Central Intelligence Agency. Electricity exports. Central Intelligence Agency. https://www.cia.gov/library/publications/the-world-factbook/fields/2044.html
25. Central Intelligence Agency. Electricity imports. *Central Intelligence Agency*. https://www.cia.gov/library/publications/the-world-factbook/fields/2043.html
26. Matthias, L., Mutschler, U.: BDEW: Germany remains electricity exporter, but imports increase significantly. In: German Energy Blog. http://www.germanenergyblog.de/?p=7218 (2011)
27. National Energy Grid: National energy grid map index. In: National Energy Grid. http://www.geni.org/globalenergy/library/national_energy_grid/index.shtml (2012)
28. Tara, P.: Germany becomes net power importer from France after atomic halt. In: Bloomberg. http://www.bloomberg.com/news/2011-05-30/germany-becomes-net-power-importer-from-france-after-atomic-halt.html (2011)
29. European Network of Transmission System Operators for Electricity: Indicative values for Net Transfer Capacities (NTC) in continental Europe Winter 2010/11, working day, peak hours (non-binding values). https://www.entsoe.eu/publications/market-and-rd-reports/ntc-values/ntc-matrix/ (2011)
30. Radio Free Europe: Tajikistan, Pakistan discuss electricity exports. In: RadioFreeEurope/RadioLiberty. http://www.rferl.org/content/tajikistan_pakistan_discuss_electricity_exports/24441274.html (2012)
31. Kelman, H.C.: Compliance, identification, and internalization: three processes of attitude change. J Conflict Resolut **2**, 51–60 (1958)
32. Wang, Y., Xiao, G., Liu, J.: Dynamics of Competing Ideas in Complex Social Systems. Cornell University Library. http://arxiv.org/abs/1112.5534 (2011)
33. Freedom House: Freedom in the World 2012: the Arab uprisings and their global repercussions. http://www.freedomhouse.org/report/freedom-world/freedom-world-2012 (2012)
34. CIA World Factbook: The World Factbook. https://www.cia.gov/library/publications/the-world-factbook/ (2012)
35. Council of Foreign Relations: Islam: governing under Sharia. http://www.cfr.org/religion/islam-governing-under-sharia/p8034 (2013)

36. Travel.State.Gov.: Country specific information. http://travel.state.gov/travel/cis_pa_tw/cis/cis_4965.html (2013)
37. USCIRF.gov.: The religion-state relationship and the right to freedom of religion or belief: a comparative textual analysis of the constitutions of majority Muslim countries and other OIC members. http://www.uscirf.gov/reports-and-briefs/special-reports/3787.html (2012)

Using Neural Network Model to Evaluate Impact of Economic Growth Rate and National Income Indices on Crude Birth Rate in Taiwan

Yi-Ti Tung and Tzu-Yi Pai

Abstract In this study, a conceptual social network, in which the artificial neural network (ANN) was used, was adopted to evaluate impact of economic growth rate (EGR) and national income indices (NII) on crude birth rate (CBR) in Taiwan. The NII included gross domestic product (GDP), GDP per capita (GDPPC), gross national product (GNP), GNP per capita (GNPPC), national income (NI), and NI per capita (NIPC). To establish the ANN model, the EGR and NII were taken as the input variables, and the CBR was taken as the output variable. The results indicated that the minimum mean absolute percentage error (MAPE), mean squared error (MSE), root mean squared error (RMSE), and maximum correlation coefficient (R) was 23.29 %, 46.02, 6.78, and 0.85, respectively when training. Those for testing were 28.93 %, 35.82, 5.99, and 0.70, respectively. The results showed that the CBR appeared to have a negative sensitivity towards three per capita indices including GDPPC (-0.0369), GNPPC (-0.1314), and NIPC (-0.3822). It suggested that the "capital dilution" would result in CBR decline. But positive EGR in a previous year should stimulate the CBR in the current year, as well as the positive macroeconomic factors including GDP, GNP, and NI. It suggested that the economic development would cause the fertility will to occur, thus increased the CBR.

Keywords Artificial neural network · Economic growth rate · National income indices · Crude birth rate · GDP · GNP

Y.-T. Tung
School of Medical Sociology and Social Work, Chung Shan Medical University,
Taichung 40201, Taiwan, Republic of China

T.-Y. Pai (✉)
Master Program of Environmental Education and Management, Department of Science Application and Dissemination, National Taichung University of Education, Taichung 40306, Taiwan, Republic of China
e-mail: bai@ms6.hinet.net

W. Pedrycz and S.-M. Chen (eds.), *Social Networks: A Framework of Computational Intelligence*, Studies in Computational Intelligence 526, DOI: 10.1007/978-3-319-02993-1_19, © Springer International Publishing Switzerland 2014

1 Introduction

The U.S. Census Bureau estimates the mid-2010 world population to be about 7 billion people and growing at 1.13 percent per year. Every second, four or five children are born, somewhere on the earth on average. In that same second, one or two other people die. Whether human populations will continue to grow at present rates and what that growth would imply for social development are among the most central and pressing questions in social science [1].

Demography encompasses essential statistics about people, such as births, deaths, and where they live, as well as total population size. Among these, fertility is vital to social evaluation and planning, especially to social policy formulation. There are two factors for describing fertility including total fertility rate and crude birth rate [1–6]. The total fertility rate is the number of children born to an average woman in a population during her entire reproductive life. Over the past 25 years, the average number of children born per woman worldwide has decreased from 6.1 to 2.6. According to the World Health Organization, 61 % of the world's 161 countries are now or bellow a replacement rate of 2.1 children per woman, including Taiwan [1]. At present, many countries actively promote the fertility will. The fertility rate reveals the number of children born per woman, but the most accessible demographic statistic of fertility is usually the crude birth rate (CBR), the number of births in a year per thousand persons.

CBR varies according to many factors such as economic factors, demographic transition, food availability, health problem, or other factors. Among these factors, economic factor was reported as the most significant factor that affected the birth rate [1–6]. Therefore, a social network in which a social architecture made up of a set of factors and a web of linkages between CBR and these factors were formed.

A social network is a social structure consists of a set of social actors (such as organizations or individuals) and the complex dyadic ties between these actors. The social network concept provides a methodic way to analyze the structure of whole social entities [7]. The research of these structures utilizes social network analysis to identify global and local patterns, locate influential entities, and test network dynamics.

Although behavior pattern between CBR and the economic factors is not fully related to theory of social networks, the social network concept can be introduced in the analysis of relationship between CBR and the economic factors. Therefore, the social network perspective offers a way to analyze the structure of CBR and economic related factors. Typically, CBR are very complex for analyzing because network interrelations between various factors result in a complicated combination of relations. In order to simplify statistical complexity and gain consistent results from the investigation data for testing CBR, the artificial neural network (ANN) is a suitable method.

The ANN is used for modeling complex relationships between inputs and outputs and has been applied in our previous works [8–13]. If the relationships between CBR and other economic factors can be determined, the local CBR patterns can be identified, the complex CBR network dynamics can be examined, and better formulation and planning for social policy can be sought.

Since CBR represents the number of births per thousand persons annually, the CBR in the current year is affected significantly by the economic factors in the previous year. However, no study using the social network concept has been done in CBR interrelation network evaluation. Additionally, no research using ANN has been proposed in this field. Therefore, a conceptual social network in which the ANN algorithm was used for interrelating different factors (actors) was proposed to evaluate the impact of economic growth rate (EGR) and national income indices (NII) on current CBR in Taiwan in this study. This chapter represents the first report of this innovative use of ANN for CBR interrelation network evaluation.

2 Materials and Methods

(1) Data set

The data for study were collected from different official annual reports. The statistical data of CBR were collected from the Statistical Yearbook of Interior from 1953 to 2011 [14]. To evaluate the effect of economic factors at previous year on current CBR, the data for EGR and other NII from 1952 to 2010 were also investigated [15]. The NII included gross domestic product (GDP), GDP per capita (GDPPC), gross national product (GNP), GNP per capita (GNPPC), national income (NI), and NI per capita (NIPC). The total number for CBR, EGR and other NII was 59. Among the total numbers of data, the numbers for training and testing were 44 and 15, respectively. To establish the ANN model, the EGR and NII were taken as the input variables, and the CBR was taken as the output variable.

(2) ANN

The ANN modeling approach, in which the important operation features of the human nervous system are simulated, attempts to solve new problems by using information gained from past experience. In order to operate analogous to a human brain, many simple computational elements called artificial neurons that are connected by variable weights are used in the ANN. With the hierarchical structure of a network of interconnected neurons, an ANN is capable of performing complex computations, although each neuron, alone, can only perform simple work. The multi-layer perceptron structure is commonly used for prediction among the many different types of structures.

A typical neural network model consisting of three independent layers: input, hidden, and output layers. Each layer is comprised of several operating neurons. Input neurons receive the values of input parameters that are fed to the network and store the scaled input values, while the calculated results in output layer are assigned by the output neurons. The hidden layer performs an interface to fully interconnect input and output layers. The pattern of hidden layer to be applied in the hierarchical network can be either multiple layers or a single layer. Each neuron is connected to every neuron in adjacent layers before being introduced as input to the neuron in the next layer by a connection weight, which determines the strength of the relationship between two connected neurons. Each neuron sums all of the inputs that it receives and the sum is converted to an output value based on a predefined activation, or transfer, function.

For prediction problems, a supervised learning algorithm is often adopted for training the network how to relate input data to output data. In recent years, the back-propagation algorithm is widely used for teaching multi-layer neural networks. Traditionally, the algorithm uses a gradient search technique (the steepest gradient descent method) to minimize a function equal to the mean square difference between the desired and the actual network outputs. The calculation of ANN was carried out using MATLAB.

2.1 Evaluation of Testing Performance

In order to evaluate the testing performance of ANN, the mean absolute percentage error (MAPE), correlation coefficient (R), mean square error (MSE), and root mean square error (RMSE) were employed and described as,

$$MAPE = \frac{1}{n}\sum_{i=1}^{n}\left|\frac{obs_i - pre_i}{obs_i}\right| \times 100\,\% \tag{1}$$

$$R = \frac{\sum_{i=1}^{n}(obs_i - \overline{obs})(pre_i - \overline{pre})}{\sqrt{\sum_{i=1}^{n}(obs_i - \overline{obs})^2 \sum_{i=1}^{n}(pre_i - \overline{pre})^2}} \tag{2}$$

$$MSE = \frac{1}{n}\sum_{i=1}^{n}(obs_i - pre_i)^2 \tag{3}$$

$$RMSE = \sqrt{\frac{1}{n}\sum_{i=1}^{n}(obs_i - pre_i)^2} \tag{4}$$

where obs_i is the observed value, pre_i is the prediction value, $\overline{obs}$ and $\overline{pre}$ are the average values of observed values and test values, respectively.

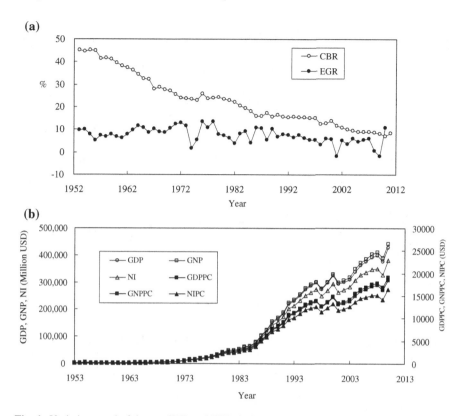

Fig. 1 Variation trend of data. **a** CBR and EGR. **b** GDP, GDPPC, GNP, GNPPC, NI, and NIPC

3 Results and Discussion

3.1 Variation Trend of Data

The number of CBR data set from 1953 to 2011 was totally 59, as shown in Fig. 1a. Among the total numbers of data, the numbers for training and testing were 44 (1953–1980 and 1996–2011) and 15 (1981–1995), respectively. The number for other data set from 1952 to 2010 was totally 59, as shown in Fig. 1a and b. The number for training and testing was 44 (1952–1979 and 1995–2010) and 15 (1980–1994), respectively. The mean, standard deviation, maximum, and minimum values of these data are shown in Table 1.

Table 1 The mean, standard deviation, maximum, and minimum values of data

Items factor	Mean	Standard deviation	Maximum	Minimum
CBR (Births per thousand persons)	22.75	11.25	45.29	7.21
EGR (%)	7.52	3.34	13.49	−1.81
GDP (Million USD)	125,105.75	143,996.57	428,186.00	1,420.00
GDPPC (USD)	5,815.95	6,343.90	18,503.00	142.00
GNP (Million USD)	127,718.97	147,474.01	441,761.00	1,420.00
GNPPC (USD)	5,934.37	6,496.49	19,090.00	141.00
NI (Million USD)	113,461.81	129,293.92	381,606.00	1,352.00
NIPC (USD)	5,280.95	5,703.73	16,491.00	133.00

Table 2 The correlation coefficients between CBR, EGR and NII factors

Factor	EGR	GDP	GDPPC	GNP	GNPPC	NI	NIPC
Correlation coefficient	0.42	−0.82	−0.83	−0.82	−0.83	−0.82	−0.83

3.2 Correlation Coefficients Between CBR and other factors

The correlation coefficients (R) between the CBR, EGR and NII factors were calculated as shown in Table 2. The R values of CBR were in the following order: EGR (0.42) > GDP, GNP, NI (−0.82) > GDPPC, GNPPC, NIPC (−0.83).

3.3 Training and Testing of ANN

To establish the ANN model, the EGR and NII at $t-1$ were taken as the input variables, and the CBR at t was taken as the output variable. The ANN consisted of three independent layers: input, hidden, and output layers. The input, hidden, and output layer was comprised of 7, 8, and 1 operating neurons, respectively, as shown in Fig. 2. The number of training was 10,000.

Figure 3a and b depict the training and testing results of CBR using ANN. All MAPE, R, MSE, and RMSE values are shown in Table 3. As shown in Table 3, when training, MAPEs between the simulated and observed values were between 9.87 and 10.86 % using ANN. When testing, the MAPEs lay between 2.90 and 5.19 %. When training, R values were about 0.96 using ANN. When testing, R values increased from 0.82−0.88 to 0.88−0.97. The MSE values were 1.12−1.29. When testing, the values were 0.21−0.84. When training, the RMSE values were 1.06−1.14 The RMSE values were 0.46−0.92 when testing.

Fig. 2 The structure of ANN

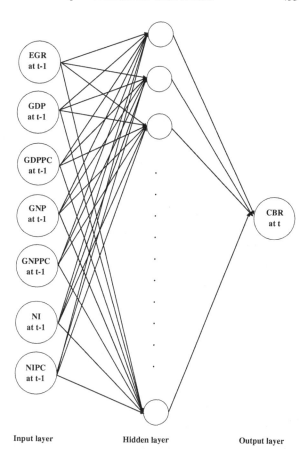

Input layer   Hidden layer   Output layer

3.4 Sensitivity Analysis Between CBR and Other Factors

The effects of the usually large uncertainties in the network factors should be taken into account by sensitivity analysis. In this study, the sensitivity of CBR for the EGR and NII factors was analyzed based on a 10 % change of the standard values.

Instead of using the traditional calculation in social network such as density, cohesion, or centrality etc., the sensitivity analysis and R were employed to represent the interrelation between CBR and other economic factors. The sensitivity (SEN) of EGR and NII factors (p) with respect to y (being CBR) was calculated by [16–19]:

$$SEN = \frac{dy/y}{dp/p} \qquad (5)$$

where dp is the change in the factor value p, and dy the change in the output y.

According to sensitivity analysis, the network in which the interrelations between various factors result in a complicated combination of relations could be

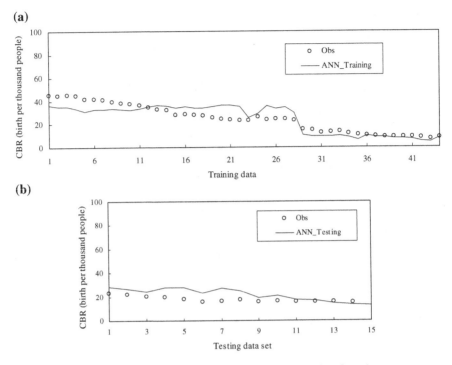

Fig. 3 The training and testing results of CBR using ANN. **a** training, **b** testing

Table 3 Training and testing performance for ANN

Performance ANN	MAPE (%)	R	MSE (%)	RMSE (%)
Training	23.28	0.85	46.02	6.78
Testing	28.93	0.70	35.82	5.99

shown in Fig. 4. The CBR values have different sensitivities towards different factors as shown in Table 4. The absolute values of CBR appeared to have a sensitivity of more than 0.1 (SEN > 0.1) towards 4 factors: EGR (0.2217), GNP (0.1091), GNPPC (−0.1314), and NIPC (−0.3822). The CBR was less sensitive towards GDP (0.0399), GDPPC (−0.0369), and NI (0.0091).

Comparable observations were also made by [2]. Brander and Dowrick [2] found that high birth rates appeared to reduce economic growth possibly through "capital dilution". They also found that birth rate declines had a strong medium-term positive impact on per capita income growth significantly. In the study proposed by [5], the results of correlations showed a negative impact of population growth on economic development in cross-country data for the 1980s.

In our study, the CBR appeared to have a negative sensitivity towards three per capita indices including GDPPC (−0.0369), GNPPC (−0.1314), and NIPC (−0.3822). The reason of the negative sensitivity also arose from the "capital

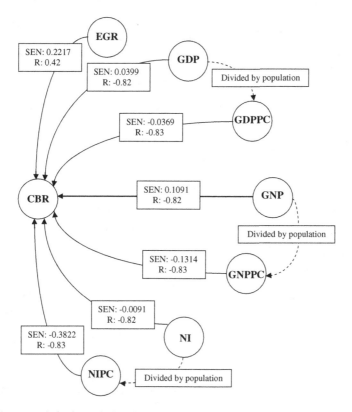

Fig. 4 The network for interrelations between CBR and other factors

Table 4 The results of sensitivity analysis

Factor	EGR	GDP	GDPPC	GNP	GNPPC	NI	NIPC
Sensitivity	0.2217	0.0399	−0.0369	0.1091	−0.1314	0.0091	−0.3822

dilution" mostly. But in our study, positive EGR at previous year would stimulate the CBR at current year (R = 0.42 and SEN = 0.2217). Meanwhile, positive macroeconomic factors including GDP, GNP, and NI would also stimulate the CBR. The results suggested that the economic development would cause the fertility will to occur, thus increased the CBR.

4 Conclusions

In this study, an ANN model was used to evaluate impact of EGR and NII on current CBR in Taiwan. The results showed that the CBR appeared to have a negative sensitivity towards three per capita indices including GDPPC (−0.0369), GNPPC (−0.1314), and NIPC (−0.3822). It suggested that the "capital dilution" would result in CBR decline. But positive EGR at previous year would stimulate the CBR at current year, as well as the positive macroeconomic factors including GDP, GNP, and NI. It suggested that the economic development would cause the fertility will to occur, thus increased the CBR. The application of social networks for this type of study could be expanded in the future.

References

1. Cunningham, W.P., Cunningham, M.A.: Principles of Environmental Science: Inquiry and Applications. McGraw-Hill Company, New York (2008)
2. Brander, J.A., Dowrick, S.: The role of fertility and population in economic growth. J. Population Econ. 7(1), 1–25 (1994)
3. Bloom, D., Canning, D., Malaney, P.: Demographic change and economic growth in Asia. Population Dev. Rev. 26, 257–290 (2000)
4. Higgins, M.: Demography, national savings, and international capital flows. Int. Econ. Rev. 39, 343–369 (1998)
5. Kelley, A.C., Schmidt, R.M.: Aggregate population and economic growth correlations: the role of the components of demographic change. Demography 32(4), 543–555 (1995)
6. Poston, D.L.: Social and economic development and the fertility transitions in mainland China and Taiwan. Popul. Dev. Rev. 26, 40–60 (2000)
7. Wasserman, S., Faust, K.: Social Network Analysis in the Social and Behavioral Sciences. Social Network Analysis: Methods and Applications. Cambridge University Press, Cambridge (1994)
8. Pai, T.Y., Tsai, Y.P., Lo, H.M., Tsai, C.H., Lin, C.Y.: Grey and neural network prediction of suspended solids and chemical oxygen demand in hospital wastewater treatment plant effluent. Comput. Chem. Eng. 31(10), 1272–1281 (2007)
9. Pai, T.Y., Chuang, S.H., Ho, H.H., Yu, L.F., Su, H.C., Hu, H.C.: Predicting performance of grey and neural network in industrial effluent using online monitoring parameters. Process Biochem. 43(2), 199–205 (2008)
10. Pai, T.Y., Chuang, S.H., Wan, T.J., Lo, H.M., Tsai, Y.P., Su, H.C., Yu, L.F., Hu, H.C., Sung, P.J.: Comparisons of grey and neural network prediction of industrial park wastewater effluent using influent quality and online monitoring parameters. Environ. Monit. Assess. 146(1–3), 51–66 (2008)
11. Pai, T.Y., Lin, K.L., Shie, J.L., Chang, T.C., Chen, B.Y.: Predicting the co-melting temperatures of municipal solid waste incinerator fly ash and sewage sludge ash using grey model and neural network. Waste Manag. Res. 29(3), 284–293 (2011)
12. Pai, T.Y., Yang, P.Y., Wang, S.C., Lo, H.M., Chiang, C.F., Kuo, J.L., Chu, H.H., Su, H.C., Yu, L.F., Hu, H.C., Chang, Y.H.: Predicting effluent from the wastewater treatment plant of industrial park based on fuzzy network and influent quality. Appl. Math. Model. 35(8), 3674–3684 (2011)
13. Pai, T.Y., Hanaki, K., Su, H.C., Yu, L.F.: A 24-h forecast of oxidant concentration in Tokyo using neural network and fuzzy learning approach. CLEAN-Soil Air Water 41(8), 729–736 (2013b)
14. Ministry of Interior, Executive Yuan, R.O.C.: Statistical Yearbook of Interior. Ministry of Interior, Executive Yuan, Taipei, R.O.C. (2010)
15. Directorate—General of Budget, Accounting and Statistics, R.O.C. National Accounts Yearbook, Directorate - General of Budget, Accounting and Statistics, Taipei, R.O.C. (2011)
16. Pai, T.Y.: Modeling nitrite and nitrate variations in A2O process under different return oxic mixed liquid using an extended model. Process Biochem. 42(6), 978–987 (2007)

17. Pai, T.Y., Chuang, S.H., Tsai, Y.P., Ouyang, C.F.: Modelling a combined anaerobic/anoxic oxide and rotating biological contactors process under dissolved oxygen variation by using an activated sludge-biofilm hybrid model. J. Environ. Eng-ASCE **130**(12), 1433–1441 (2004)
18. Pai, T.Y., Chang, H.Y., Wan, T.J., Chuang, S.H., Tsai, Y.P.: Using an extended activated sludge model to simulate nitrite and nitrate variations in TNCU2 process. Appl. Math. Model. **33**(11), 4259–4268 (2009)
19. Pai, T.Y., Shyu, G.S., Chen, L., Lo, H.M., Chang, D.H., Lai, W.J., Yang, P.Y., Chen, C.Y., Liao, Y.C., Tseng, S.C.: Modelling transportation and transformation of nitrogen compounds at different influent concentrations in sewer pipe. Appl. Math. Model. **37**(3), 1553–1563 (2013)

Index

W. Pedrycz and S.-M. Chen (eds.), *Social Networks: A Framework*
of Computational Intelligence, Studies in Computational Intelligence 526,
DOI: 10.1007/978-3-319-02993-1, © Springer International Publishing Switzerland 2014

Printed in the United States
By Bookmasters